Solar System Update

Philippe Blondel and John W. Mason (Editors)

Solar System Update

Published in association with
Praxis Publishing
Chichester, UK

Editors
Dr Philippe Blondel, CGeol FGS PhD MSc
Senior Scientist
Department of Physics
University of Bath
Bath
UK

Dr John W. Mason
Olympus Mons
51 Orchard Way
Barnham
West Sussex PO22 0HX
UK

SPRINGER–PRAXIS BOOKS IN ASTRONOMY AND PLANETARY SCIENCES
SUBJECT *ADVISORY EDITORS*: Dr. Philippe Blondel, C.Geol., F.G.S., Ph.D., M.Sc., Senior Scientist, Department of Physics, University of Bath, Bath, UK; Dr. John Mason B.Sc., M.Sc., Ph.D.

ISBN 3-540-26056-0 Springer-Verlag Berlin Heidelberg New York

Springer is part of Springer-Science + Business Media (springeronline.com)

Bibliographic information published by Die Deutsche Bibliothek

Die Deutsche Bibliothek lists this publication in the Deutsche Nationalbibliografie; detailed bibliographic data are available from the Internet at http://dnb.ddb.de

Library of Congress Control Number: 2005936392

Printed in Germany

Cover design: Jim Wilkie
Completed in LaTex: EDV-Beratung, Germany

Printed on acid-free paper

Latest Advances in Solar System Research

Solar System science is advancing faster than ever before, and even professional researchers in the field find it hard to keep up with developments. As this volume goes to press, the MESSENGER spacecraft is en route to Mercury, the first spacecraft to visit the planet in over 30 years; Venus Express and Mars Reconnaissance Orbiter have recently entered orbit around Venus and Mars, respectively; the Cassini-Huygens mission continues to produce stunning scientific results at Saturn; New Horizons has begun its nine-and-a-half year journey to Pluto; and the Rosetta spacecraft is on its way to a rendezvous with Comet 67P/Churyumov-Gerasimenko in 2014. In addition, ground-based telescopes and the Hubble Space Telescope continue to make important discoveries, observing a wide range of Solar System objects, including those such as Uranus and Neptune where no spacecraft is currently active or planned.

This first volume of *Solar System Update* comprises a collection of twelve topical reviews, each presented as a separate chapter, covering important areas of Solar System science. The contributions have been written by scientists at the forefront of research in the selected areas, in a style which, we hope, will be accessible not only to advanced undergraduate students and beginning graduate students, but also to professional astronomers, planetary scientists and physicists, helping them to keep abreast of the latest developments in related fields. We have attempted to highlight a diverse range of topics, from recent studies of the Sun and 'space weather', to research on the major planets and planetary satellites, including the very latest results from the Cassini mission at Saturn, and extending to the recent exciting discoveries in the outer Solar System beyond Neptune.

As the hub of our planetary system, it is appropriate that Chapter 1 of this volume should focus on the Sun. Volker Bothmer examines early ideas about the solar wind, influences of solar activity on the Earth's magnetic field and geomagnetic storms. He shows how space-borne instruments, observing particularly in the EUV and soft X-ray regions of the EM-spectrum, have given us new insights into solar physics and revealed details of changing solar activity as never before. Observations of the dynamic solar atmosphere, acquired over many years, by the Yohkoh, SoHO and TRACE spacecraft are comprehensively presented. Taken in combination with observations from near-Earth satellites, these studies have revealed in detail how solar eruptions affect interplanetary and near-Earth space. Discoveries about the origin and evolution of the most intense solar eruptions, the coronal mass ejections, and of the slow and fast solar wind streams, have provided a firm basis for space weather forecasts in near-Earth space. The author also looks ahead to probable future developments in this field.

In many respects, Mercury has been the forgotten planet of the past three decades, having been visited by just one spacecraft, Mariner 10, in 1974/75. But as Robert Strom and Ann Sprague show in Chapter 2, Mercury may provide answers to many important questions regarding the formation and evolution of our Solar System. They carefully examine what has been learned about Mercury from the Mariner 10 flybys and from 30 years of ground-based telescopic observations. The authors describe the planet's exosphere, the exciting discovery of possible water ice deposits near Mercury's poles, the unique internal structure of the planet and models to explain its significant magnetic field. Only 45% percent of Mercury's surface were imaged by Mariner 10, but Earth-based images show some features on the side not imaged by Mariner. Important features of Mercury's surface geology, ground-based spectroscopic observations which have provided clues to Mercury's surface composition, and features relating to the tectonic, thermal and geologic history of the planet are all reviewed. Some interesting aspects of Mercury's origin and orbital history are described and the author's complete their chapter with a discussion of the main objectives of the MESSENGER mission due to make its first flybys of Mercury in 2008 and to go into orbit around the planet in 2011.

Although the surface of Venus was mapped in detail by the Magellan spacecraft in the early 1990s, the planet has been relatively neglected for the past 15 years, in spite of the fact that a great many questions remain about its atmosphere, surface and interior. Answering such questions is important in the context of the evolution and stability of terrestrial planet environments and has a relevance to global change problems on the Earth. Now, some of these questions are set to be addressed by two new space missions, Europe's Venus Express and Japan's Planet-C. In Chapter 3, Dmitri Titov, Hakan Svedhem and Fred Taylor examine our current knowledge of the atmosphere of Venus, and highlight areas where our knowledge is incomplete or where there is contradictory evidence. They discuss the temperature structure, composition and chemistry, cloud layers, radiative effects and atmospheric dynamics, as well as interactions of the atmosphere with the solar wind flow and with the planet's surface. The probable evolution of the planet's atmosphere and climate are also examined. The authors conclude by describing important aspects of the Venus Express and Planet-C missions and make some predictions for the results which can be expected.

A lively debate over the origin and evolution of the Moon continues more than 30 years after the departure of the last human visitors. It is therefore fitting that Chapter 4 has been provided by Apollo 17 astronaut-geologist Harrison Schmitt, the only author in this volume to have actually visited the body they are writing about! A sizeable consensus exists that a few tens of millions of years after the formation of the Earth, the Moon belatedly came into existence as a consequence of a "giant impact" between the very young Earth and a planetoid having about 11–14% of Earth's mass. Harrison Schmitt reviews the main tenets of this giant impact hypothesis and its consequences. He then goes on to explain why, in his opinion, such ideas deserve intense questioning, since the giant impact scenario is based on computer modelling and simulations rather than also being objectively tested against all the geochemical data available from the collected lunar samples. The author subsequently describes the major problems with the giant impact hypothesis, which lie with geochemical and geophysical information we have about the interior of the Moon below about 500 km. The author subsequently provides some alternative ideas about the origin of the Moon and its subsequent evolution, and he reviews the various episodes of large basin formation early in the Moon's history.

Mars has been the subject of intense scrutiny in recent years and it will come as no surprise that two chapters in this volume have been devoted to the Red Planet. In Chapter 5, Stephen Lewis and Peter Read discuss evidence for climate change on Mars. Although the atmosphere and surface of Mars are extremely dry at present, spacecraft imagery has revealed surface features apparently caused by flowing water on the surface in the past, leading to speculation that the planet was once relatively warm and wet. Recent space missions have provided a detailed picture of the present day climate of Mars, including the discovery of water ice in the permanent polar ice caps as well as in the upper layers of the surface across much of the planet. The Mars Exploration Rovers have also found *in situ* evidence for ancient water in the nearby rocks and landscape including, at the Opportunity site, rocks which appear to be a mixture of sediments formed during episodic events of shallow surface water. All the evidence indicates that, in certain regions, the Martian climate must at times have been substantially different to the present day. The authors discuss the oscillations in Mars' obliquity, orbital eccentricity and solar insolation that would lead to changes in climate, and whether such cycles may be associated with the observed polar layered deposits. Both long-term and short-term climate changes are described, including observed variations on very short timescales. Finally, numerical modelling and simulation are discussed as tools for understanding the past, present and future climates of Mars, and the results of some experiments performed with a Mars Global Circulation Model are presented.

The spacecraft observations which are more clearly defining the history of water on Mars appear to indicate that significant amounts of water were present on the surface of Mars in the past. If so, the possibilities for habitable environments on Mars in the past were substantial and diverse, as discussed by Thomas McCollom in Chapter 6. Whether or not life might exist, or has once existed, on Mars has been the subject of scientific debate for over 100 years, and continues vigorously today. The author discusses in detail the evidence that Mars appears to posses all the critical elements for life as we know it, including the required chemical elements, energy sources and the presence of water. He also describes the range of physical environments known to be tolerated by micro-organisms on Earth and relates these to Mars. The inhospitable nature of the martian surface at present indicates that potential habitats for extant life will be very limited, but habitable environments might well exist at or near the surface during more favourable climatic epochs and certainly long ago, if Mars did indeed have a warmer, wetter past.

The giant planet Jupiter and its satellites are currently not under scrutiny by any spacecraft, although new missions are planned. Nevertheless, as Patrick Irwin describes in Chapter 7, there is continuing great interest in the giant planets of our Solar System, particularly now that so many Jupiter-sized planets are being discovered in orbit around other stars. The question of whether our Solar System is normal, unusual or perhaps even unique is an intriguing one, but the discovery of more and more extrasolar planetary systems is now making such comparisons possible. After summarising the gross characteristics of Jupiter, Saturn, Uranus and Neptune, the author introduces how exoplanets are discovered through radial velocity and transit detection methods. He reviews the distributions of mass, orbital radius and orbital eccentricity for known exoplanets and shows that Extrasolar Giant Planets (EGPs) may be grouped into five classes ranging from Class I, Jupiter-like planets, to Class V, the so-called 'Hot Jupiters'. The implications of the discoveries for Solar System formation models are discussed in detail, and a wide range of further methods for detecting exoplanets and the implications of such discoveries are described.

Our current knowledge of the icy moons of Jupiter is reviewed by Richard Greenberg in Chapter 8. The author discusses in detail how the initial compositions of the Galilean satellites were determined by conditions in the circum-jovian nebula, and how they subsequently evolved under the complex interplay of orbital and geophysical processes, including orbital resonances, tides, internal differentiation and heat. He shows how the three satellites with large water-ice components – Europa, Ganymede and Callisto – are very different from one another as a consequence of the way in which these complex processes have played out. The evidence for liquid water is strongest for Europa, where a deep ocean lies just below the surface. On Ganymede, much of the surface has been modified by tectonics, with limited evidence for fluid exposure and flow. Callisto is the most primitive of the icy satellites with an ancient, heavily-cratered surface, and an interior only partially differentiated by density. The recent discovery that the tiny satellite Amalthea is also icy has raised new challenges to models of the formation and evolution of the jovian system, as the author describes.

The ongoing Cassini-Huygens mission at Saturn continues to produce a steady stream of stunning images, surprising discoveries and outstanding science. In Chapter 9, Ellis Miner, Dennis Matson and Linda Spilker summarise the primary science results from both the Cassini orbiter and Huygens Titan probe up to the publication of this volume. The initial results reviewed include images and other data from the close encounter with the outer retrograde satellite Phoebe prior to Saturn orbit insertion, from ring studies and magnetospheric studies during orbit insertion and the first year in orbit, from studies of Titan during the first several flybys and the Huygens probe mission, and from close flybys of several of Saturn's icy satellites. The results include discovery of several new satellites and a new radiation belt between the rings and Saturn's atmosphere, remarkable observations of wave phenomena in the rings, numerous discoveries relating to the icy satellites and, perhaps most surprising of all, the discovery of fountain-like sources of tiny water ice particles spraying material high above the surface of the geologically active moon Enceladus. As this volume goes to press, the Cassini observations continue unabated, and are steadily rewriting our knowledge of satellite characteristics in general and of the planet Saturn, its ring system and satellites, as the orbital tour continues.

No spacecraft has visited the ice giants Uranus and Neptune since Voyager 2 provided 'snapshots' of these planets in 1986 and 1989, respectively. However, as Heidi Hammel describes in Chapter 10 our understanding of these two planets has evolved rapidly in recent years due to a combination of improvements in instrumentation and techniques, and to continued observation by both ground-based telescopes and the Hubble Space Telescope. Many years of painstaking telescopic observation have been necessary to reveal intrinsic changes in the outer giants, and this should not be surprising for planets where the length of each season is measured in decades rather than months. The changing appearance of the atmosphere of Uranus and its current cloud patterns and zonal winds are reviewed. Recent observations of the uranian rings, satellites and ionosphere are presented. For Neptune, the rapidly evolving atmosphere is assessed in the context of long-term records, and advances in characterizing the planet's clumpy ring system and remarkable changes in the atmosphere of Neptune's moon Triton are described. Concluding remarks include a synopsis of future exploration of these dynamic planetary systems.

The outer regions of the Solar System beyond the planets are also producing their share of exciting developments, as Audrey Delsanti and David Jewitt describe in Chapter 11. Beyond the orbit of Neptune lies the Kuiper belt, a vast number of objects in a flattened ring, collisionally-processed relics from the formation of the Sun and planets. The Kuiper belt is the source of the Centaurs and Jupiter-family comets and studies of its objects may reveal much about early conditions in the Solar System. The authors describe the various known dynamical sub-types of Kuiper belt objects (KBOs). While the majority of KBOs are small, about a dozen have diameters greater than 1000 km, including Pluto and the recently discovered (and possibly larger) giant KBOs 2003 UB_{313} (a binary system), 2003 EL_{61} (a triple system) and 2005 FY_9. The properties of Pluto and the other large KBOs are described, and a large number of binary and multiple systems are discussed. The observed structure of the Kuiper belt, including the inclination distribution, velocity dispersion, radial extent, mass and size distribution and the probable origins of this structure are all carefully reviewed. The surface properties of KBOs, including albedo, colour, origins of the colour diversity, surface processes and spectra are also discussed.

Comets are the icy nomads of our planetary system, thought to have been around, relatively unchanged, since the Solar System's formation, and often considered to be 'time capsules' of valuable information for scientists interested in learning about the early history of the Sun and planets. But, as David Hughes explains in Chapter 12, although our knowledge of comets has advanced hugely in recent years, our understanding is still very much 'skin deep'. The mass and density of the nucleus, its internal structure and strength are all poorly understood. Whether the physical and chemical properties vary from place to place in a specific cometary nucleus, or from comet to comet, is a matter of conjecture. The author summarises our knowledge of the physical properties and chemical composition of the cometary nucleus, based on the four comets, Halley, Borrelly, Wild-2 and Tempel-1, which have had their nuclei imaged by spacecraft. The rate at which the cometary nucleus decays, its effect on cometary activity and the thickness of the all-encompassing surface dust layer are also discussed. The results of recent cometary space missions are summarised, special attention being paid to the Deep Impact mission to comet Tempel-1.

This book has benefited from the support and assistance of a large number of people. We would like to offer our sincere thanks to all of the contributing authors for their considerable efforts, perseverance and enthusiasm for this project. We are also most grateful to Frank Herweg of Springer, Heidelberg for his invaluable assistance and advice in the preparation of the LaTeX files for this book, including his work on many of the illustrations prior to publication. Finally, we are also indebted to Sue Peterkin and Romy Blott of Praxis Publishing for their very considerable assistance at all stages in the organization and coordination of this project, and to Clive Horwood, Publisher, for his encouragement, advice and patience throughout.

April 2006 Philippe Blondel and John W. Mason

Contents

List of Contributors

Volker Bothmer
Institute for Astrophysics
University of Göttingen
Friedrich-Hund-Pl.1
37077 Göttingen
Germany
bothmer@astro.physik.uni-goettingen.de

Audrey Delsanti
Institute for Astronomy
University of Hawaii
2680 Woodlawn Drive
Honolulu
HI 96822
USA
delsanti@ifa.hawaii.edu

Richard Greenberg
Department of Planetary Sciences and
Lunar and Planetary Laboratory
University of Arizona
1629 E University Blvd
Tucson
AZ 85721-0092
USA
greenberg@lpl.arizona.edu

Heidi B Hammel
Space Science Institute
4750 Walnut Street, Suite 205
Boulder
CO 80301
USA
hbh@alum.mit.edu

David W Hughes
Department of Physics and Astronomy
Hicks Building
University of Sheffield
Sheffield S3 7RH
UK
d.hughes@sheffield.ac.uk

Patrick G J Irwin
Atmospheric, Oceanic and Planetary
Physics
Clarendon Laboratory
Department of Physics
University of Oxford
Parks Road
Oxford OX1 3PU
UK
irwin@atm.ox.ac.uk

David Jewitt
Institute for Astronomy
University of Hawaii
2680 Woodlawn Drive
Honolulu
HI 96822
USA
jewitt@ifa.hawaii.edu

Stephen R Lewis
Department of Physics and Astronomy
The Open University
Walton Hall
Milton Keynes MK7 6AA
UK
S.R.Lewis@open.ac.uk

Dennis L Matson
Jet Propulsion Laboratory
Caltech
4800 Oak Grove Drive
Pasadena
CA 91109-8099
USA
Dennis.L.Matson@jpl.nasa.gov

Thomas M McCollom
Center for Astrobiology
University of Colorado
Boulder
CO 80309
USA
tom.mccollom@lasp.colorado.edu

Ellis D Miner
Jet Propulsion Laboratory (Retired)
11335 Sunburst Street
Lake View Terrace
CA 91342-7342
USA
ellis.d.miner@jpl.nasa.gov

Peter L Read
Atmospheric, Oceanic &
Planetary Physics
Clarendon Laboratory
Department of Physics
University of Oxford
Parks Road
Oxford OX1 3PU
UK
p.read1@physics.ox.ac.uk

Harrison H Schmitt
University of Wisconsin-Madison
P O Box 90730
Albuquerque
NM 97199
USA
schmitt@engr.wisc.edu

Linda J Spilker
Jet Propulsion Laboratory
MS 230-205
Caltech
4800 Oak Grove Drive
Pasadena
CA 91109-8099
USA
Linda.J.Spilker@jpl.nasa.gov

Ann L Sprague
Lunar and Planetary Laboratory
University of Arizona
Tucson
AZ 85721
USA
sprague@lpl.arizona.edu

Robert G Strom
Department of Planetary Sciences
University of Arizona
Tucson
AZ 85721
USA
rstrom@jupiter.lpl.arizona.edu

Håkan Svedhem
Research and Scientific Support
Department European Space
Agency/ESTEC PB 299
NL-2200AG Noordwijk
The Netherlands
H.Svedhem@esa.int

Fred W. Taylor
Atmospheric, Oceanic and Planetary Physics
Clarendon Laboratory
Department of Physics
University of Oxford
Parks Road
Oxford OX1 3PU
UK
fwt@atm.ox.ac.uk

Dmitri V. Titov
Max Planck Institute for Solar System Research
Max-Planck-Straße 2
37191 Katlenburg-Lindau
Germany
titov@mps.mpg.de

1 The Solar Atmosphere and Space Weather

Volker Bothmer

Abstract. First ideas about possible physical influences of the Sun on Earth other than by electromagnetic (EM) radiation were scientifically discussed more seriously after Richard Carrington's famous observation of a spectacular white-light flare in 1859 and the subsequent conclusion that this flash of EM radiation was connected with the origin of strong perturbations of the Earth's outer magnetic field, commonly referred to as geomagnetic storms, which were recorded about 24 hours after the solar flare. Tentatively significant correlations of the number of geomagnetic storms and aurorae with the varying number of sunspots seen on the visible solar disk were found in the long-term with respect to the roughly 11-year periodicity of the solar activity cycle. Although theories of sporadic solar eruptions were postulated soon after the Carrington observations, the physical mechanism of the transfer of energy from the Sun to the Earth remained unknown. Early in the 20th century Chapman and Ferraro proposed the concept of huge clouds of charged particles emitted by the Sun as the triggers of geomagnetic storms. Based on the inference of the existence of a solar magnetic field, magnetized plasma clouds were subsequently introduced. Eugene Parker derived theoretical evidence for a continuous stream of ionized particles, the solar wind, leading to continuous convection of the Sun's magnetic field into interplanetary space. The existence of the solar wind was confirmed soon after the launch of the first satellites. Since then the Sun is known to be a permanent source of particles filling interplanetary space. However, it was still thought that the Sun's outer atmosphere, the solar corona, is a static rather than a dynamic object, undergoing only long-term structural changes in phase with the Sun's activity cycle. This view completely changed after space borne telescopes provided extended series of solar images in the EUV and soft X-ray range of the EM spectrum, invisible to ground-based observers. The remote-sensing observations undertaken by Yohkoh, followed by multi-wavelength movies from SoHO (Solar Heliospheric Observatory) and high resolution EUV imaging by TRACE (Transition Region and Coronal Explorer) have revealed to date that the Sun's atmosphere is highly dynamic and never at rest. Solar eruptions have been tracked into space in unprecedented detail. In combination with near-Earth satellites, their interplanetary and geo-space effects could be investigated in depth, having provided the roots for space weather forecasts. This chapter summarizes the discoveries about the origin and evolution of solar storms and their space weather effects, providing a comprehensive picture of the most important links in the Sun-Earth system. It finally provides an outlook to future research in the field of space weather.

1.1 Early Concepts of Solar-Terrestrial Physics

Shortly after the development of a sensitive variation compass or declinometer by George Graham in 1722, and his discovery that the compass needle is continually in motion, Anders Celsius and Olof Hiorter recognized a relationship between the appearances of the aurora and short-time magnetic fluctuations of the direction of the Earth's magnetic field (Hiorter 1747). Jean Jacques d'Ortous de Marian (1733) published a textbook devoted to the phenomenon in which he proposed a similar existence of aurora in the southern hemisphere of the Earth, and also its correlation with sunspot frequencies. In the first decade of the 19th century Alexander von Humboldt (1808) studied magnetic disturbances in Berlin and termed them "Magnetische Ungewitter", later on referred to as geomagnetic storms (e.g., Bartels 1932). Around 1830 Carl Friedrich Gauss performed systematic measurements of the fluctuations of the Earth's magnetic field in the "Göttinger Magnetische Verein" (Gauss 1839). He stated, that unravelling the origin of the "Rätselhafte Hieroglyphenschrift der Natur" will become a triumph of modern science. Almost at the same time, Heinrich Schwabe reported a 10-year periodicity in sunspot counts (Schwabe 1844). Interestingly, systematic observations of sunspots had been taken already for about 200 years, e.g., in 1611 by Johannes and David Fabricius and Christoph Scheiner (Fabricius 1611). In 1852 Edward Sabine claimed that the sunspot periodicity is evident also in geomagnetic activity records and that there might be a physical impact of the Sun's activity on the Earth.

Nobody had ever recognized a direct link between solar and terrestrial phenomena, but on September 1, 1859, Richard Carrington observed together with Robert Hodgson an outstandingly intense "white-light flare" – a sporadic electromagnetic emission (EM) on the Sun's visible surface, lasting for some minutes. "I was suddenly surprised at the appearance of a very brilliant star of light, much brighter than the Sun's surface, most dazzling to the protected eye ..." reported Hodgson in the "Monthly Notices of the Royal Astronomical Society" (Carrington 1860). The flare observed by Carrington and Hodgson was followed by a "Magnetisches Ungewitter" about 18 hours later and Carrington claimed that the solar flare might have been the indicator of solar processes that subsequently led to the geomagnetic storm.

Figure 1.1 shows the fluctuations of the horizontal (H) component of the Earth's magnetic field observed at the Colaba station in Bombay (Lakhina 2004). The magnitude of the magnetic field variations is typically a couple of thousands of nano Tesla (nT) during large magnetic storms (e.g., Schlegel 2000). The onset of a large magnetic storm is usually characterized by a rapid increase in the field magnitude, called the sudden storm commencement (SSC), which as we shall see in Sect. 1.5, is caused by compression of the Earth's outer magnetic field, its magnetosphere (Gold 1959) through the impact of a fast transient corpuscular stream from the Sun. During large magnetic storms aurora can be seen at latitudes as far south as Hawaii or Italy (e.g., Schlegel 2000). It has been reported that Tiberius, Caesar around 34 A.D., sent troops to Ostia in order to extinguish a fire in the harbour as noticed from the red patchy sky – the aurora seen from some distance (Seneca ~60 A.C.). The various types of aurora and their excitation mechanisms are described in detail, e.g., in Prölss (2004).

Around the same time that Richard Carrington discovered a solar flare, Angelo Pietro Secchi demonstrated spectroscopically, that solar prominences are real solar features

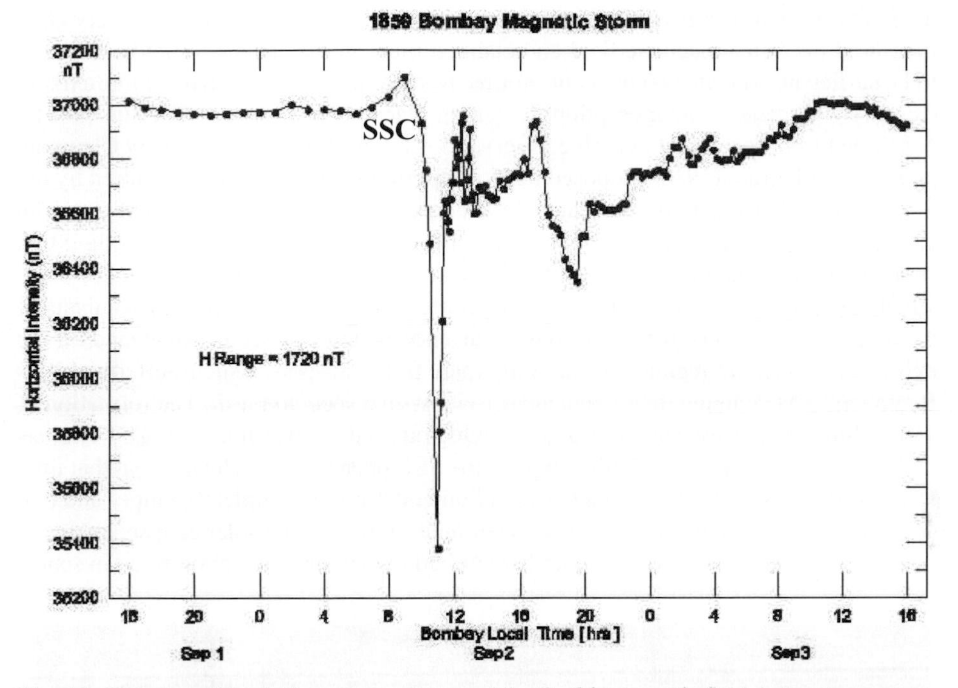

Fig. 1.1. Variation of the horizontal component of the Earth's magnetic field during the geomagnetic storm on September 2–3, 1859 as recorded by the Colaba station in Bombay, India. The sudden storm commencement (SSC) preceeding the storm is indicated, Adapted from Lakhina et al. (2004).

which sometimes rise from the solar surface upwards with speeds of several hundreds of km/s, hence overcoming the Sun's gravitational field (Secchi 1870).

The discussion of ideas on how energy could be transferred from Sun to Earth besides EM-radiation were stimulated in the aftermath of Carrington's observation, e.g., by Antoine Henry Becquerel (1880) and Eugen Goldstein (1880) in order to explain such "delayed" geomagnetic storms. It should be noted that Charles Young (1872) noticed that small geomagnetic fluctuations appear also directly at the onset of a flare and today it is known that these non-delayed fluctuations are caused by the flare's EM-radiation which produces an atmospheric change in electric conductivity at Earth, exciting a current inducing the magnetic fluctuations.

George Francis Fitzgerald (1892) proposed that "a sunspot is a source from which some emanation like a comet's tail is projected from the Sun ..." and asked "is it possible then that matter starting from the Sun with the explosive velocities we know possible there, and subject to an acceleration of several times solar gravitation, could reach the Earth in a couple of days?". Oliver Lodge (1900) suggested that magnetic storms are caused by " ... a torrent or flying cloud of charged atoms or ions". In the same year, Edward Walter Maunder (1904) noted that the appearances of large sunspot groups were associated with the appearances of large geomagnetic storms. He also noticed a 27-day recurrence rate persistent in the time-series of geomagnetic activity records, evident for the relationship of geomagnetic activity with the solar rotation rate. George Ellery Hale measured the Sun's magnetic field and claimed that sunspot magnetic fields seem not to be sufficient to create geomagnetic storms. Rather, the storms seem to be more likely caused by processes – solar eruptions? – taking place in areas surrounding sunspots he concluded from spectroheliographic observations (Hale 1908, 1919). Sydney Chapman and Vincent Ferraro (1930) proposed the concept of huge neutral clouds emitted by the Sun as possible triggers of the magnetic sudden storm commencements (SSCs). The clouds were thought to cause compressions of the Earth's magnetosphere, leading to a short increase in the magnetic field intensity at the beginnings of the storms at times when they impinged upon the Earth. Julius Bartels (1932) derived from the analysis of geomagnetic variations that the 27-day recurrence periodicity seems to depend on the existence of specific regions on the Sun, apart from sunspots, which he termed "M-Regionen" – M denoting their magnetic activity with respect to Earth. The formation of a travelling shock wave caused by a fast moving rarefied gas was introduced by Thomas Gold (1955) in order to help further explain the SSC phenomenon. However, at that time it was still believed that the space between Sun and Earth is completely empty and that the solar corona, which is visible for some minutes during a total solar eclipse, as shown in Fig. 1.2, is a quasi-static object purely confined to the Sun's neighbourhood in space.

Fig. 1.2. The solar corona as seen during the total solar eclipse on July 11, 1991. Courtesy: High Altitude Observatory (HAO), Boulder, USA.

1.2 Discovery of the Solar Wind, its Source Regions and Geomagnetic Effects

From cometary observations, Ludwig Biermann (1951) deduced that the Sun seems to be a continous source of particles that blow through the solar system and past its objects, creating the anti-sunward directed tails of comets. The emitted particles he referred to as solar "corpuscular radiation" or "solar wind". Eugene Parker (1958) supported Biermann's suggestion by his hydrodynamic calculations that first showed theoretically that the solar corona cannot exist in thermal equilibrium but has to expand, finally giving rise to the solar wind and its embedded outward convected solar magnetic field, the so-called interplanetary magnetic field (IMF).

Eugene Parker's discovery was scientifically contradictive at that time and his first submitted summary of his results was rejected by the reviewers.

Konstantin Gringauz in 1959 flew "ion traps" on the Soviet Lunik 2 and 3 missions, instruments measuring the total electric charge of ions. He reported that the signal fluctuated as the spacecraft spun around its axis, suggesting an ion flow entering the instrument whenever it faced the Sun (Gringauz 1961). Analysis of the data however failed to find the necessary signal as well as another ion trap flown on NASA's Explorer 10 mission. Finally in 1962 when Mariner 2 flew towards Venus, it provided the first detailed in-situ measurements of the solar wind and its time-dependent structure over the period of investigation. The principal investigator of this instrument was Marcia Neugebauer, and the principal investigator of the magnetometer verifying the existence of the IMF was Charles Sonett (Neugebauer et al. 1966, Sonett & Abrahams 1963).

The solar wind stream structure observed was dominated by slow and fast solar wind streams in a 27-day recurrent manner, suggesting their sources rotated with the Sun as would be the case in the presence of Bartels M-regions. However, the measurements also showed that this basic pattern was frequently disrupted by strong irregularities. Table 1.1 summarizes the basic properties of fast and slow solar wind streams at Earth's orbit.

With the advent of the space age instruments could be flown out of the Earth's shielding atmosphere, allowing the first observations of the Sun at X-ray and EUV-wavelengths (see Fig. 1.3) invisible to ground-based observers. The first extended space borne observations of the solar atmosphere in X-rays were taken during the Skylab mission in 1973 (e.g., Bohlin 1978).

Table 1.1. Basic solar wind characteristics near Earth's orbit. Adapted from Schwenn (1990).

	Fast Wind	Slow Wind
→	$>\sim 400$ km/s	$<\sim 400$ km/s
→	np $\sim 3\,\mathrm{cm}^{-3}$	np $\sim 8\,\mathrm{cm}^{-3}$
→	∼95% H, 4% He, minor ions and same number of electrons	∼94% H, ∼5% He, minor ions and same number of electrons – great variability
→	B ~ 5 nT	$B \sim 4$ nT
→	Alfvénic Fluctuations	Density Fluctuations
→	Origin in Coronal Holes	Origin 'Above' Coronal Steamers

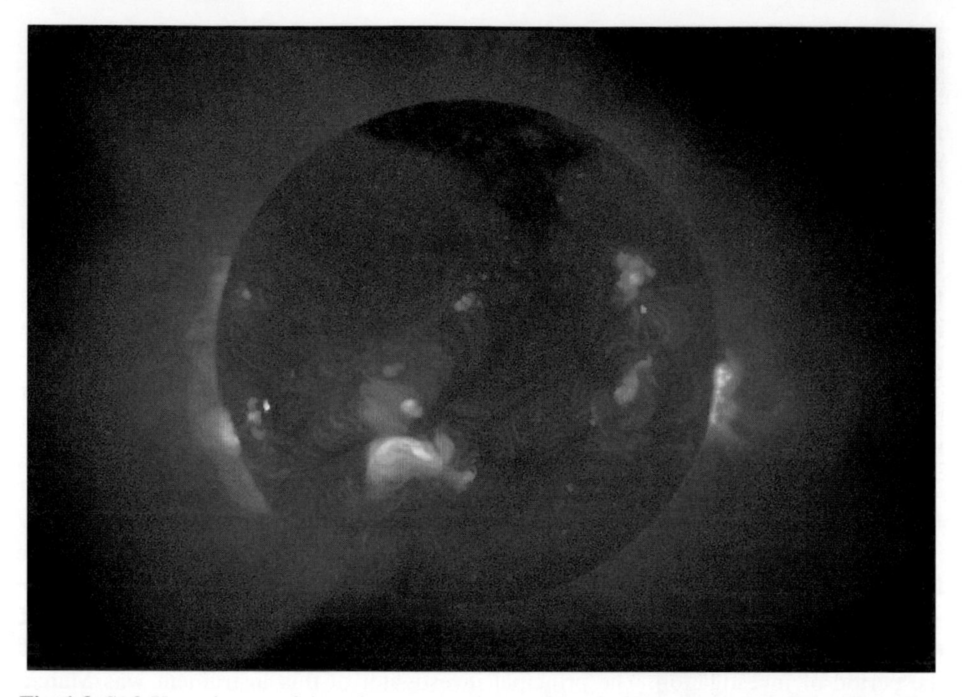

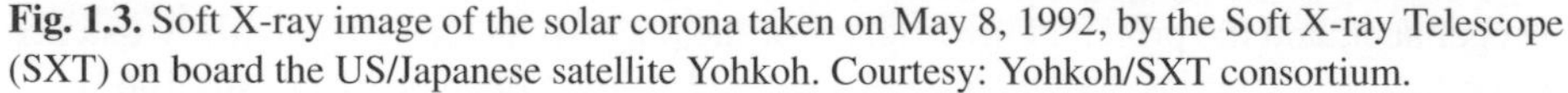

Fig. 1.3. Soft X-ray image of the solar corona taken on May 8, 1992, by the Soft X-ray Telescope (SXT) on board the US/Japanese satellite Yohkoh. Courtesy: Yohkoh/SXT consortium.

The Solar and Heliospheric observatory (SoHO) was launched on December 2, 1995. The joint ESA/NASA mission started its scientific observations in February 1996 in the L1 orbit of the Sun-Earth system, ~1.5 million km in the sunward direction ahead of Earth. SoHO carries 12 high-tech scientific instruments which measure solar oscillations, photospheric fields, chromospheric and coronal plasma properties and in-situ flows of solar particles (Fleck, Domino & Poland 1995). SoHO has been successfully operated, except for the short phase (today known as the SoHO holidays) during its temporary loss in 1998, for ten years now and has provided stunning new insights into the physics of the Sun and its heliospheric impacts (e.g., Fleck & Svetska 1997, Lang 2000).

Figure 1.4 shows EUV-observations of the solar corona at 195 Å corresponding to temperatures of about 1.5 million K emitted from FeXII ions as imaged by the Extreme-ultraviolet Imaging Telescope (EIT) on board SoHO together with in-situ solar wind and geomagnetic data measured in August/September 1996. The right EUV image was taken ten days after the first one so that the coronal hole had already disappeared behind the Sun's western limb. The active bright shining loops of the active region seen before in the Sun's eastern hemisphere can be nicely recognized at the western limb. The highest solar wind speed measured during this time period at Earth's orbit was related to the passage of the low latitude extension of the northern coronal hole. It takes about three days for the solar wind at a speed of 600 km/s to reach the distance of Earth. The peak of the solar wind speed on August 27, 1996 (day 243) is consistent in time with what is expected from the remote sensing observations. The intensity-time profile of the Ap-

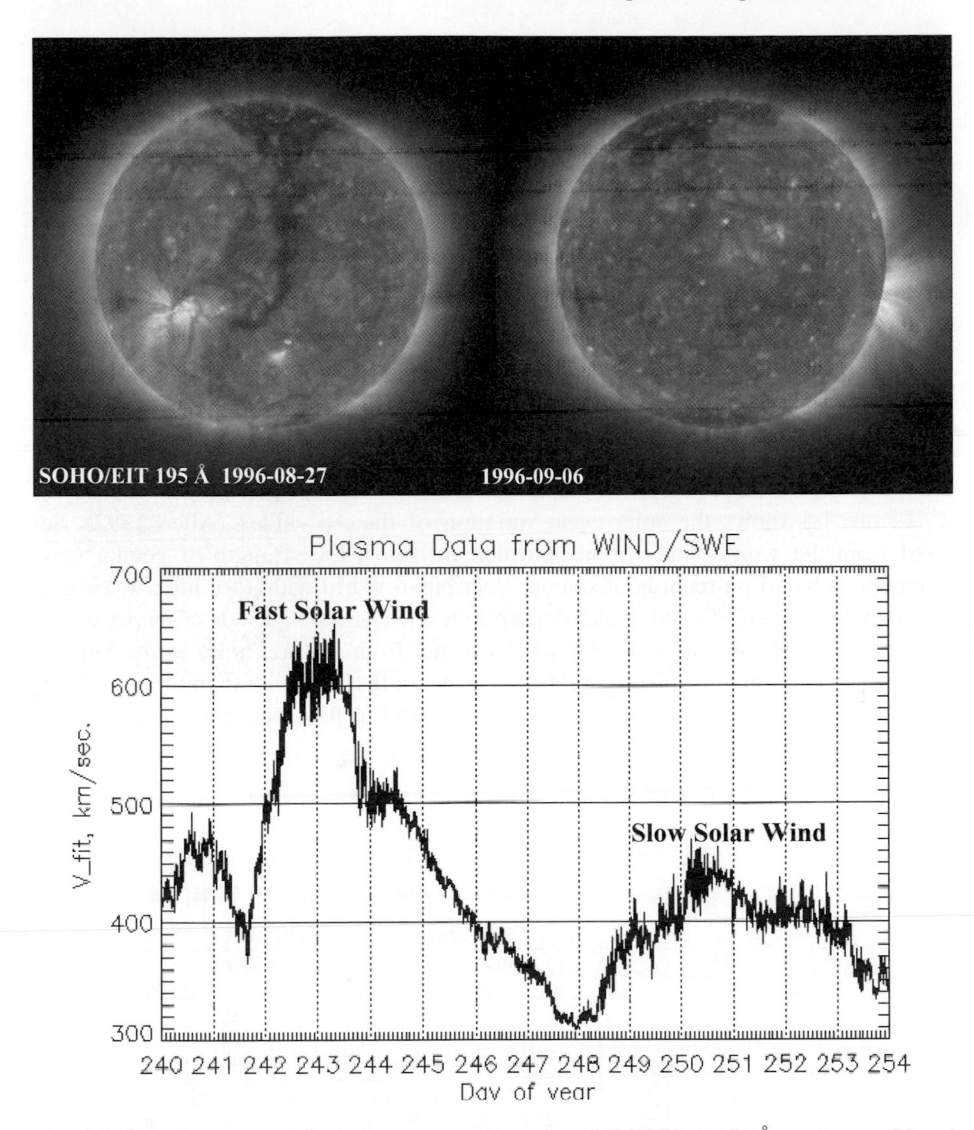

Fig. 1.4. Top: Structure of the solar corona as seen by SOHO/EIT at 195 Å on August 27 and September 6, 1996. Bottom: Solar wind speed measured from August 27 until September 9, 1996. The fast (~620 km/s) solar wind stream measured by the Solar Wind Experiment on the WIND spacecraft at 1 AU on August 29–30 corresponds to central meridian passage of the low-latitudinal extension of the northern polar coronal hole. Geomagnetic activity levels are provided through the Ap index.

index, a world-wide geomagnetic activity index described in more detail further on in this section, is indicating the level of the magnetic fluctuations observed on Earth. In case the low-latitudinal extension of the coronal would increase in heliolongitude, the time interval of the high speed flow at 1 AU would increases correspondingly.

Observations of the corona taken simultaneously with solar wind data obtained by near-Earth satellites have revealed that recurrent (the period of solar rotation is 25.4 days and hence 27.3 days as seen from Earth) geomagnetic storms are caused by these fast solar wind streams with typical speeds in the range of 500–800 km/s (Burlaga & Lepping 1977; Crooker & Siscoe 1986; Tsurutani 2001). Contrary to Bartels' earlier belief that recurrent storms are due to magnetic active regions at the Sun, we now know that they stem from coronal holes, i.e. rather magnetic "quiescent" solar regions. Recurrent geomagnetic storms are dominant especially in the declining phase of sunspot cycles (e.g., Richardson, Cliver & Cane 2001) because at these times the large polar coronal holes exhibit persistent low latitude extensions over time periods of several months (e.g., Timothy, Krieger & Vaiana 1975).

Figure 1.5 shows the solar cycle variation of the Ap*-index (Allen 1992), derived from the world-wide Ap-index which measures the strength of geomagnetic fluctuations based on records of stations distributed world-wide (see http://www.gfz-potsdam.de/pb2/pb23/GeoMag/niemegk/kp_index/). The time periods of increased geomagnetic activity due to fast solar wind streams from coronal holes at the Sun are primarily evident in the Ap* time-profile as peaks in the declining to minimum phases of the sunspot cycle (e.g., Richardson et al. 2001). In Fig. 1.5, the geomagnetic activity

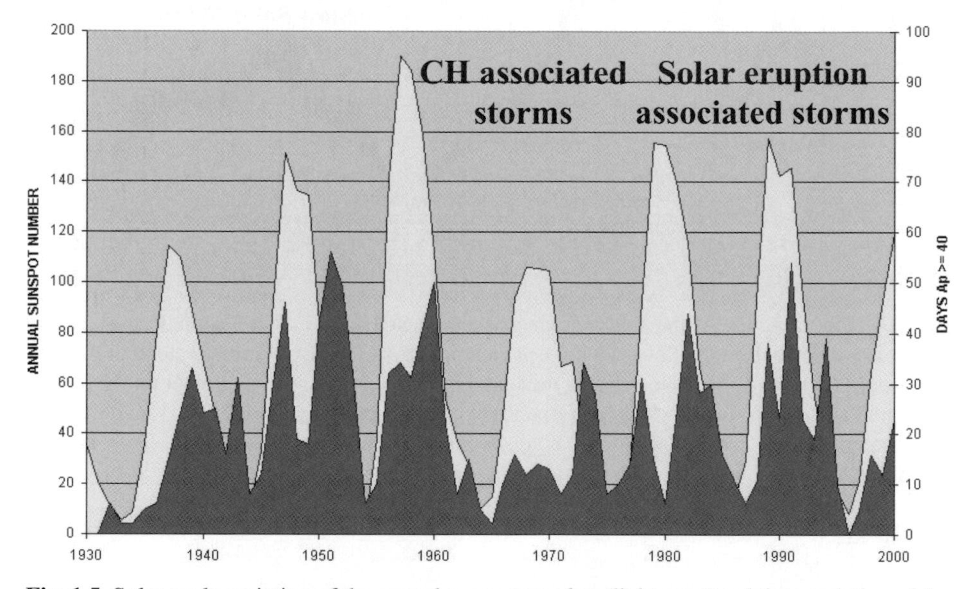

Fig. 1.5. Solar cycle variation of the annual sunspot number (light grey) and time variation of the number of geomagnetic disturbed days with Ap* > 40 nT (dark grey). Courtesy: Allen, http://www.ngdc.noaa.gov/stp/GEOMAG/image/APStar_2000sm.gif

phases during the solar cycle that were dominated by phases of transient solar eruptions are marked for comparison in cycles 21 and 22 accordingly.

The physical origins of geomagnetic activity are induction currents caused by the solar wind's electric field impacting the Earth's magnetosphere. The solar wind plasma and the outward convected magnetic field as IMF impose the electric field $E = -v \times B$ on the Earth's magnetosphere that results in a complex system of magnetospheric and ionospheric current systems (e.g., McPherron 1995). Its prime components in both hemispheres are the ionospheric polar electrojets and the near equatorial magnetospheric ring current measured through the AE and Dst indices (e.g., Bartels & Veldkamp 1949; Siebert 1971; Rostoker 1972; Mayaud 1980). The current systems are schematically shown in Fig. 1.6.

The electrojet currents are primarily driven by electrons which spiral along magnetic field lines into the ionosphere where the current density j is given by $j = \sigma E$, with σ being the conduction strength and E being the electric field that drives the current. The electric field E is of the order of several 10^{-3} V/m at times of weak energy input by the solar wind and reaches magnitudes of several 0.1 V/m during large storms (e.g., Schlegel 2000).

The electrojets and the ring current are the major drivers of the fluctuations of the Earth's magnetic field. Comparable to the "Richter-Skala" characterizing earthquake-magnitudes, the overall geomagnetic activity index Kp ("Kennziffer Planetarisch") is a

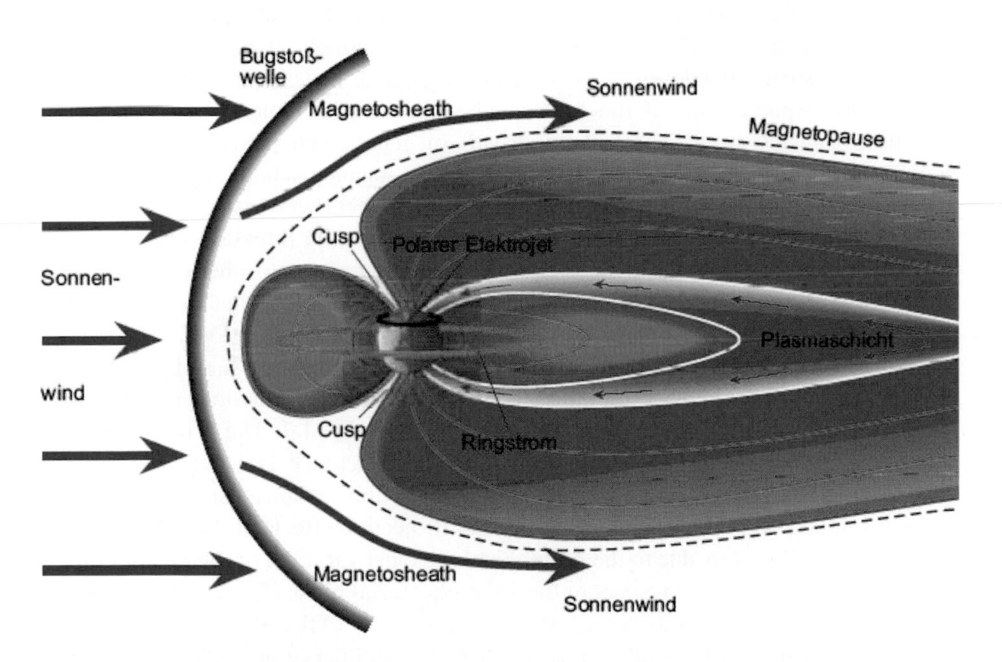

Fig. 1.6. Schematic representation of the solar wind flow impinging the Earth's magnetosphere and the main induced electric current systems in the terrestrial magnetosphere and ionosphere. Note that an electrojet current is also flowing in the southern auroral oval. Courtesy: K. Schlegel (2001).

quasi-logarithmic parameter, derived as daily 3 hour values from 0 to 9, from magnetograms recorded by 13 stations all over the world (Bartels & Veldkamp 1949; Siebert 1971). It is important to note, that Kp is not only made up out of the intensity of the two current systems, but that it also depends on the spatial position of the polar electrojets, i.e. large pressure pulses driven by solar wind flows causing stronger compressions of the Earth's magnetosphere and subsequent movement of the electrojets to lower latitudes, lead also to enhanced Kp-values, but will appear less pronounced in the ring current index Dst. Hence a difficulty in an exact global determination of the magnitude of a given magnetic storm is the superposition of compressional effects and current strengths. The Kp index and the magnetospheric ring current Dst are both measured in units of nano Tesla (nT). The 3-hour ap values and the daily Ap index represent the linear counterparts of Kp; a description of the conversion method between Kp and ap can be found at http://www.gfz-potsdam.de/pb2/pb23/GeoMag/niemegk/kp_index.

The subsolar point up to which the magnetosphere stretches out in space in the sunward direction can be calculated from the pressure balance between the solar wind's dynamic pressure and the pressure provided by the Earth's magnetic field (e.g., Kivelson & Russell 1995, Parks 2003) as $2nmV^2 = B^2(r_{\mathrm{MP}})/2\mu_0$, with n, V being the density and velocity of the solar wind, m being the proton mass and B being the strength of the Earth's magnetic field. Any pressure variation of the solar wind causes the magnetosphere to flutter in space like a flag as shown e.g. in the simulations by Goodrich et al. (1998). Any variations of the solar wind electric field $E = -v \times B$ leads in turn to fluctuations of the geomagnetic field, either caused through variation of the solar wind speed or its magnetic field magnitude. Note, that for simplicity dynamic pressure variations due to solar wind density variations are neglected. A detailed overview on solar wind-magnetosphere coupling functions can be found, e.g. in Gonzalez (1990).

The solar wind speed is one of the drivers of geomagnetic activity, but it is well known now that the energy transfer into the Earth's magnetosphere depends crucially upon whether the IMF has a southward-directed component with respect to the ecliptic plane, more precisely whether it is directed opposite to the direction of the Earth's magnetic field as measured in the GSM (Geocentric Solar Magnetospheric Coordinates) system (e.g., Russell & McPherron 1973). The main driver of geomagnetic activity is referred to for short simply as the southward ($-Bz$)-component of the IMF. A systematic introduction to common coordinate systems has been provided by Russell (1971).

At times near solar activity minimum the inner heliosphere is dominated by fast solar wind streams from the polar coronal holes as shown in Fig. 1.7. The solar wind speed is a key factor parameter in modulating geomagnetic activity. However, the strength of the southward component of the IMF and its direction remains also important even without southward IMF-components because of projection effects of the IMF and magnetospheric field vectors due to the diurnal and seasonal variation of the position of the Earth's magnetic axis with respect to the solar equatorial plane (Russell & McPherron 1973, Crooker & Siscoe 1986, Ness 2001). The variability of the -Bz component caused by these geometrical effects often exceeds that of V and hence depending on the time variation of V and B complicated seasonal and diurnal variations are caused (Russell & McPherron 1973, Phillips et al. 1992, Schreiber 1998, Richardson, Cliver & Cane 2001). To understand the solar wind structure of the inner heliosphere at any given moment is thus crucial in order to determine the status of the magnetosphere.

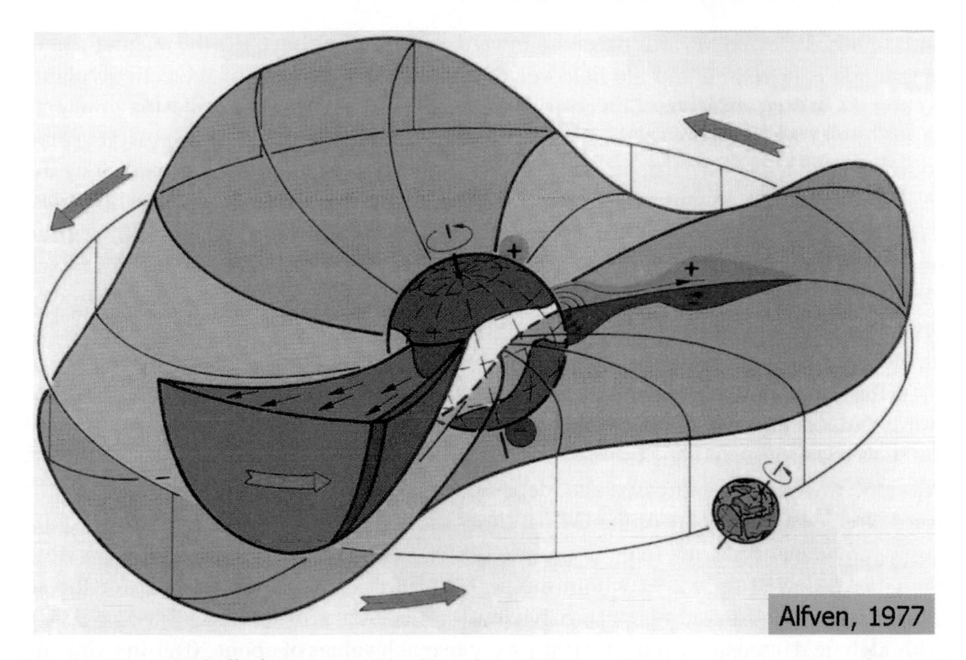

Fig. 1.7. The solar ballerina proposed by H. Alfvén (1977) – schematic picture of the structure of the inner heliosphere near solar minimum. The Sun's poles are covered with large coronal holes of opposite magnetic field polarity. Co-rotating solar wind streams of opposite magnetic polarity with a heliospheric current sheet (HCS) in between them dominate the flow pattern. The situation shown here is typical for odd cycles.

1.3 Structure of the Heliosphere

The structure of the inner heliosphere near solar minimum resembles the skirt of a "ballerina" as proposed by Alfven in (1977) and sketched in Fig. 1.7. At some solar radii distance from the Sun the solar wind in both hemispheres is comprised of the two flows from the large polar coronal holes with opposite magnetic field polarity, yielding a so called two sector structure of the IMF. An observer, e.g., at Earth orbit, would pass through the upper and lower parts of the ballerina skirt during one solar rotation, i.e. would record phases of opposite polarity of the IMF. For a typical solar wind speed around 450 km/s, based on the rotation period of the sun, the interplanetary magnetic field lines form the shape of an Archimedean (Parker) spiral with an average direction of 135° as measured counterclockwise with respect to the sunward direction. A detailed review on the physics of the inner heliosphere has been provided by Schwenn & Marsch (1990, 1991).

Another dominant feature of the heliosphere near solar minimum associated with the fast/slow stream solar wind pattern is the long-lasting formation of co-rotating interaction regions (CIRs). CIRs are caused by fast solar wind streams catching up slower solar wind streams originating in solar longitude westward of the fast streams. In the stream interaction process, the slow speed wind is compressed in its trailing edge and deflected in the sense of solar rotation, whilst the fast speed wind is compressed in its leading portion and slightly deflected towards the anti-sunward direction. Within CIRs the magnetic field magnitude is increased and the field vector may be deflected out of the ecliptic plane. At some distances from the Sun co-rotating shock waves form if the plasma gradients of the interacting streams exceed the critical thresholds in terms of the Alfvén- and sound-speeds. Charged particles are accelerated by CIR shocks and can stream along the magnetic field lines to heliospheric distances far away from the local acceleration sites. CIR shocks typically form at distances around 2 AU from the Sun, but some have been observed by the Helios spacecraft at distances as close as 0.3 AU (Schwenn & Marsch 1990). The physics of CIRs in the 3-D heliosphere has been explicitely summarized in reviews on these topics by Gosling & Pizzo (1999), Balogh et al. (1999) and Crooker et al. (1999).

CIRs are important triggers of geomagnetic storms because they can cause southward components of the IMF at times when the solar wind speed rises, i.e. at entry into the high speed stream (e.g., Tsurutani 2001). In the leading edge of high speed streams Alfvénic waves are compressed and, depending on the IMF-polarity, significant southward directed components of the IMF at times of an increased wind speed, excite quite large geomagnetic storms (e.g., Tsurutani 2001). The compression of the IMF within the CIR, followed by waves within the high speed stream are the reasons for the so called typically observed two step behaviour in recurrent storms (e.g., Burlaga 1977). Although field intensities of the IMF at 1 AU can reach values of about 20 nT in CIRs, the variability of the IMF is still larger in the aftermath of transient solar events as described in Sect. 1.5. This is the reason why geomagnetic storms caused by CIRs usually do not exceed Kp-values of 7+ (Bothmer & Schwenn 1995).

Ulysses, launched in 1990, was the first spacecraft to explore the uncharted third-dimension of the heliosphere. It orbits the Sun nearly normal to the ecliptic plane at distances between 1 and 5 AU (Balogh, Marsden & Smith 2001; Marsden 2001). The measurements of the solar wind speed and IMF-direction taken during Ulysses first and second orbit around the Sun, which included a gravity assist maneuver at Jupiter to reach the higher heliographic latitudes, are displayed in Fig. 1.8. At higher latitudes outside the ecliptic, only fast solar wind with speeds of about 750 km/s was encountered by Ulysses. Slow solar wind with speeds less than 450 km/s was confined to heliospheric regions in a narrow belt reaching from about 20° N to 20° S with respect to the heliographic equator, i.e. close to the ecliptic plane. It is important to point out that the origin of the slow speed solar wind is not known, but it is believed that it stems from intermitted leakage from solar regions with previously closed loops, including the loop tops of overlying coronal streamers (Wang et al. 1998).

The dominant magnetic polarity observed by Ulysses in both heliospheric hemispheres is consistent with the expected global magnetic polarity of the Sun during odd cycles: Positive in the N, whilst negative in the S (Fig. 1.9). Short-term intervals with faster streams seen at lower latitudes can be usually attributed to either transient streams which will be described in detail in section 1.5, or to streams from equatorward extensions of polar coronal holes as shown in Figure 1.4 or to low latitude short-lived coronal holes that occur near solar maximum at equatorial latitudes as can be seen in Fig. 1.10. As described earlier, the intermittent flow structure of slow and fast solar wind streams gives rise to stream-stream interactions and respective CIRs. The structure of the heliosphere near solar maximum gets increasingly complex as the level of solar activity

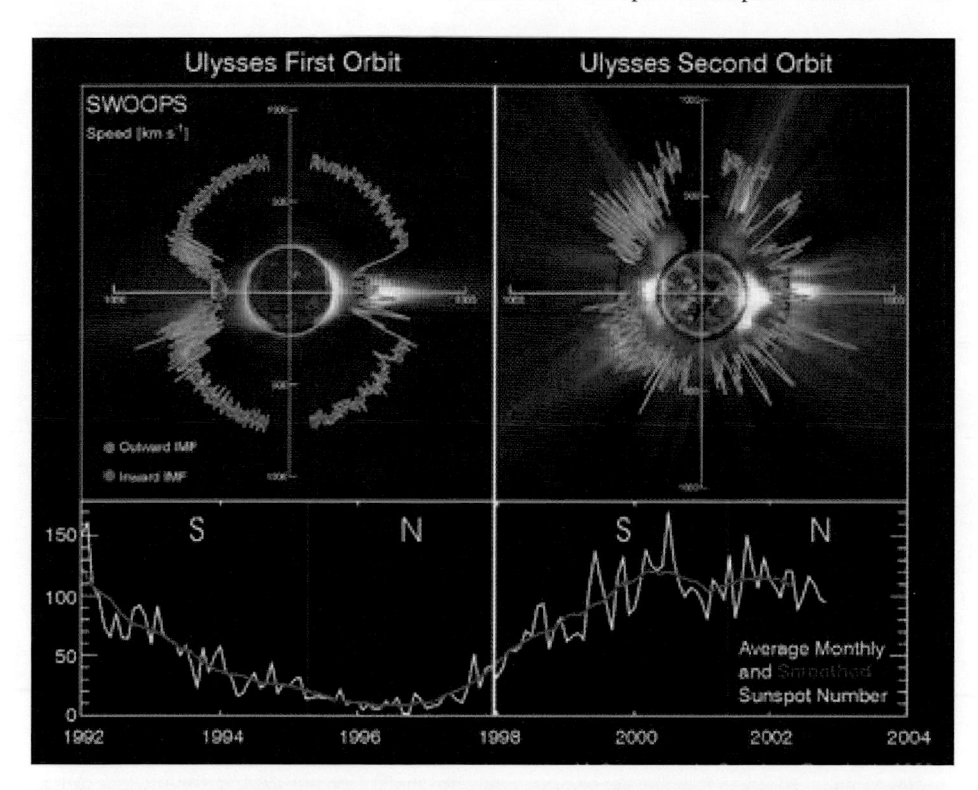

Fig. 1.8. Solar wind structure in the 3-D heliosphere near solar minimum (left diagram) and solar maximum (right diagram) as measured with the Ulysses spacecraft at distances to the Sun between 1–5 AU during the first two orbits around the Sun in 1992–2003. The solar wind speed is provided by the radial distance from the "Sun's center" and the blue and red colors denote the magnetic polarity of the IMF, with blue labeling inward directed fields (– polarity) and red labeling outward directed fields (plus polarity). Ulysses SWOOPS solar wind data are superposed on composite solar images obtained with the SOHO EIT and LASCO C2 instruments and with the Mauna Loa K-coronameter. From McComas et al. (2003).

rises. The reason for these changes are directly connected to the variability of the solar magnetic field and the processes of its reversal.

1.4 The Changing Solar Magnetic Field and Dynamic Corona

With increasing solar activity the relatively simple structure of the solar corona as shown in Fig. 1.9 changes drastically. The basic structure of the solar magnetic field and corona comprised of open fields at polar latitudes and closed fields distributed at equatorial latitudes vanishes. Figure 1.10 shows the changing structure of the Sun's EUV corona at 195 Å in different images, taken by SoHO during 1996 near solar minimum, during the increasing phase of the solar cycle in 1998 and finally in 1999 close to solar maximum.

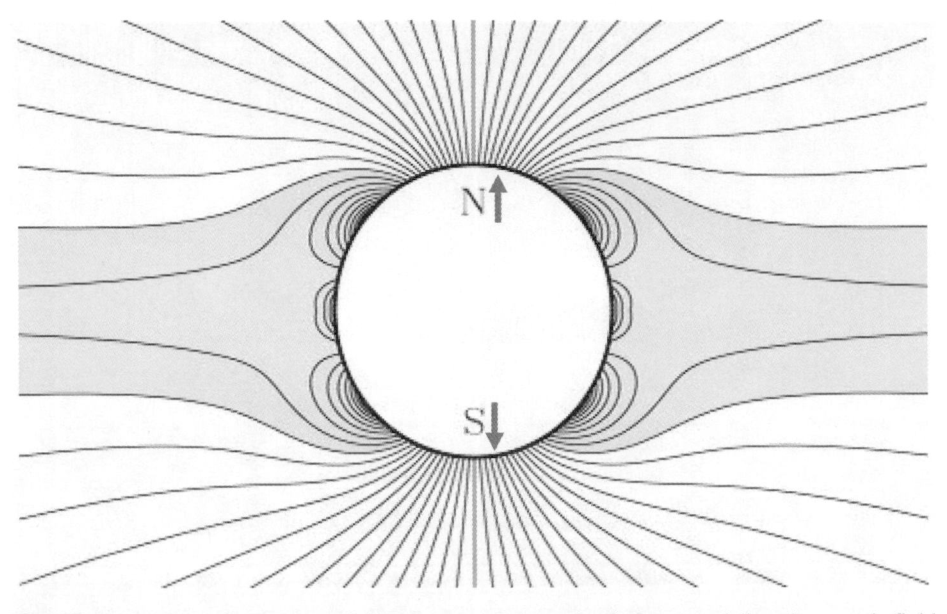

Fig. 1.9. Basic magnetic structure of the solar corona comprised of open and closed magnetic field lines. The white areas are filled with fast solar wind from the polar coronal holes, the grey shaded areas with slow solar wind released in a variable manner from closed field regions. The global magnetic polarity is indicated for odd cycles. Courtesy: M. Aschwanden (2004).

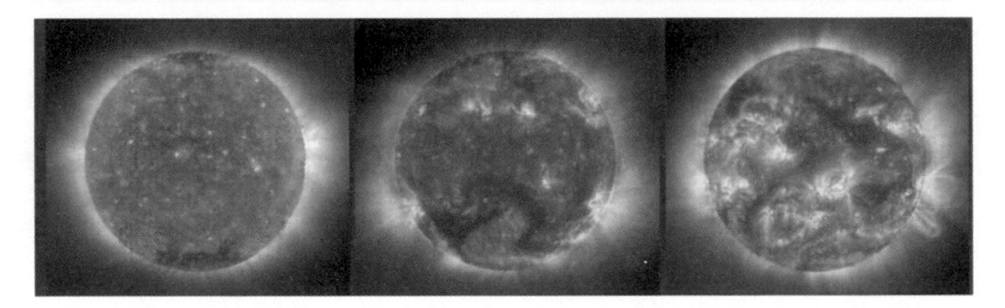

Fig. 1.10. Three EUV-images of the solar corona taken by SoHO/EIT at 195 Å in 1996, 1998 and 1999. Solar maximum was in 2000. Note the strong change in coronal structure, the increase of the number of bright shining coronal regions with stronger underlying photospheric magnetic fields and the appearances of dark coronal regions with underlying open magnetic field structure at lower latitudes. Courtesy: SOHO/EIT consortium.

The US/Japanese satellite Yohkoh (the Japanese word for sun-beam) was launched in October 1991 and has provided for almost 10 years stunning images of the Sun's X-ray variability (e.g. Uchida 1994). The X-ray variability and the photospheric magnetic field varies strongly over the course of the 11-year solar cycle as can be seen from the collection of images in Fig. 1.11. The individual X-ray images were taken about every half year from 1991 until 2001 shown together with the evolution of the longitudinal component of the photospheric magnetic field as observed by the US Kitt Peak National

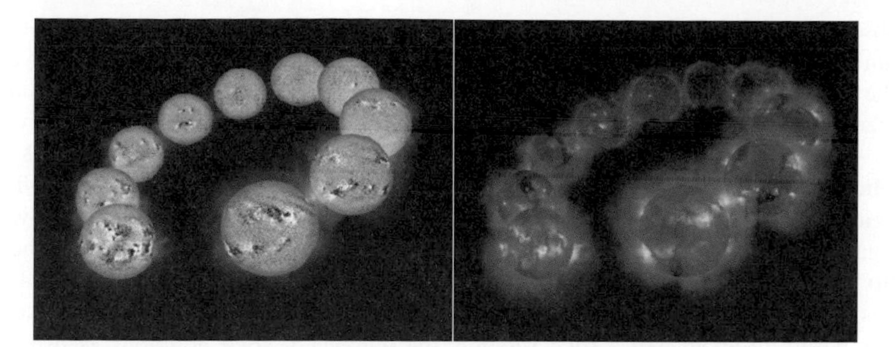

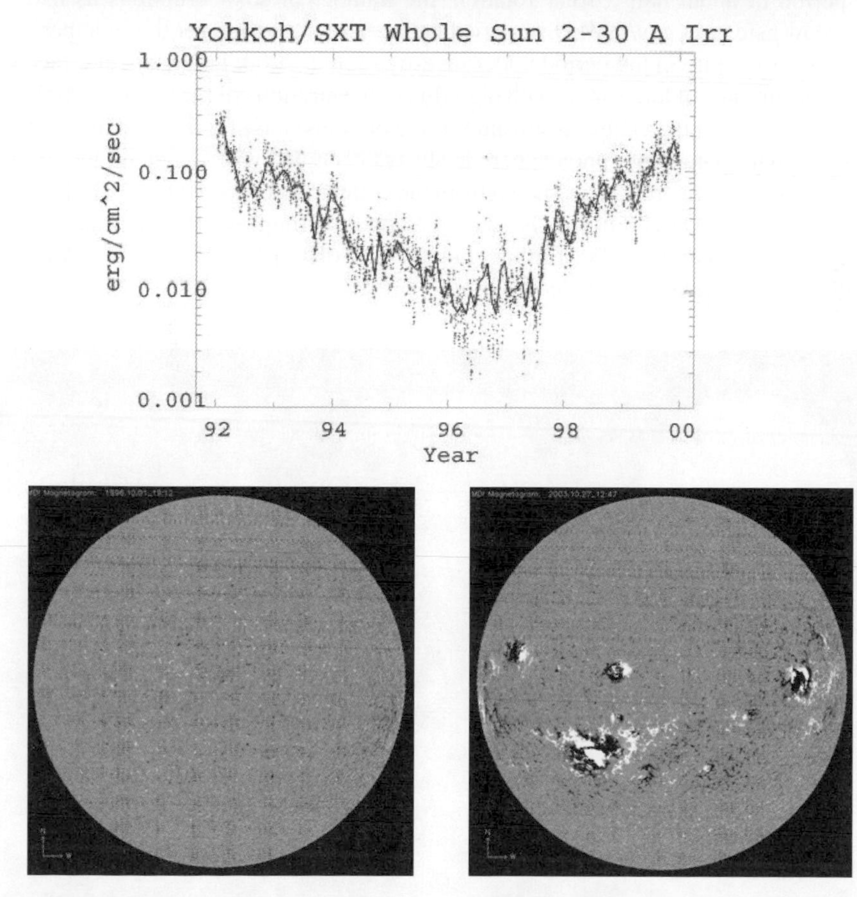

Fig. 1.11. Top left: Variation of the longitudinal component of the photospheric magnetic field over the course of solar cycle 23 as observed by KPNSO. Top right: Variability of the Sun's X-ray corona as observed by Yohkoh. Center plot: X-ray intensity-time profile as measured by Yohkoh SXT from minimum to maximum in cycle 23. The points are daily average irradiance values. The curve is a plot for average values over the period of a solar rotation. Courtesy: Loren Acton, Montana State University. Bottom: SoHO/MDI magnetograms taken in October 1996 (left) and in October 2000 (right). Dark regions correspond to negative magnetic polarities in the photosphere, white areas to positive magnetic polarities. Courtesy: KPNO, Yohkoh and SoHO/MDI consortia.

Observatory (KPNO) in Tucson, Arizona, USA. Such observations have shown that the Sun's X-ray radiation varies by about a factor of 10^3 over the course of the solar cycle. The varying solar magnetic field driven by dynamo processes acting within the Sun's interior and causing the 11 year magnetic polarity reversals is the driver of solar activity.

At the bottom of Fig. 1.11 two full disk magnetograms taken by the SOHO/MDI (Michelson Doppler imager) instrument in October 1996 and 2000 are shown. The white regions on the disk denote positive magnetic polarity and the dark ones regions with negative magnetic polarity. There is a remarkable difference obvious in the photospheric flux and hence in the number of active magnetic regions that radiate in EUV and X-rays. Seven days before and after the day the magnetograms were taken, i.e. over a time period of about half a solar rotation, the number of solar eruptions as listed at http://cdaw.gsfc.nasa.gov/CME_list/ varied between less than ten for the time period in 1996 to over seventy in the period under investigation in 2000, i.e. by about a factor of 10. The occurrence of large areas with high flux concentrations of the magnetic field, the most intense ones being visible as sunspots, is the obvious cause for the dramatic increase in solar activity. The Sun's photosphere is always occupied with magnetic fields, even when no sunspots are present, as shown from the comparison of SoHO/MDI/EIT images in Fig. 1.12 with GONG (Global Oscillation Network Group) white-light images taken from ground observatories. For further comparison SoHO/EIT 195, 284, 304 Å images and a Catania H_α-image are added.

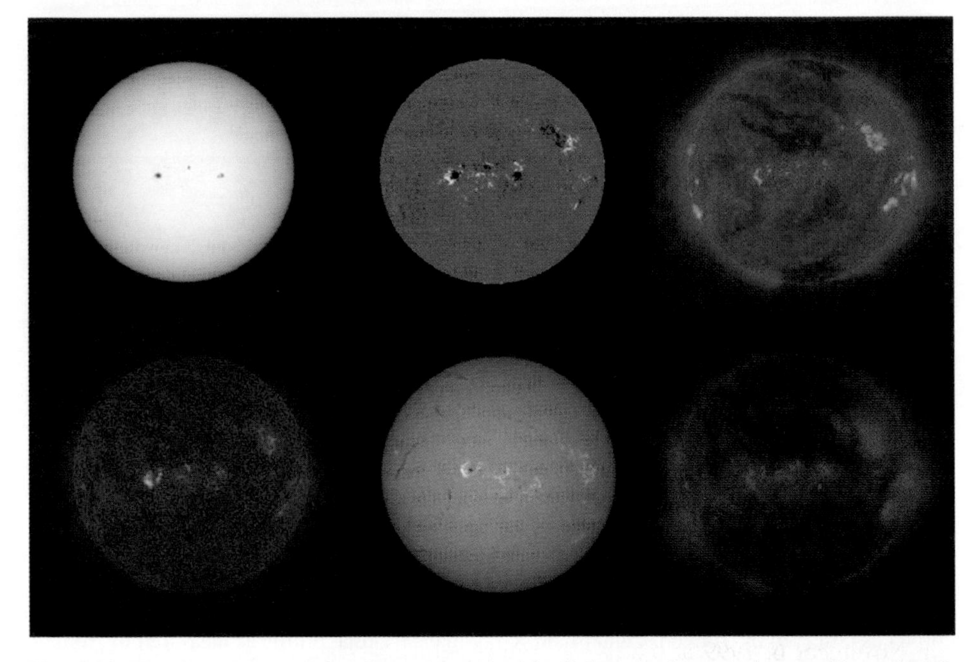

Fig. 1.12. Top from left to right: Ground-based white-light solar image taken on November 9, 2005 taken by the GONG project; SoHO/MDI magnetogram; SoHO/EIT 195 Å image. Bottom from left to right: SoHO 284 image; ground-based H-alpha image from Catania; SoHO 195 image. Courtesy: SoHO/EIT/mDI consortium and Solar Weather Browser (Copyright © 2004, 2005 The Royal Observatory of Belgium) at http://sidc.be/SWB.

Fig. 1.13. SOHO/MDI/EIT illustration of the magnetic carpet which may heat the solar corona. In this scenario the overlying field lines in the solar corona are connected to the underlying photospheric fields which "refresh" in a 40 hour time period. Courtesy: SOHO/MDI/EIT consortium.

Even at times when the photosphere does not reveal any larger magnetic flux concentrations as apparent from the absence of sunspots, permanently small-scale changes take place in the photosphere due to the rise of hot plasma cells from the Sun's interior which cool at the solar surface and subsequently sink. Magnetic bipoles (e.g., Wang & Sheeley 1989) of various spatial scales, the smaller ones with low intensities replenishing itself in time periods of about 40 hours are known to give rise to the black and white (salt and pepper) pattern in the MDI magnetograms, a solar phenomenon termed the magnetic carpet (http://umbra.nascom.nasa.gov/ssu/magnetic_carpet.html). Figure 1.13 shows how the small-scale magnetic field of the photospheric carpet connects the network on the spatial scale of supergranulation cells, while the larger scale magnetic fields extend up into the corona (Aschwanden, 2004).

Figure 1.14 taken by the TRACE (Transition Region And Coronal Explorer) mission shows fine structures of the solar corona in unprecedented spatial resolution. The image (with another color table on the right) of coronal loops over the eastern limb of the Sun was taken in the TRACE 171 Å pass band characteristic of a plasma at 1 MK temperature on November 6, 1999 at 02:30 UT. The image was rotated over +90 degrees in the clockwise direction.

On the base of the new observations Fig. 1.15 shows how the views of the physics of the solar corona have changed in time from a quasi-static simple gravitationally stratified solar atmosphere to a complex, highly time-variable system made up out of of small-scale

Fig. 1.14. TRACE observations show unprecedented fine structures of corona features. The image (with another color table on the right) of coronal loops over the eastern limb of the Sun was taken in the TRACE 171 Å pass band on November 6, 1999, at 02:30 UT. Courtesy: TRACE consortium.

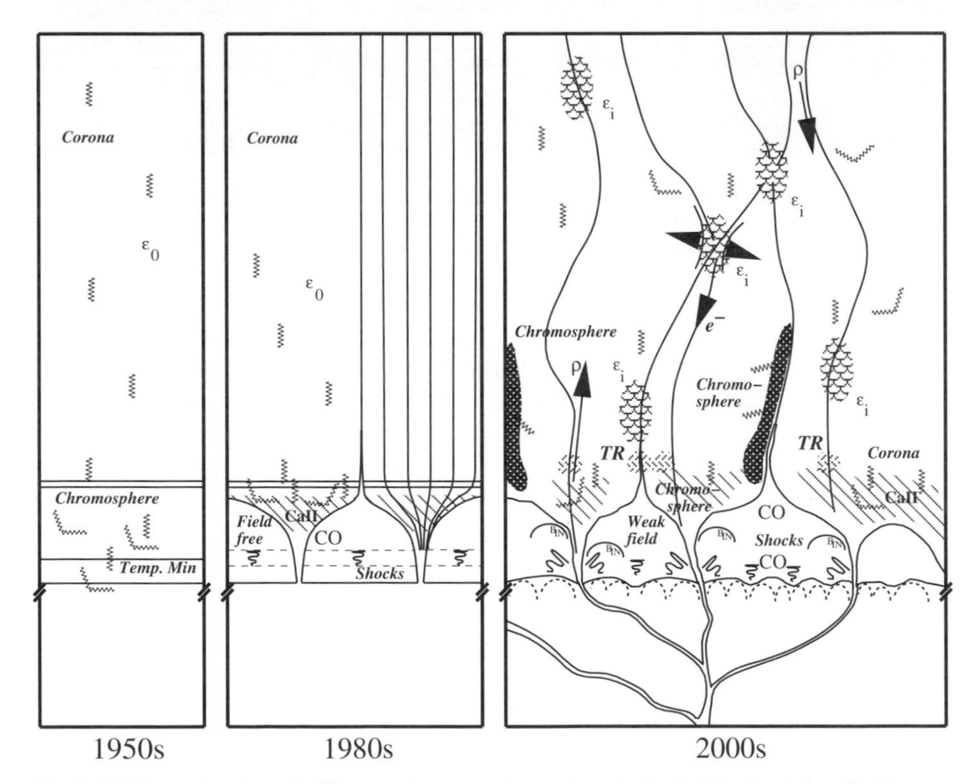

Fig. 1.15. Three sketches of different solar coronal concepts. Left: Gravitational stratified layers; middle: vertical fluxtubes with chromospheric canopies; right: fully inhomogenous mixing of photospheric, chromospheric and coronal zones by dynamic processes such as heated upflows, cooling downflows, intermittent heating, non-thermal electron beams, field line motions and reconnections, emission from hot plasma, absorption and scattering in cool plasma, acoustic waves, and shocks. Courtesy: Aschwanden (2004), Schrijver (2001).

magnetic networks (from Aschwanden 2004). However, the heating of the solar corona, i.e. the existence of a million degree hot corona above the 5800 K "cool" photosphere is still one of the unsolved mysteries in natural science.

Although, the Sun's atmospheric layers are ever changing on small scales most of the time the interplanetary medium seems practically unaffected at the distance of Earth by transient particle beams or radiation from reconnection processes associated with microflare activity so that the main role is left to the solar wind from the open regions of the sun's magnetic field, i.e. coronal hole flows.

1.5 The Explosive Corona – Coronal Mass Ejections

After the revolutionary observation of the white-light solar flare in September 1859 by Richard Carrington (see Sect. 1.1) and his conclusion that the flare might have been indicative for solar processes triggering about 17 hours later the major magnetic storm at Earth, scientists focused their research for decades in order to obtain conclusive physical relationships between these two phenomena, but without obtaining unambiguous results. Today we refer to these studies in the context of "the solar flare myth" (Gosling 1993). The primary reason for the long undiscovered true physical links in solar-terrestrial physics is primarily related to the faintness of the solar corona, being 10^6 times less bright in intensity than the visible solar disk, i.e. the photosphere. Observations of the corona remained illusive until Bernhard Lyot (1939) invented the coronagraph, the first telescope able to detect the faint corona from Earth apart from during total solar eclipses. A coronagraph detects essentially photospheric light scattered from free electrons in the outer solar atmosphere. The electrons had escaped their parent atoms when they became highly ionized in the hot corona. This polarized light is also referred to as the Thomson-scattered light of the K-corona, with K denoting the German word "kontinuierlich". The continuum corona is the prime ingredient of the white-light features visible in coronagraph images. Other major components of coronal brightness are from photospheric light scattered at dust particles, the so-called F (Fraunhofer)-corona, and from a number of emission lines of highly ionized atoms, e.g., Fe^{+12}. A detailed introduction into the physics of the solar corona has been given by Aschwanden (2004). A scientifically highly important feature of the solar corona is that the plasma-β is typically less than 1, i.e. the thermal pressure of the plasma is much smaller than its magnetic pressure and hence the ionized atoms and electrons are structured by the Sun's magnetic field.

In the early 1970s for the first time coronagraphs were developed for space missions and successfully flown on the OSO (Orbiting Solar Observatory) 7 mission and some years later on board of Skylab, subsequently with the P78-1 and Solar Maximum missions (SMM) and currently with SoHO (e.g., St. Cyr et al. 2000). The first observations of the solar corona at time cadences of several tens of minutes recorded by space borne coronagraphs yielded a big surprise: The frequent appearance of large coronal "bubbles", exceeding greatly the Sun's size at some solar radii distance, that were propagating outward into space with speeds of several hundreds of km/s in the telescope's fields of view from about 2 to 6 solar radii (Koomen et al. 1974, Hildner et al. 1976, Howard et al. 1985, Sheeley et al. 1986, Burkepile & St. Cyr 1993, Howard et al. 1997, St. Cyr et

Fig. 1.16. A coronal mass ejection (CME) at the Sun's East-limb observed by SoHO/LASCO/C3 coronagraph on August 5, 1999. The field of view is ~6–30 RS. The speed of the CME was about 700 km/s but it was accelerated to about 1000 km/s when it reached 10 solar radii (http://cdaw.gsfc.nasa.gov/CME_list/). Courtesy: SoHO/LASCO consortium.

al. 1999, Yashiro et al. 2004). These large-scale coronal transients are today commonly referred to as coronal mass ejections or CMEs.

Figure 1.16 shows a typical CME observed by SoHO on August 5, 1999. The speed of the CME was initially about 700 km/s, but in this case it was accelerated to about 1000 km/s during its outward motion up to distances of at least 10 solar radii as derived from the measured velocity increase in the field of view (~2–30 R_S) of the SoHO LASCO (Large Angle Spectrometric Coronagraph) instruments (Brueckner et al. 1995).

SoHO has recorded so far more than 10,000 CMEs (http://cdaw.gsfc.nasa.gov/CME_list/) with unprecedented resolution in space and time, allowing very detailed studies of their white-light structures, origins and kinematics (e.g., Chen et al. 2000, Cremades & Bothmer 2004). CMEs carry roughly 5×10^{12} to 5×10^{13} kg of solar matter into space (e.g., Howard et al. 1985, Vourlidas et al. 2002). Their speeds are often fairly constant over the first solar radii, with the prime acceleration taking place commonly just within the first solar radius (e.g., St. Cyr et al. 1999). However some CMEs are accelerated significantly longer as was the case in the sample event shown in Figure 1.16. The average speed of CMEs is at the order of 400 km/s, though some are substantially slower and others reach extreme high speeds of almost 3000 km/s (e.g., Yashiro et al. 2004). On average the kinetic energy of a CME is around 10^{23} to 10^{24} J (e.g., Vourlidas et al. 2002) which is comparable to the energy of large solar flares. Their angular widths are in the range of 24° to 72° (e.g., Yashiro et al. 2004). During low solar activity CMEs occur at low heliographic latitudes, but almost all around the Sun at times near solar maximum. While near solar minimum the daily average CME rate is ~1, while it is ~4 near solar maximum (Yashiro et al. 2004). Table 1.2 summarizes the basic characteristics of CMEs.

CMEs are best associated with eruptive prominences (disappearing filaments) as shown by Webb & Hundhausen 1987 and to a lesser extent with solar flares, though

Table 1.2. Basic properties of CMEs.

Basic characteristica of CMEs	
Speed	<300–3000 km/s
Mass	5×10^{12}–5×10^{13} kg
Kinetic Energy	10^{23}–10^{24} J
Angular Width	~24°–72°
Occurrence Frequency	~1–~4 (sol. min.- sol. max.)

each individual phenomenon may occur without the others (e.g., Subramanian & Dere 2001). Gopalswamy et al. (2003) found more than 70% of the SoHO/LASCO CMEs during the years 1996–2002 to be associated with prominence eruptions. Flare associated CMEs seem to be strongly connected to magnetic active regions as evident from their brightness in low coronal EUV observations and from the enhanced underlying photospheric magnetic flux. This finding seems plausible as one can easily imagine that stronger changing photospheric flux gives rise to coronal heating and flaring processes in active regions occurring at preferential lower heliographic latitudes in the course of the solar cycle, following the well known butterfly pattern of sunspots (http://science.msfc.nasa.gov/ssl/pad/solar/images/bfly.gif).

In a recent study Zhang et al. (2001) have analyzed for a number of events the temporal and physical relationship between coronal mass ejections and flares in detail. In these cases the CMEs did slowly evolve through the fields of view of the LASCO coronagraphs. According to their results the kinematic evolution of flare associated CMEs shows a three-phase development: an initiation, an impulsive acceleration and a propagation phase. In the initiation phase the CME slowly rises for a time period of several tens of minutes followed by the onset of the X-ray flare and the impulsive acceleration phase of the CME until finally the acceleration ceases and the CME starts propagating further out with constant speed as shown in the velocity-time diagram in Fig. 1.17 derived for the CME observed on June 11, 1998. Certainly future high time-cadence solar observations will shed more light on the physical details of the onset of CMEs.

CMEs develop rapidly into large-scale objects with diameters multiple the size of the Sun itself as shown in Fig. 1.18 from the study of structured CMEs performed by Cremades & Bothmer 2004. The typical three-part structure of the CMEs evident in Figs. 1.16 and 1.18, consisting of a bright leading edge, a dark void and a bright trailing core, is evident already in the SoHO/EIT 195 Å images of the low corona shown at the top of Fig. 1.18. Cremades & Bothmer (2004) pointed out, that CMEs originate from magnetic loop/flux rope systems that already existed in the low corona at heights below about 1 solar radii and that they expand in a self-similar manner into the field of view (~2–6 R_S) of the LASCO/C2 coronagraph. In the bottom right image of Fig. 1.18 the identified source region of the CME is located in a composite SoHO/EIT/MDI image. The purple and blue colors denote regions of opposite magnetic field polarity in the photosphere. Bothmer & Cremades (2004) found that bipolar regions in the photosphere are generally the underlying source regions of CMEs, independent of whether these were active regions or ones in which the magnetic flux was already decaying and that had persisted considerably longer in time. At the higher latitudes the source regions of CMEs were typically more spatially extended and associated with prominences.

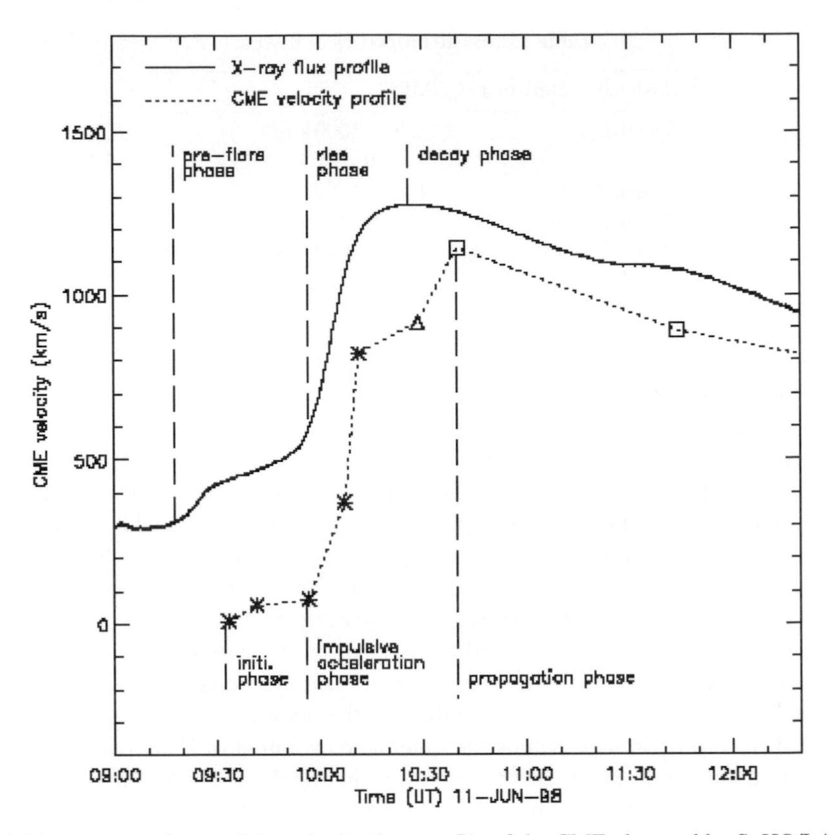

Fig. 1.17. The three phases of the velocity-time profile of the CME observed by SoHO/LASCO on June 11, 1998. Courtesy: Zhang et al. 2001.

In a systematic study of the source region properties of the underlying bipolar regions Bothmer & Cremades (2004) found that the 3-D topology of structured CMEs observed in the field of view of LASCO/C2 can be classified according to a basic scheme in which the fundamental parameters are the heliographic position and orientation of the source region's neutral line separating the opposite magnetic polarities. If one assumes that the average orientation of the neutral lines, separating bipolar regions as CME sources, follows Joy's law, the characteristic white-light shape of a CME seen in the FOV of a coronagraph can be explained naturally through the basic scheme presented in Fig. 1.19. CMEs originating from the visible solar disk are seen at the East-limb in cross-section and sideways at the West-limb. The scheme reverses for CMEs originating at the backside of the Sun as viewed from the position of the observer assumed in Fig. 1.19. Howard et al. (2005) have successfully reproduced the white-light pattern for the CMEs shown in Fig. 1.19 through a graduated cylindrical shell (GCS) model, hence supporting the findings by Cremades & Bothmer (2004) on the 3-D structure of CMEs.

The apparent profile of an individual CME may differ more or less from the basic scheme presented in Fig. 1.19 because of the solar variability of the fundamental under-

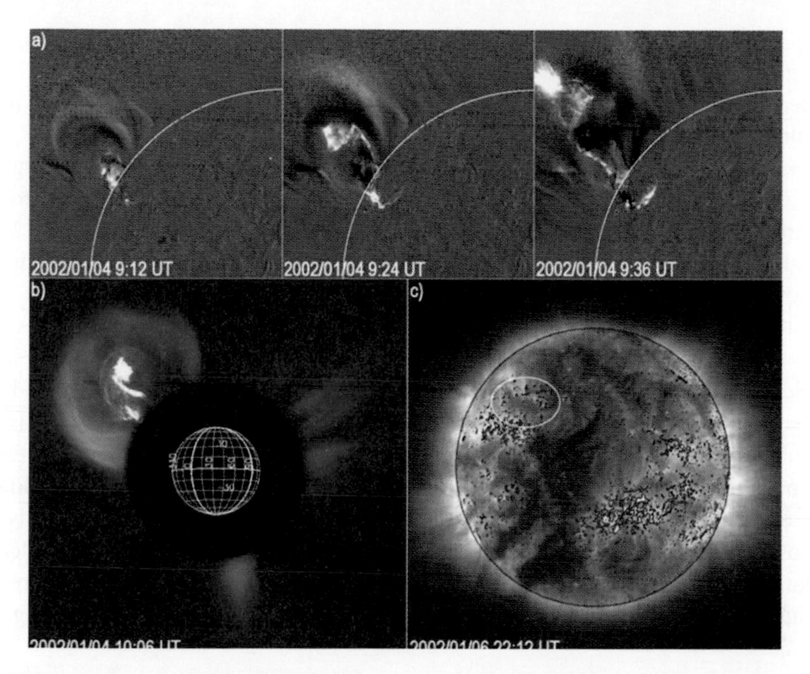

Fig. 1.18. Evolution of the East-limb CME observed by SoHO/EIT/LASCO on January 4, 2002. Top: Running intensity difference images showing the prominence eruption in the low corona at 195 Å. Bottom left: The CME as observed by LASCO/C2. Bottom right: Contours of a SoHO/MDI magnetogram superposed onto an EIT 195 Å image. The circle denotes the identified CME source region in the low corona and photosphere. From Cremades and Bothmer (2004).

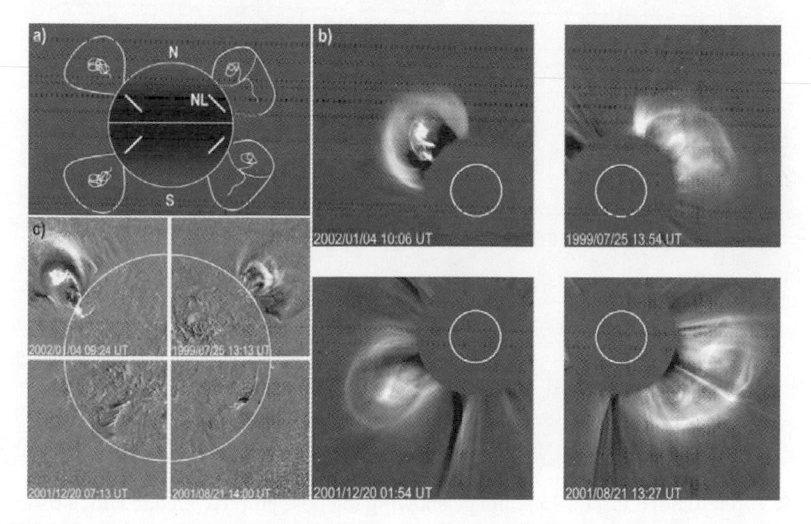

Fig. 1.19. a) Idealized scheme showing the extreme cases of CME projection for front-side events. NL denotes the photospheric neutral line. b) Four projected CMEs observed by SoHO/LASCO/C2 matching the scheme. c) Source region 195 Å signatures for the four CMEs. For the northern events eruptive features are selected while for the southern ones post-eruptive features are shown. From Cremades and Bothmer (2004).

lying parameters, e.g., many neutral lines are not straight but have rather complicated topologies, especially in active regions. The degree of correspondence with the scheme also depends on the absolute values of the source region lengths which will impose difficulties for small values typically found in compact active regions.

Contrary to the white-light structure of the CMEs shown in the scheme in Fig. 1.19, events originating from near the center of the solar disk appear as unstructured halos (Howard et al. 1982) as shown in the bottom sequence of images in Fig. 1.20. Middle and right images at the top show the disappearance of a filament in H_α, the left image shows a post-eruptive EUV arcade that developed after the CME's onset in its low coronal source region (Tripathi, Bothmer & Cremades 2004). Multi-wavelengths observations of the CME's source region are shown in Fig. 1.21 based on EUV 195 Å images from SoHO/EIT, soft X-ray observations from Yohkoh and H_α-images from the French Observatory at Paris/Meudon (http://bass2000.obspm.fr/home.php). The view is complemented by SoHO/MDI magnetograms. From Fig. 1.20 it is obvious that CMEs stem from localized spatial source regions on the Sun. In this case the front-side halo CME did several days later pass Earth's orbit as evident from the WIND and ACE (Advanced Composition Explorer) satellite data (Bothmer, 2003, Yurchyshin 2001). The calculated orientation of the CME was found to lie almost normal to the ecliptic plane in agreement with the expected orientation of the filament in the CME's source region (see Fig. 1.20)

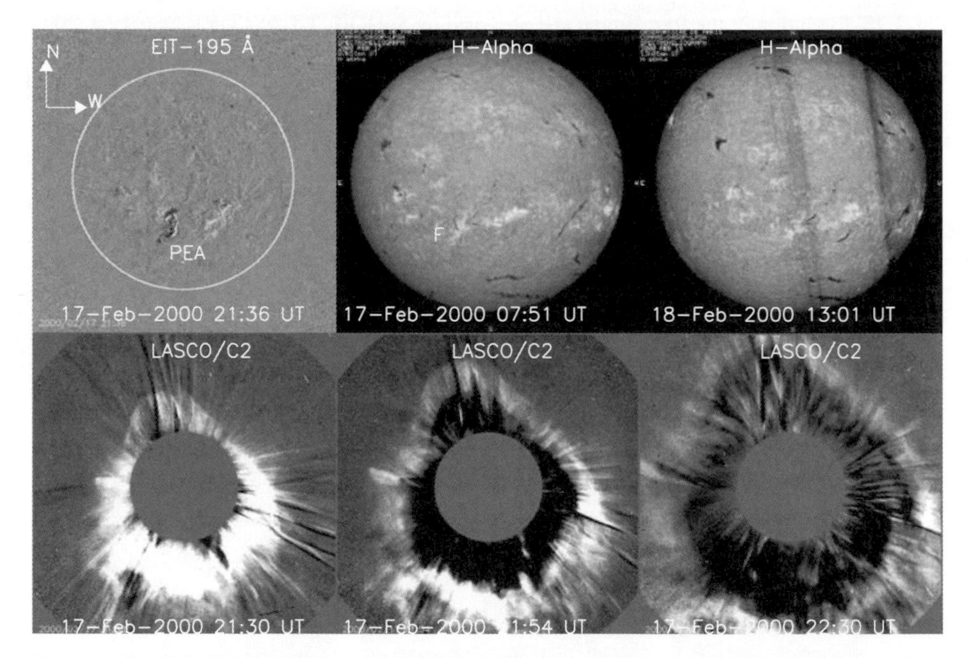

Fig. 1.20. Top left: SoHO/EIT 195 A image showing the post-eruptive arcade which formed after the front-side halo CME observed by LASCO/C2 on February 17, 2000. Middle and right images: H_α-images from the Paris/Meudon Observatory showing the disappearance of the associated filament. Bottom images: SoHO/LASCO/C2 images showing the near-Sun development of the halo CME. The speed of the CME was about 600 km/s. Note the asymmetry of the halo in the NE to SW direction. From Tripathi, Bothmer & Cremades 2004.

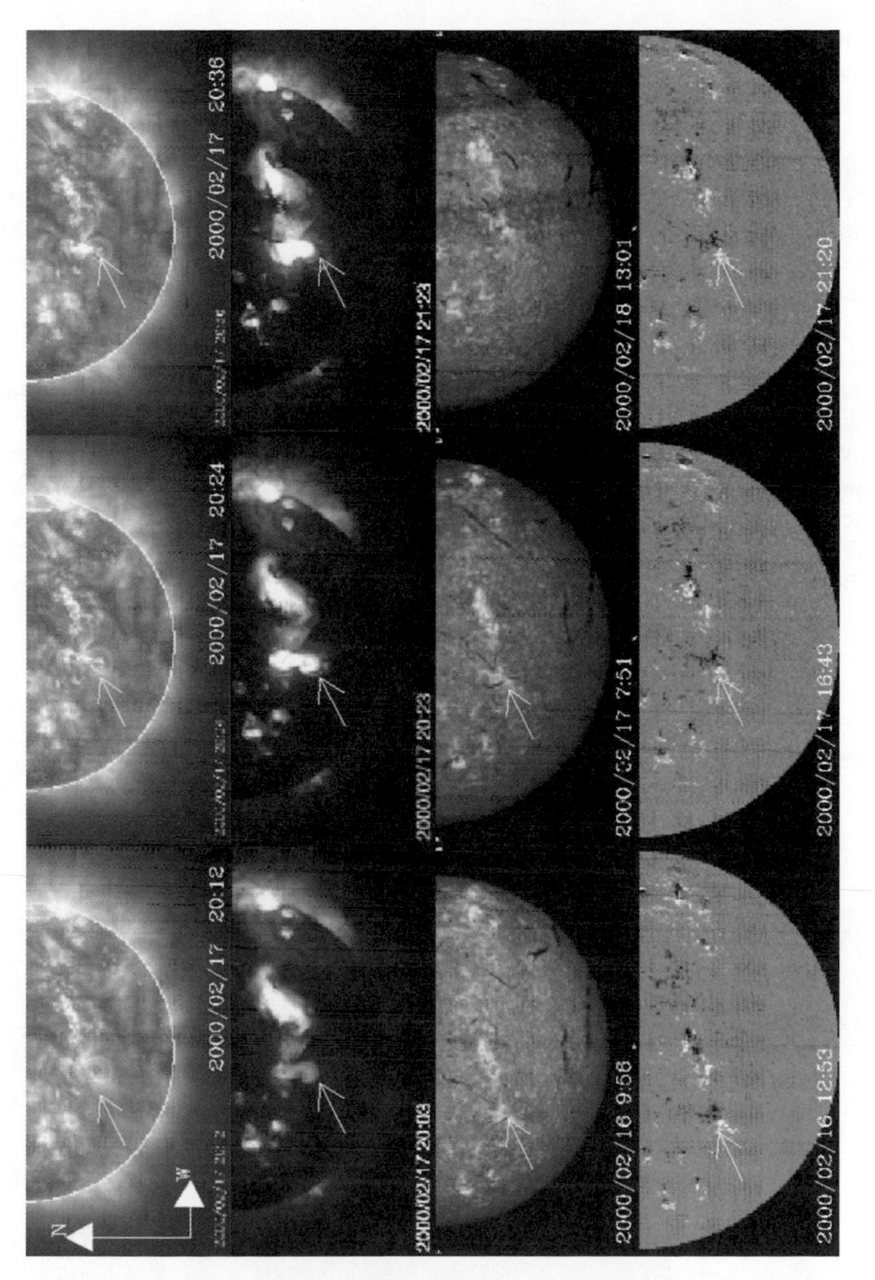

Fig. 1.21. Multi-wavelengths observations showing the source region of the halo CME observed by SoHO/LASCO on February 17, 2000. Right images: SoHO/EIT 195 A images showing the development of a post-eruptive arcade. Second panel from right: Yohkoh soft X-ray images showing a bright sigmoidal structure in the CME's source region. Third panel from right: H_α-images showing the disappearance of the associated filament. Left panel: SoHO/MDI observations of the magnetic field structure in the underlying photspheric source region of the CME. From Tripathi, Bothmer & Cremades 2004.

and the magnetic field configuration was consistent with the expected one from the MDI observations (Bothmer 2003).

The EUV signatures CMEs leave after their onset in the low corona on the solar disk can be used to discriminate whether they are front-sided or back-sided events (e.g., Triapthi, Bothmer & Cremades, 2004, Zhukov & Auchere 2004). These features include "EIT waves" and "dimmings" as shown in Fig. 1.22. The coronal waves seen by EIT typically propagate with speeds of several hundreds of km/s but are not seen for all CMEs (Thompson 2001; Wang 2000; Klassen et al. 2000). According to Tripathi, Bothmer & Creamades EUV post-eruptive arcades seen for a couple of hours after the onset of a CME are a definitive CME proxy even without the availability of simultaneous coronagraph observations. Intensity brightening in soft X-rays near the onset time of CMEs, as seen in the Yohkoh observations shown in Fig. 1.21, are other good CME proxies (Canfield, Hudson & McKenzie (1999).

While solar flares emit EM-radiation, CMEs are responsible for the convection of magnetized solar plasma into interplanetary space, with the fastest (>1000 km/s) CMEs typically causing the most intense interplanetary disturbances and in case of the presence of a southward Bz at Earth's orbit also the strongest magnetic storms (e.g., Bothmer 2004; Bothmer & Schwenn 1995; Gosling 1993; Tsurutani 2001). Depending on the speed of

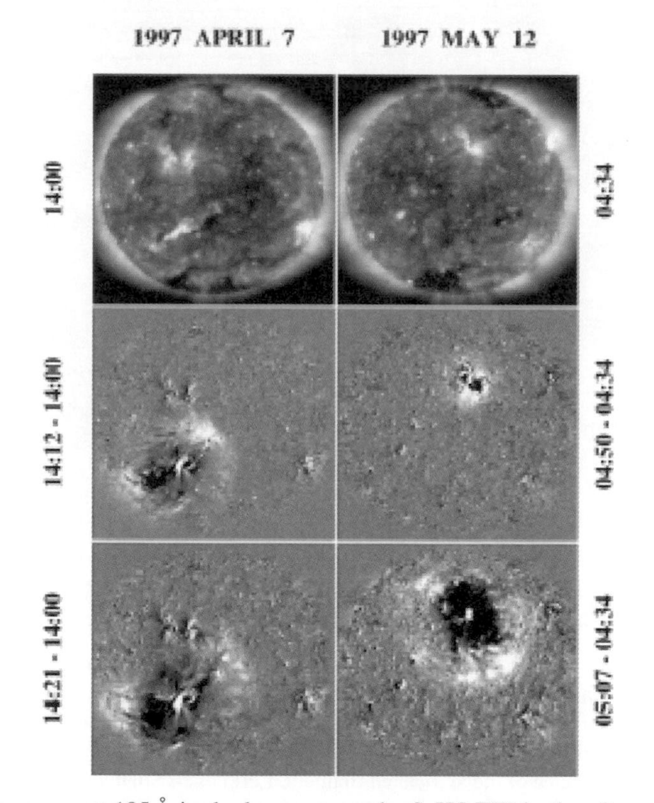

Fig. 1.22. Waves seen at 195 Å in the lower corona by SoHO/EIT in the aftermath of the two CMEs on April 7 and May 12, 1997. From Wang 2000.

the CME, the delay of the geomagnetic storm with respect to the solar event is in the time range from less than a day to several days (e.g., Brueckner et al. 1998).

1.6 Space Storms and Space Weather Effects

The first in-situ measurements of the solar wind already showed that apart from slow and fast solar wind streams, the interplanetary medium is frequently disrupted by transient flows often associated with interplanetary shock waves, discernable as strong discontinuities in which all plasma parameters (velocity, density, temperature) and the magnetic field intensity abruptly increase by some orders of magnitude (e.g., Gosling et al. 1968). Figure 1.23 shows an interplanetary shock wave detected by Helios 1 on May 13, 1981 (Sheeley et al., 1985). The plasma speed abruptly increases from 600 to over 1200 km/s. It was commonly assumed that solar flares are the causes of such interplanetary shocks (e.g., Chao & Lepping, 1974).

The simultaneous operations of the Solwind coronagraph onboard the P-78/1 satellite and the German/US sun-orbiting spacecraft Helios 1, which explored the in-situ characteristics of the inner heliosphere over the range 0.3–1 AU in the ecliptic-plane together with its sister spacecraft Helios 2, allowed for the first time during the years 1979–1982 a direct study of the interplanetary effects of CMEs (Sheeley et al. 1985, Bothmer & Schwenn 1996). Sheeley et al. (1985) found that 72% of the interplanetary shock waves detected by Helios 1 were associated with large, low-latitude mass ejections on the nearby limb, with most of the associated CMEs having had speeds in excess

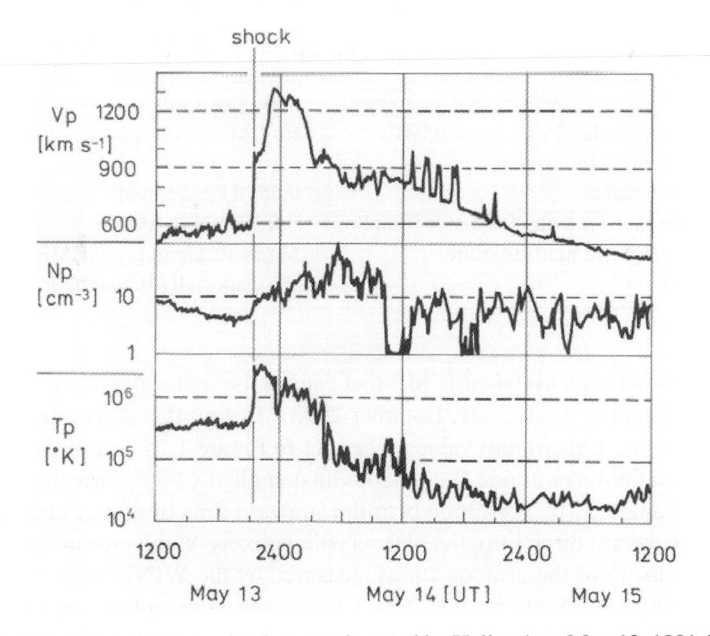

Fig. 1.23. A strong interplanetary shock wave observed by Helios 1 on May 13, 1981. The plasma parameters plotted from top to bottom are: Proton speed, density and temperature. From Sheeley et al. 1985.

of 500 km/s, some even having had speeds in excess of 1000 km/s. These observations clarified that CMEs are the sources of interplanetary shock waves and not solar flares. Today it is well known that flares and CMEs can occur without each other, although the most intense events commonly occur jointly, and the concrete physical relationships between the two physical phenomena is a subject of ongoing research (e.g., Harrison 1986, Zhang et al. 2001).

Often an interplanetary shock wave was found to be followed several hours later by a transient solar wind stream with unusual plasma and magnetic field signatures, likely being the driver of the shock wave (Burlaga et al. 1981, Bothmer & Schwenn 1996). Systematic analyses of the wealth of satellite data obtained since the beginning of the space age has made it possible to establish reliable identification criteria of transient magnetized plasma flows as interplanetary consequences of CMEs (e.g., Gosling 1990). To distinguish these CMEs in the solar wind from their solar counterparts they are termed interplanetary coronal mas ejections or ICMEs (e.g., Cane & Richardson 2003).

The classic plasma, magnetic field and suprathermal particle signatures of ICMEs at 1 AU are helium abundance enhancements, unusual ion and electron temperatures an ionization states (e.g., He+, Fe16+), higher than average magnetic field strengths (>5 nT), low variance of the magnetic field, smooth rotations of the magnetic field direction over time periods of several hours, bi-directional suprathermal (>40 eV) electron streaming, bi-directional suprathermal ion flows and plasma composition anomalies (e.g., Gosling 1993 and references therein, Henke et al. 1998, 2001). Table 1.3 summarizes basic characteristics of ICMEs at 1 AU.

Table 1.3. Characteristics of interplanetary coronal mass ejections (ICMEs) at 1 AU.

Characteristics of ICMEs at 1 AU	
Speed	300–3000 km/s
Interplanetary shocks ahead of ICMEs	for CMEs with speeds >400 km/s
Magnetic Field Intensities	<10–>100 nT
Radial extension	0.25 AU ≈ 24 hours
Radial expansion with distance R from Sun	$\sim R^{+0.8}$ (R in [AU])
Helical magnetic field structure	Magnetic cloud type ICMEs (1/3)
Plasma-β	<1 (especially in magnetic clouds)

Figure 1.24 shows a classical ICME that caused the major magnetic storm on July 15/16, 2000 (Lepping et al. 2001, Bothmer 2003). During this storm the Ap and Kp-indices reached their maximum values. The plot in Figure 1.24 shows the time-profile of the magnetic field magnitude B, the latitudinal angle θ (+90° correspond to North), the azimuthal angle φ (0° corresponds to the sunward direction, measured positive in the counterclockward direction), the plasma bulk velocity V, the proton density NP and the thermal velocity of the protons vth as measured by the WIND spacecraft (courtesy D. Berdychevsky, from Lepping et al. 2001). Data gaps were replaced by measurements from the GEOTAIL satellite (http://www-spof.gsfc.nasa.gov/istp/geotail/geotail. html).

On July 15, the plasma parameters abruptly rise and the plasma speed increases to about 1100 km/s. This corresponds to the arrival of the interplanetary shock wave at

Table 1.4. Interplanetary causes of geomagnetic storms during 1997-2001 and average Ap-levels. The acronyms are: CIR (co-rotating interaction region); ICME (interplanetary coronal mass ejection); MICMEs (Multiple interplanetary coronal mass ejections). From Bothmer 2004.

Cause of Storm	Number of Ap > 20 days	Number of events	Typical Ap-Range
Slow wind	8	8	<30
CIR/CH	90	55	<60
combined ICME/CIR	18	11	<150
ICME	101	81	<170
MICMEs	38	30	<200
Total Number	255	185	

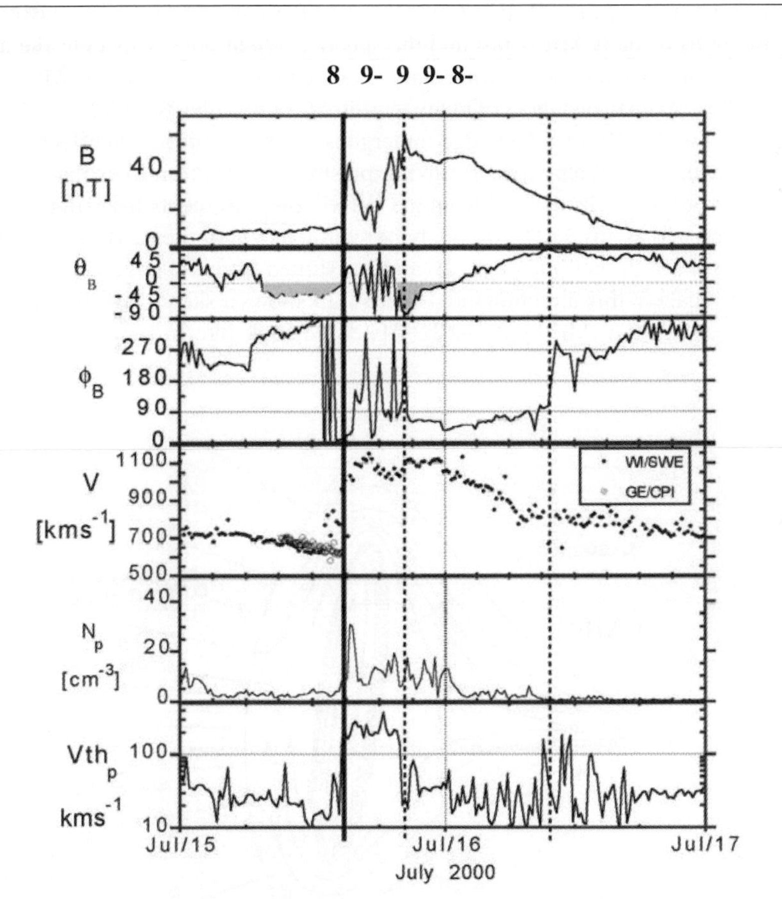

Fig. 1.24. A magnetic cloud type ICME observed by the WIND spaceraft on July 15/16, 2000. Data gaps are filled with GEOTAIL data. The arrival time of the interplanetary shock is marked by the solid line, the boundaries of the ICME are labeled with dashed lines. Parameters displayed from top to botoom are: Magnetic field magnitude B and the polar (θ) and azimuthal angle (φ), bulk velocity of the plasma, proton density and proton thermal speed. Courtesy: D. Berdychevsky, from Lepping et al. 2001.

14:35 UT, labelled with a solid thick line. Several hours later on the same day at around 19 UT it was followed by the ICME as identified from the unusual high magnetic field strength and smooth southward to northward rotation of the magnetic field direction with respect to the ecliptic plane. The rotation started on July 15 at about 19 UT and lasted until July 16, 09 UT. At the top of Fig. 1.24, Kp-values greater than 7+ are listed in the corresponding 3-hour intervals. The time-period of the strongest magnetic disturbances measured at Earth corresponds with the time interval of the occurrences of the strong southward magnetic field components caused by the ICME, even without taking details of the WIND orbit into account. The magnetic storm starts with the arrival of the shock wave triggering turbulent IMF fluctuations, followed by its main phase at times of the southward directed magnetic field at the leading edge of the ICME. The magnetic field strength at the nose of an ICME is typically considerably enhanced (e.g., Bothmer & Schwenn 1998) if the ICME is fast and the speed gradient with respect to the ambient solar wind is large so that the interplanetary shock can be driven by the ICME far out in the heliosphere, even to distances of many tens of AU or its outer boundaries (Richardson et al. 2005). The ICME's leading edge undergoes strong compression effects due to its interaction with the ambient slower moving plasma ahead. From the solar wind speed and duration of the ICME derived from the WIND measurements the radial size of the ICME is estimated as ~0.3 AU. Although the spatial extent of the ICME is not known in the direction out of the ecliptic plane, it can be assumed from the large radial size, that it is very large-scale in this direction too. Figure 1.25 shows a sketch how the propagation and the effects on the IMF by a fast ICME could look like in the inner heliosphere

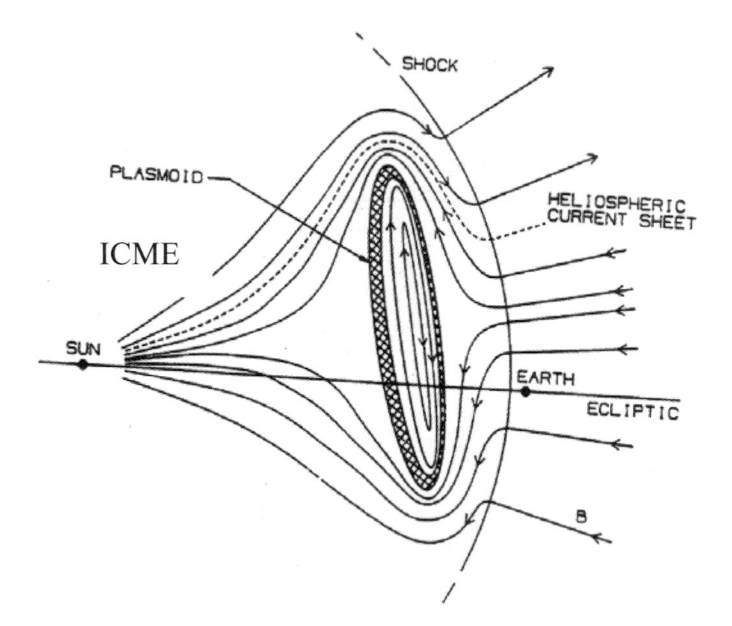

Fig. 1.25. view of a fast ICME propagating through the inner heliosphere viewed in a meridional cross-section. The deflections of the IMF and HCS between the driven shock and the ICME are sketched. From Gosling 1990.

as viewed from a meridional perspective (Gosling, 1990). The deflections of the IMF in the region between the shock wave and the leading edge of the ICME, in which a high turbulent plasma regions forms, is termed "draping of the IMF in the ICME sheath (McComas et al. 1988).

ICMEs that exhibit large-scale helical internal magnetic field configurations as the one shown in Fig. 1.24 are termed magnetic clouds (Klein & Burlaga 1982). In magnetic cloud type ICMEs the plasma-β is typically much smaller than 1 independent of the distance to the Sun in the inner heliosphere (Bothmer & Schwenn 1998). According to Bothmer & Schwenn (1994, 1998) and Bothmer & Rust (1997) magnetic cloud type ICMEs can possess four different orientations of the magnetic field as characterized by systematic rotations with respect to the ecliptic plane: SEN, SWN, NES, NWS, with N,S,E,W denoting subsequently the direction in which the magnetic field is directed at the leading edge of the ICME, its center and its trailing edge. These orientations characterize the helicity or chirality of a large-scale cylindrical magnetic flux tube in agreement with a self-consistent MHD-model as shown in Fig. 1.26 (Goldstein 1983; Bothmer & Rust 1997; Bothmer & Schwenn 1998). If one considers that the fingers of the observer's hand point towards the direction of the magnetic field in the leading edge of the ICME and the thumb would point along the magnetic field direction at the center of the ICME (the axis), and finally one takes into account that the direction at the trailing edge is opposite to the direction at its leading edge, one can classify SEN and NWS type ICMEs as left-handed helical structures and SWN and NES type ICMEs as right-handed helical structures (Bothmer & Rust 1997). According to this classification, the ICME on July 15/16, 2000 is of type SEN, i.e. it possesses left-handed magnetic chirality.

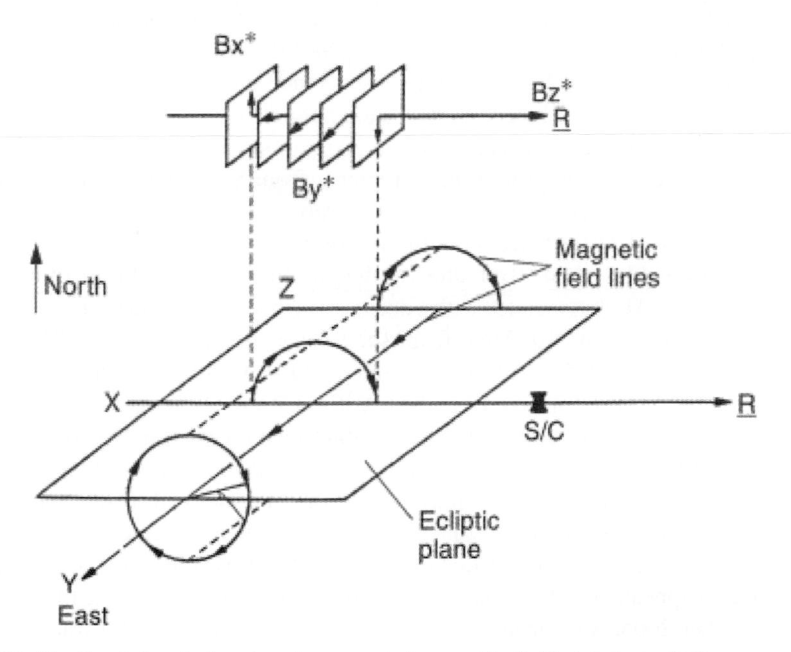

Fig. 1.26. Idealized sketch showing the expected magnetic field signatures during passage of a large-scale cylindrical magnetic flux tube over a spacecraft. From Bothmer & Schwenn 1998.

Magnetic cloud type ICMEs are of specific importance in the context of geomagnetic storms because their internal magnetic field structure can lead to large southward components of the IMF at 1 AU (e.g., Zhang and Burlaga 1988). From a study of magnetic clouds in solar cycle 23 observed by the WIND and ACE satellites, Bothmer (2003) concluded that magnetic clouds trigger geomagnetic storms basically in two different ways or in combinations of these two ways: 1. through their specific internal magnetic field configuration – not only by SN or NS rotations, but also through the cloud's axis orientation when it is highly inclined and possessing a southward field direction (see also Mulligan, Russell & Luhmann 1998), and 2. through draping of the ambient IMF, especially in case of fast ICMEs driving shock waves ahead. The energy transfer from the solar wind into the magnetosphere is most sufficient if ICMEs are associated with long-lasting (several hours) strong (<-10 nT) components of the magnetic field at 1 AU (e.g., Gonzalez and Tsurutani 1987; Tsurutani 2001; Bothmer & Schwenn 1995) and activity is further amplified by ICME with high speeds (especially those >1000 km/s). It should be noted that contrary, in case of northward (+Bz)-values associated with passage of an ICME, a decrease in geomagnetic activity occurs (Veselovsky et al. 2005).

The potential of ICMEs, i.e. CMEs at the Sun, to cause high V and low $-$Bz-values at Earth's orbit is the prime reason why they are the drivers of all major geomagnetic storms with Kp $>$ 7+ (Bothmer & Schwenn 1995). Bothmer (2004) analyzed in the framework of a European Union project (INTAS 99-727) the causes of all geomagnetic storms with disturbed days with Ap-values $>$ 20 nT during the years 1997–2001 based on an unprecedented coverage of solar wind measurements provided by near Earth satellites. The results of this investigation are summarized in Table 1.3 which confirms that ICMEs are the prime drivers of major storms as proposed by Bothmer & Schwenn (1996) in an earlier study of all storms with Kp-values greater than 8-, and also that super-intense storms are often triggered by multiple interacting ICMEs.

Fast ICMEs driving shock waves from close to the Sun out into the heliosphere are also capable of accelerating charged particles to energies of MeV or GeV (e.g., Reames 1999). Figure 1.27 shows a mosaic of solar observations for the July 14, 2000 CME including MeV electron and proton measurements from SoHO/COSTEP. COSTEP is the COmprehensive SupraThermal and Energetic Particle analyzer (e.g., Bothmer 1997). The time delay at 1 AU to the CME onset time at the Sun is typically less than 15 minutes for the electrons and less than 30 minutes for the protons. The arrival time of the ICME (CME passage) at 1 AU is associated with a decrease in the electron and proton intensities. Protons of energies around 100 MeV from such solar energetic particle (SEP) events cause "particle snowstorms" in the images from the SoHO optical telescopes when they pass through the CCDs, as can be seen in the LASCO C2 and C3 images in Fig. 1.26. The effects of the MeV protons on the efficiency of the SoHO solar panels is shown in Fig. 1.27. Intense SEP events represent a serious threat for manned space flights to the Moon and Mars and during extravehicular activities on the International Space Station (e.g., Reitz 1997, NASA "Foundations of solar particle event risk management" 1996). One of the most recent space weather effect due to a series of fast CMEs/ICMEs associated with the unexpected appearance of a strong active region in the declining phase of solar cycle 23 in September 2005, was the about one week malfunction of the star trackers of the Rosetta spacecraft (http://sci.esa.int/sciencee/www/object/index.cfm?fobjectid=37921) which was on its orbit to comet 67P/Churyumov-Gerasimenko.

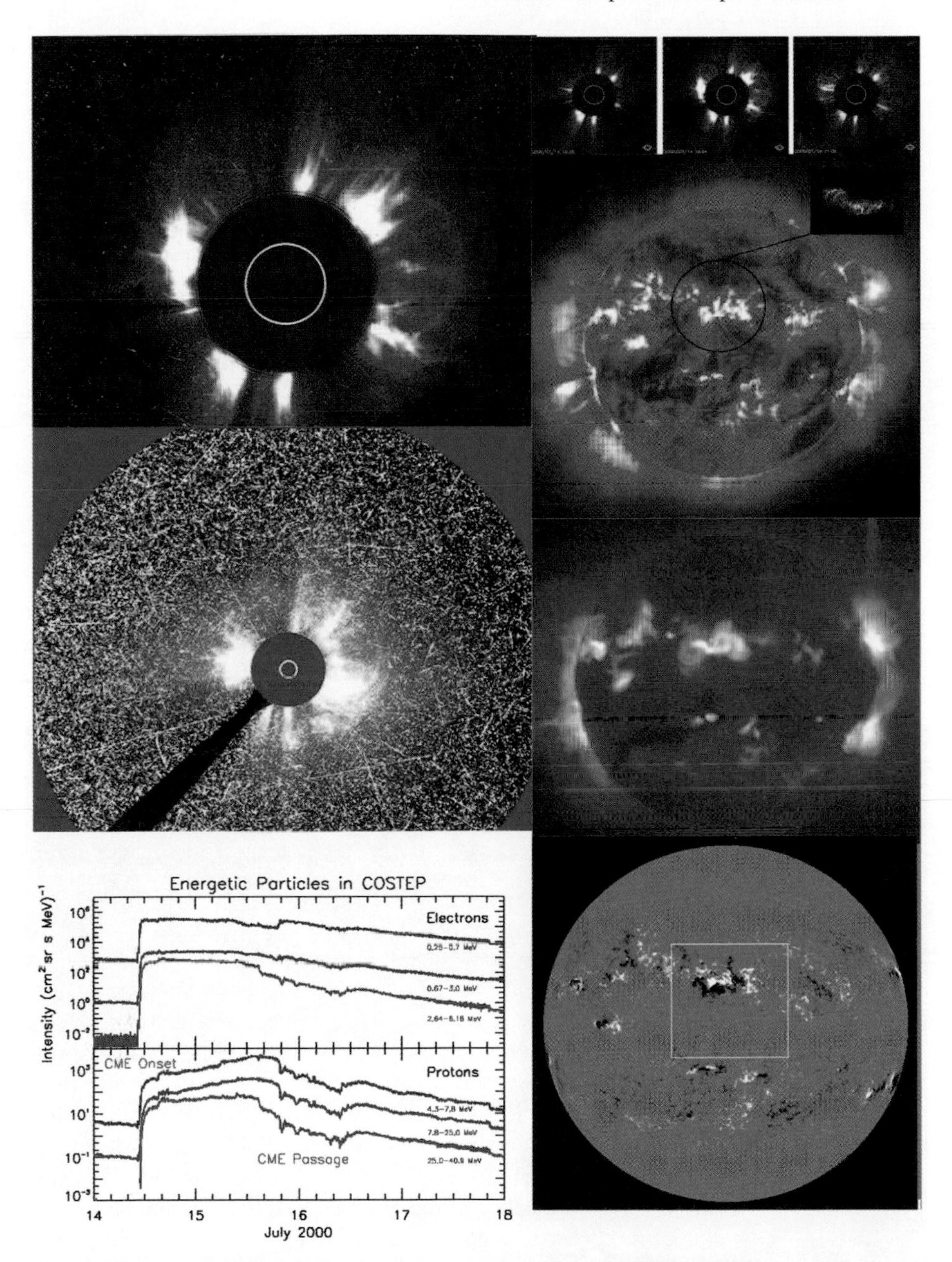

Fig. 1.27. Mosaic of solar observations showing the CME on July 14, 2000 observed by the LASCO C2 and C3 coronagraphs (left, top and middle images and top row of right image sequnece). The large image to the top right shows a SoHO/EIT 195 Å image of the CME's low coronal source region and an inset of a TRACE image at the same wave-length showing the associated posterupticve arcade. Middle and bottom images show the Sun in soft X-rays as observed by Yohkoh and the structure of the photospheric magnetic field as observed by SoHO/MDI. The left lower plot shows MeV electrons and protons measured by SoHO/COSTEP from the SEP event caused by the CME. Courtesy: SoHO, Yohkoh, TRACE consortia.

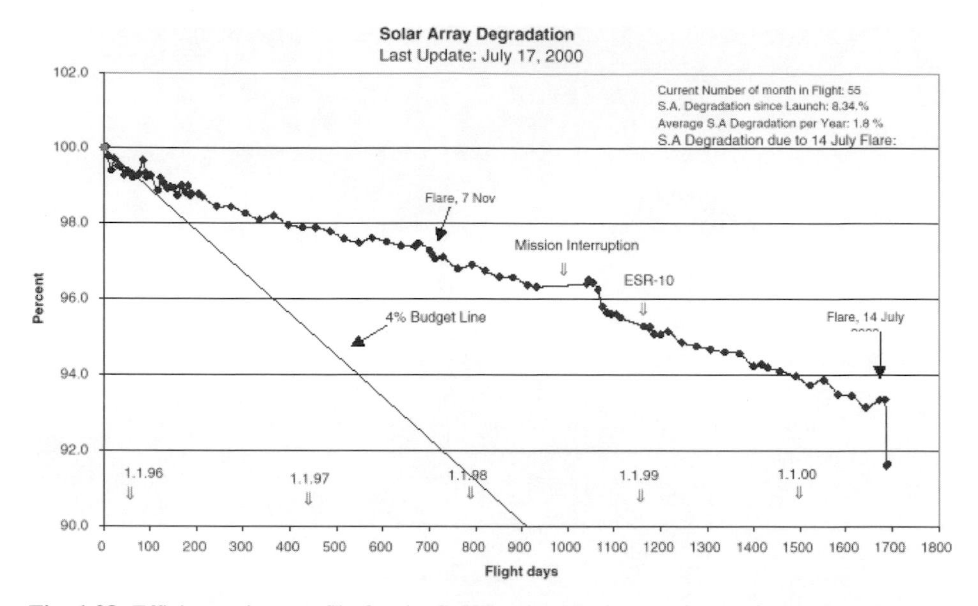

Fig. 1.28. Efficiency-time profile for the SoHO solar panels from launch in 1995 until end of 2000. The July 14, 2000 solar particle event caused a decrease of the efficiency at the order of a few percent. Courtesy: ESA/NASA SoHO project.

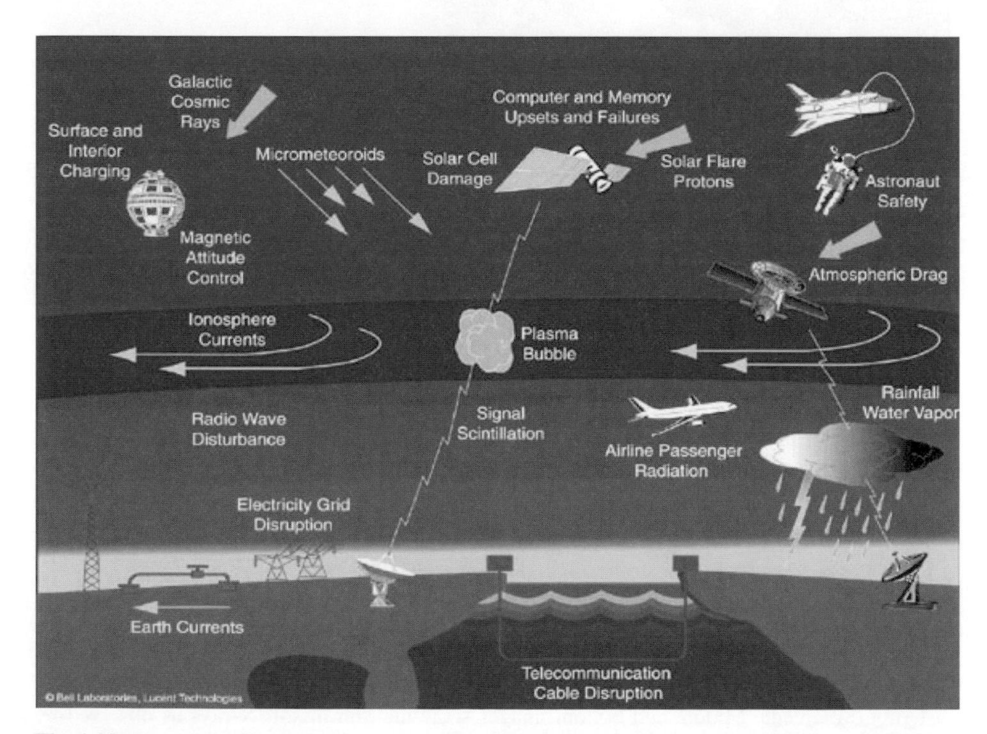

Fig. 1.29. Summary of known effects caused by enhanced solar activity (flares, CMEs and ICMEs). Adapted from Lanzerotti 1997.

CMEs/ICMES as prime drivers of space storms further cause directly and indirectly a series of serious space weather hazards (Figure 1.29): Surface charging on geo-synchronous telecommunication satellites, spacecraft communication problems and even losses, black-out of electric power systems like in the case of the Hydro-Quebec (Canada) power grid in March 1989, corrosion of oil pipelines and even affect our atmosphere and climate.

1.7 The Future – Forecasting Space Weather

The new scientific discoveries obtained since the advent of the space age have provided us with a new view on the physics of the Sun and its dynamic atmosphere. Besides the solar EM-radiation, the Sun permanently emits a magnetized plasma, either in a quasi steady form of slow and fast solar wind streams or, occasionally in an explosive manner in form of coronal mass ejections (CMEs) that evolve in the interplanetary medium as ICMEs (interplanetary coronal mass ejections). To know at any given moment of time the energy state of the Earth's magnetosphere and the electric currenmts flowing in it, i.e. to provide reliable forecasts of the magnetosphere's energy state on time-scales of several days in advance, continous monitoring of the evolution of the solar photospheric magnetic flux and the global structure of the corona seems a crucial task for which additionally a deep interdisciplinary understanding of the physical processes at the Sun, in the interplanetary medium and of the Earth's magnetosphere and atmosphere is required. It cannot be overemphasized in this context that besides solar physics, the interplanetary evolution of the solar wind streams and of CMEs/ICMEs is of extreme importance to help establish reliable space weather predictions and that, also the varying structure of the Earth's magnetic field has to be taken into account, i.e. all parts of the combined system.

The first approximation to help model the solar wind stream structure in the inner heliosphere may be provided through the so-called potential field source surface (PFSS) model in which the coronal magnetic field is assumed to be current free (rot $\boldsymbol{B} = 0$) (e.g., Hoeksema 1984, Wang & Sheeley 1992, Schrijver & de Rosa 2003). The data input for these models are ground- and space-based magnetograms. New magnetohydrodynamic (MHD) codes are also able to take to some extent contributions from active regions into account. The structure of the global solar corona has been calculated, e.g., based on the Magnetohydrodynamics Around a Sphere (MAS) model developed by the SAIC (Science International Corporation) group for the range 1–30 $\mathrm{R_S}$ based on the strength of the radial magnetic field $\mathrm{B_r}(\theta, \varphi)$ as a function of solar latitude (θ) and longitude (φ) provided through full disk synoptic, i.e. the data cover the time period of a full solar rotation) magnetograms and temperature $\mathrm{T_e}(\theta, \varphi)$ and density $\mathrm{n_e}(\theta, \varphi)$ values, e.g., obtained through SoHO/EIT data, of the properties of the coronal base. This model has been used to estimate the solar wind speeds at the orbit of the Ulysses spacecraft during the Whole Sun Month Campaign in 1996 as shown in Fig. 1.30.

Wang, Hawley and Sheeley (1997) have modeled the solar wind expansion in the heliosphere by taking into account the locations, areal sizes, rotation and solar-cycle evolution of coronal holes. The solar wind flows from the coronal holes can then be can be reproduced by applying extrapolation techniques to the measurements of the

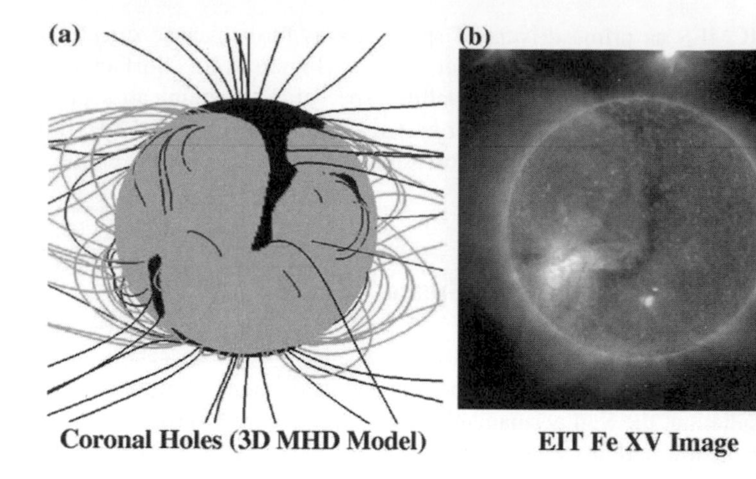

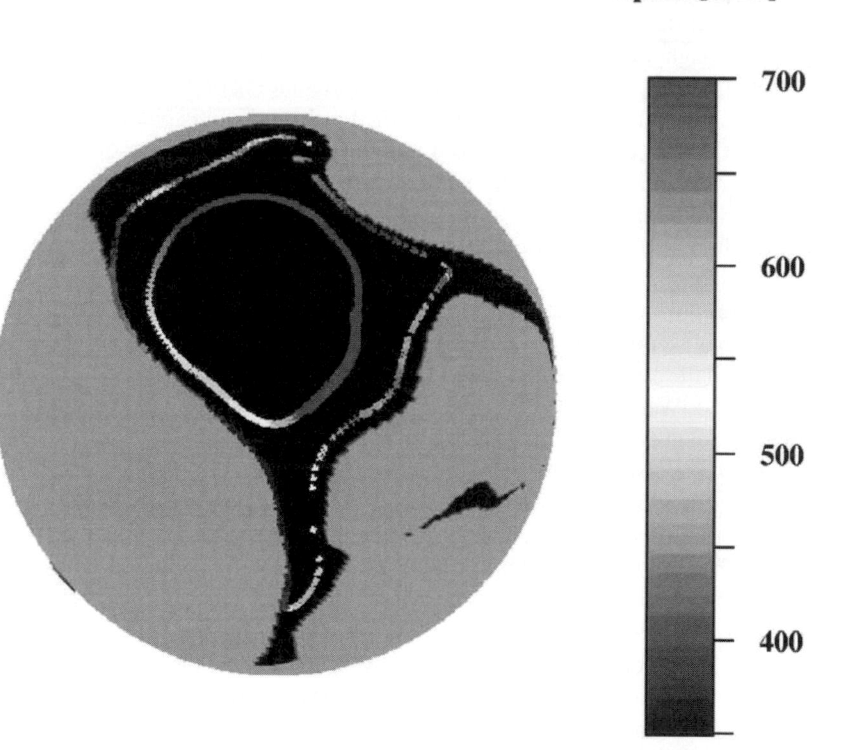

Fig. 1.30. Top: (a) Coronal holes (open field regions, colored black) from an MHD model of the solar corona for August 27, 1996. Closed field regions are colored gray. (b) EIT FeXV image for the same day. Bottom: Mapping of solar wind velocity measurements back to the Sun. Red marks highest solar wind speeds. Coronal hole boundaries are modeled as for the top left image but shown as viewed from above the north pole of the Sun. From Balogh & Bothmer et al. 1999.

photospheric magnetic field and its expansion to estimate the bulk speed, mass and energy densities of the solar wind plasma in the heliosphere. Odstrcil et al. (2002, 2003) have developed a time-dependent solar wind model, the so-called ENLIL code, for the range 21.5 R_S to 1.6 AU for a heliographic latitude range of $\sim 60°$ in the inner heliosphere. The model is also based on solar magnetograms and uses further the MAS or WSA (Wang-Sheeley-Arge) code to locate the inner radial boundary conditions. A detailed summary on the current modeling efforts has been given by Aschwanden et al. (2006). It should be noted, that unfortunately the magnetograms obtained currently are not providing exact values of the photospheric magnetic field at heliographic latitudes beyond about 60°. Figure 1.31 shows how the method by Wang, Sheeley & Arge is

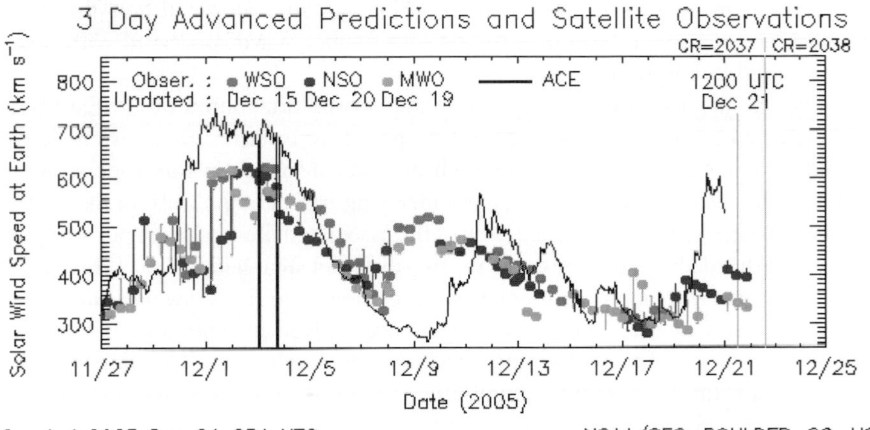

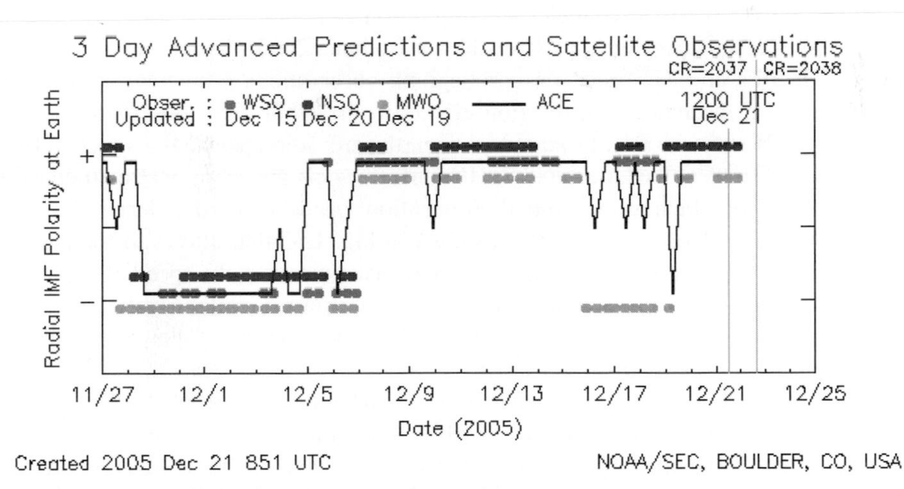

Fig. 1.31. Comparison of predicted and measured solar wind velocities (top) and IMF polarities (bottom) at 1 AU in December 2005. In-situ data are provided by the ACE satellite. WSO, NSO, MWO denote the different magnetic field data used as input for the model calculations. Courtesy: NOAA/SEC. Bottom:

applied by the NOAA National Space Environment Laboratory, Boulder, CO, USA to forecast solar wind speed and IMF polarity at Earth's orbit and to compare them with in-situ data from the ACE satellite (http://www.sec.noaa.gov/ws/).

The modeling and prediction of ICMEs at Earth's orbit is probably the most challenging task today in terms of space weather forecasts. CMEs originate from localized bipolar regions separating opposite polarity magnetic fields on the Sun as shown by Cremades & Bothmer (2004). Bipolar regions appear in various spatial sizes and with a variety of intensities, sometimes within a few hours, and to date the physical characteristica of the magnetic regions that produce CMEs are poorly understood. The top plot of Fig. 1.32, from Cremades, Bothmer & Tripathi (2005), shows the source regions of CMEs during 1996–2002 together with the evolution of the longitudinal component of the photospheric magnetic field. The bottom part shows an example of how the small-scale emergence of flux in the source region of a filament/CME observed with SoHO may have caused its eruption.

The internal magnetic field configuration and spatial orientation of CMEs/ICMEs may be predicted according to the scenario proposed by Bothmer & Schwenn (1994, 1998) and Bothmer & Rust (1997) in which the helical magnetic flux rope structure of CMEs/ICMEs can be inferred from the underlying magnetic polarity in the CME's photospheric source regions, the orientation of the associated filaments and post-eruptive arcades and the handedness preference in the two different solar hemispheres (Fig. 1.33). The study of events with opposite orientations as compared to Joy's law indicates that in these cases the hemispheric-handedness dependence is reversed. This is an encouraging result in terms that it seems to become possible in the near future to reliably predict the internal magnetic structure and spatial orientation of ICMEs from the Sun to 1 AU. Future studies should help establish quantitative forecasts for the field strength, diameter and arrival time of ICMEs. It should be noted, that the effects of the global structure of the corona should be taken into account since CMEs are deflected in their direction of propagation with respect to their low coronal source regions in the presence of polar coronal holes as shown in Fig. 1.34 (Cremades & Bothmer 2005). To determine the arrival times of ICMEs at 1 AU based on coronagraph observations is another challenging task because of the inherent projection effects in the white-light observations (e.g., Cremades & Bothmer 2004). Front-side, i.e. earthward, -directed CMEs watched from the position of Earth or the L1 orbit of SoHO, unfortunately appear as unstructured halos (see Fig. 1.20) for which reliable radial propagation speeds are hard to derive. DalLago et al. (2005) have developed the model shown in Fig. 1.35 that allows to forecast the arrival time of CMEs at Earth's orbit based on measurements of the speed of expansion and propagation for CMEs originating from different source locations at the Sun (limb, near limb, halo events). The travel time (T_{tr}) to 1 AU was found to follow the following expression: $T_{tr} = 203 - 20.77 \cdot \ln(V_{exp})$.

Future research is certainly needed to understand better the onset of CMEs, their 3-D structure and interplanetary evolution as ICMEs. A new mission dedicated to help unravel these questions and help clarify the space weather effects of CMEs/ICMEs is currently under preparation for launch in spring 2006. The NASA STEREO (Solar TErrestrial RElations Observatory) consists of two suitably, nearly identical spacecraft, equipped with optical telescopes (coronagraphs, EUV imagers and interplanetary cameras) that allow for the first time to study the 3-D structure of CMEs and their evolution from

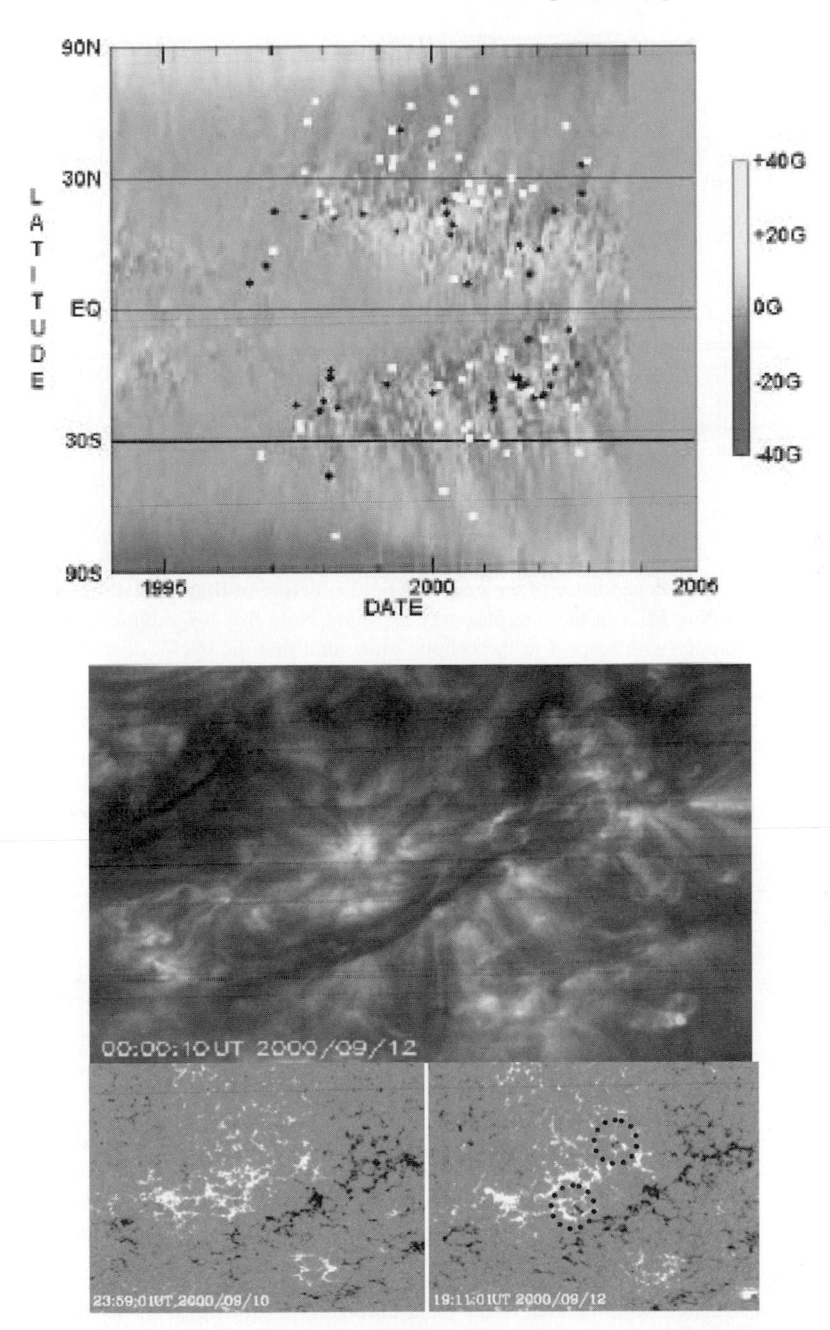

Fig. 1.32. Top: Source regions of structured CMEs in 1996–2002 and the longitudinal evolution of the photospheric magnetic field. Active regions are marked dark blue, decaying ones are labeled bright blue. From Cremades, Bothmer & Tripathi 2005. Bottom: A filament before its eruption, it was associated with a CME with a speed of ∼1600 km/s, as imaged by SoHO/EIT at 195 Å and the evolution of the photospheric magnetic flux in its source region. The two circles mark the locations where new magnetic flux emerged. Courtesy: Tripathi & Bothmer (2006).

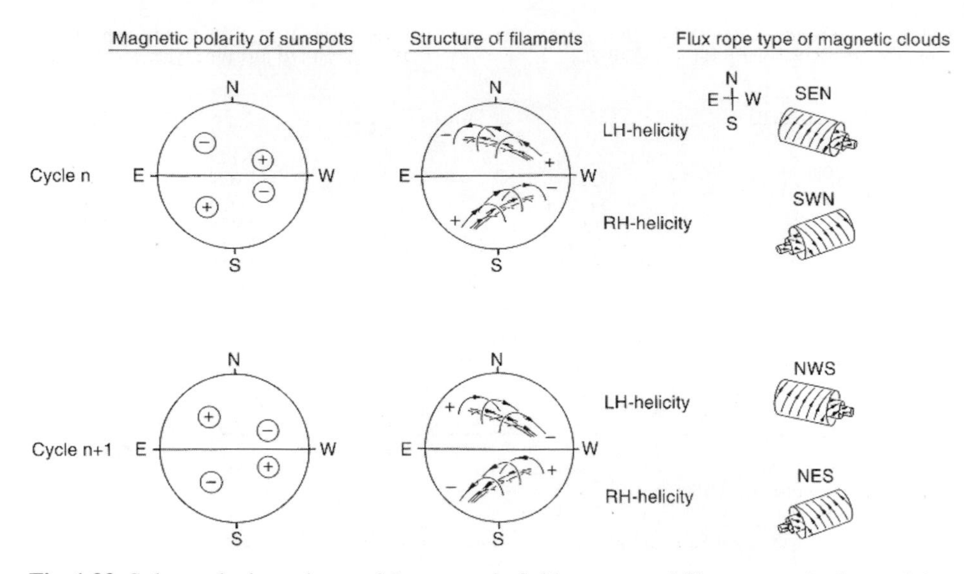

Fig. 1.33. Solar cycle dependence of the magnetic field structure of filaments at the Sun and that of the corresponding MCs in the interplanetary medium. Note that for simplicity the MCs are oriented horizontally with respect to the ecliptic plane and that the cycles do not indicate any overlaps. From Bothmer & Rust (1998).

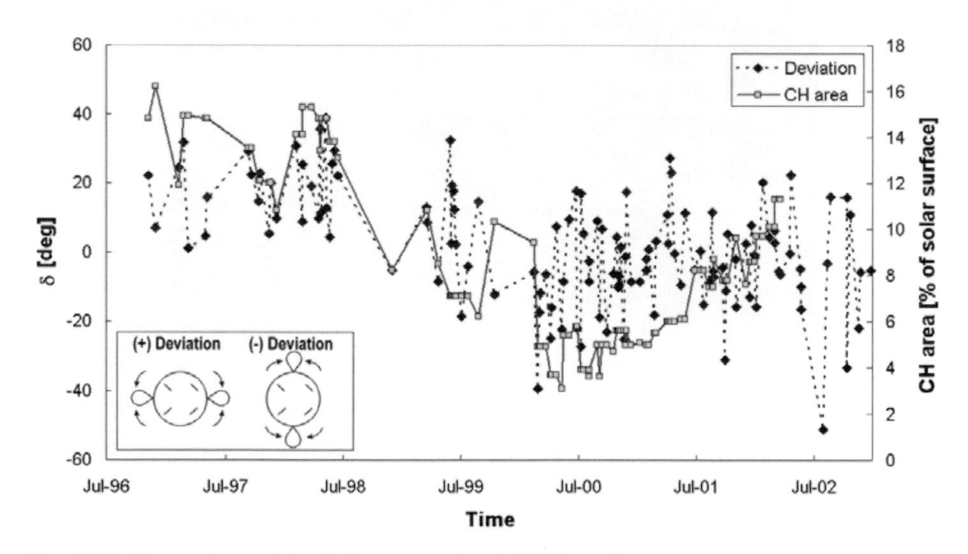

Fig. 1.34. Comparison of the deflection angle of CMEs as measured with respect to their low coronal source regions with the area of the polar coronal holes for the years 1996-2002. From Cremades & Bothmer (2004).

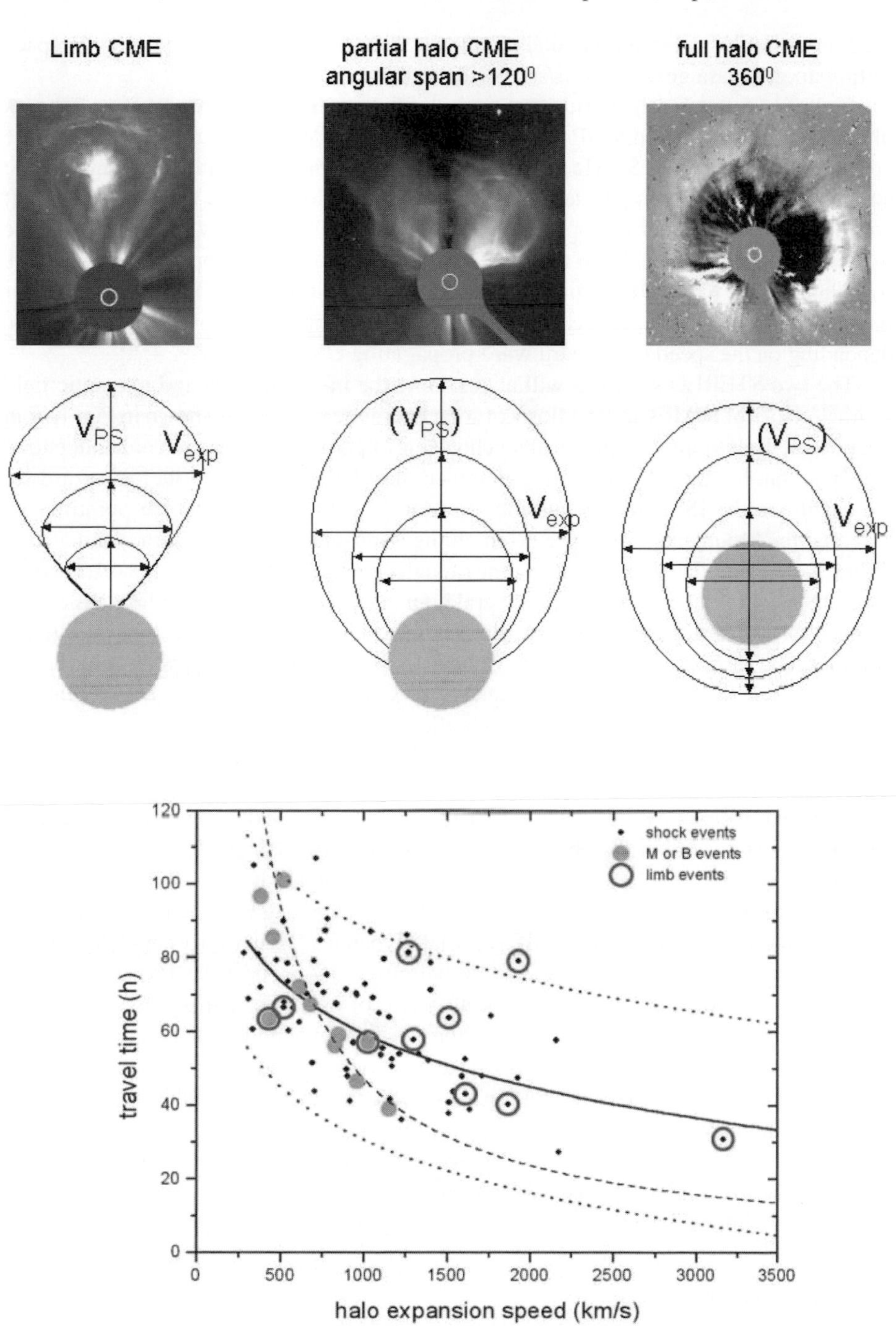

Fig. 1.35. Top images: Schematic scheme showing the direction of the propagation speed (V_{PS}) and that of the expansion speed (V_{exp}) for CME at the limb, near the limb and for halos events. Bottom: Comparison of the travel time of CMEs to 1 AU with the calculated halo expansion speeds. From Dal Lago et al. (2005).

the Sun to Earth's orbit and beyond simultaneously from new vantage points in space (http://stp.gsfc.nasa.gov/missions/stereo/stereo.htm).

Figure 1.36 shows schematically the orbit of the two STEREO spacecraft and the set of near Sun imagers called SCIP (Sun Centered Imaging Package) including the entrance aperture mechanisms SESAMe (SECCHI Experiment Sun Aperture Mechanisms, the command send for opening the telescope doors is "SESAMe open"). The twin spacecraft will drift by 22° per year in opposite directions in near 1 AU orbits with respect to the Sun-Earth line. Scientists hope that STEREO will approximately operate until 5 years from launch, i.e. until 2011 which is the next solar maximum. STEREO will allow for the first time real-time space weather predictions with forecast times from $\sim$1 to 5 days, depending on the speed of the earthward propagating CMEs/ICMEs.

The two STEREO satellites will also sample the in-situ plasma and magnetic field characteristics of ICMEs and the flows of solar energetic particles as shown in Fig. 1.27 in Sect. 1.6 at the bottom. The physical mechanisms of particle acceleration are understood poorly so far, but they are of high importance in order to forecast radiation hazards to astronauts on the ISS and for missions to Mars and the Moon – worth pointing out here that the Moon and Mars have no shielding atmosphere like the Earth – and also to airline passengers. Though the travel time of high energy particles to Earth's orbit is only several tens of minutes, long-time – several hours to days – exposures of humans can be avoided through real-time CME/flare alerts. Unfortunately the prediction of solar flares and energetic particle events much more ahead in time can presently only be achieved by monitoring the Sun's photospheric magnetic flux, with sunspot observations being the simplest means. Helioseismology methods allow one to detect the appearances of newly emerging magnetic flux on the solar surface at the back side of the Sun (Fig. 1.37) which appears a promising tool for advanced warning capabilities, however intense magnetic flux can also emerge on time-scales of just a few hours, e.g., as observed during the large solar storms at the end of October 2003 (e.g., Veselovsky et al. 2005).

Upcoming missions like STEREO and Solar-B, which will provide soft X-ray observations of the solar corona and high resolution vector magnetic field measurements of photospheric magnetic fields, or SDO (Solar Dynamics Observatory), which will provide high resolution photospheric and coronal observations, or Solar Probe, a mission designed to visit for the first time the atmosphere of our Sun, will provide new exciting scientific observations and will strengthen our understanding of the physics of the Sun and that of the complex Sun-Earth system as a whole and hence help establish the next steps to enable real-time reliable space weather predictions which are required for mankinds future evolution. National projects all over the world, either provided through local centers of scientific excellence or by small individual national satellite missions will contribute to this challenging path with the goal to unravel the physics of the Sun-Earth system through the physics of Space Weather (e.g., Bothmer & Daglis 2006).

Acknowledgements

The project Stereo/Corona, supported by the German "Bundesministerium für Bildung und Forschung" through the "Deutsche Zentrum für Luft- und Raumfahrt e.V." (DLR, German Space Agency) is a science and hardware contribution to the optical imaging package SECCHI, currently being developed for the NASA STEREO mission to be

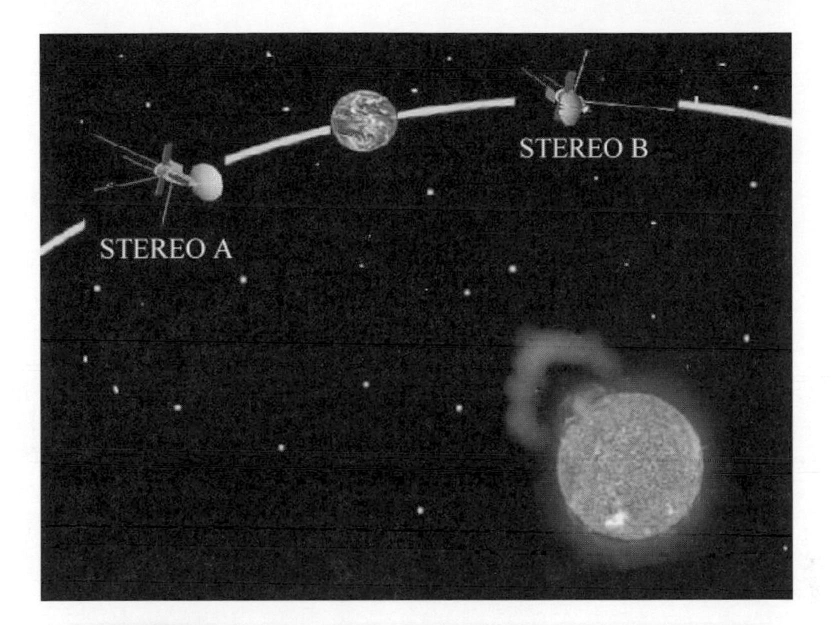

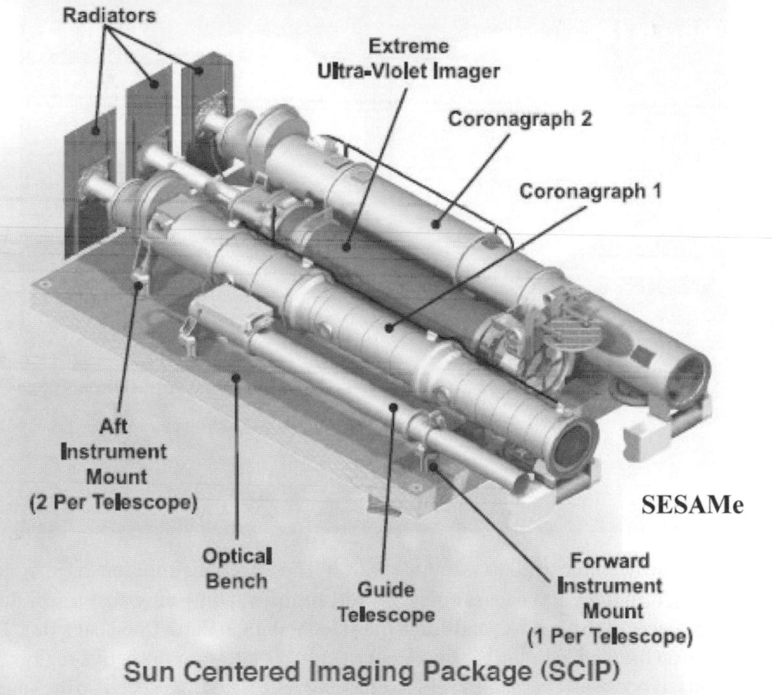

Fig. 1.36. Top: Schematic representation of the orbit of the STEREO spacecraft's A (ahead) and B (behind). Courtesy: NASA STEREO consortium. Bottom: The SCIP (Sun Centered Imaging Package) of the SECCHI (Sun Earth Connection Coronal and Heliospheric Investigation) imaging package for the NASA STEREO mission.

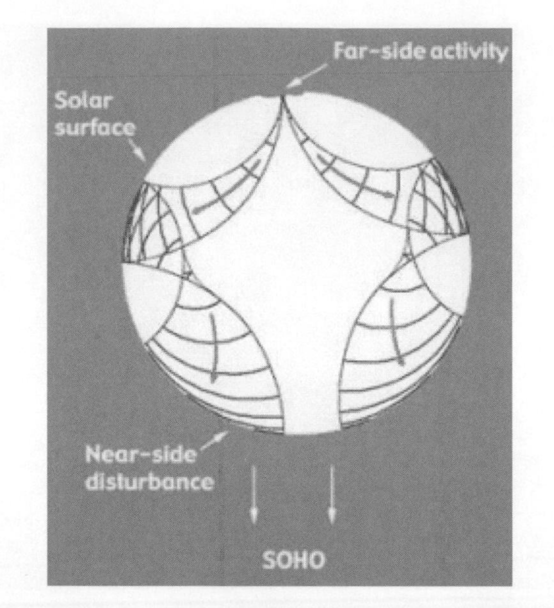

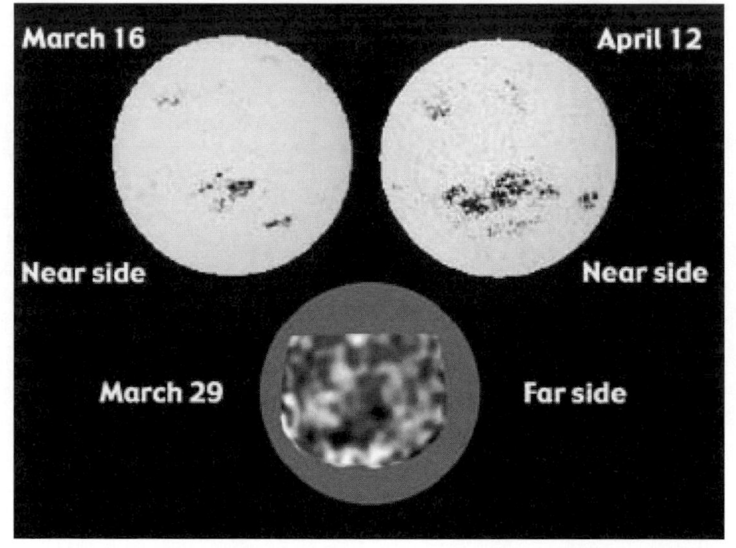

Fig. 1.37. nasa.gov/headlines/y2000/ast09mar_1.htm Top: An active region on the side of the Sun facing away from the Earth causes sound waves, represented by arcs, that travel through the interior, bounce once off the surface, and reach the side facing the Earth (the near side). The waves generate ripples on the near side surface and are reflected back toward the active region. An active region reveals itself because it possesses very strong magnetic fields that speed up the sound waves. Waves that pass through an active region have a round trip travel time about twelve seconds shorter than the average of 6 hours. The difference becomes evident when sound waves shuttling back and forth get out of step with one another. Bottom: Three-panel image, showing absolute magnetic field strength of the same feature (upper left) one-half solar rotation before, and (upper right) one-half solar rotation after the (below, center) holographically imaged farside region. Courtesy: NASA and the European Space Agency. http://science.nasa.gov/headlines/y2000/ast09mar_1.htm

launched in 2006. Further information can be found at http://stp.gsfc.nasa.gov/missions/stereo/stereo.htm. I thank all the members of the SoHO/LASCO/EIT/MDI consortium who built the instruments and provided the data used in this paper. LASCO, EIT and MDI images are courtesy of SOHO consortium. SoHO is a project of international co-operation between ESA and NASA. The CME Catalogue is generated and maintained by NASA and The Catholic University of America in cooperation with the Naval Research Laboratory. I acknowledge the use of H_α data from the Observatory of Meudon. NSO/Kitt Peak data shown are produced cooperatively by NSF/NO AO, NASA/GSFC and NOAA/SEL. I also like to thank the Yohkoh and TRACE consortia in the same way as the SoHO consortium. Thanks go further to Dr. Hebe Cremades, Dr. Durgesh Kumar Tripathi and Ralf Kutulla who helped me solving technical issues and/or provided results from their work. I finally like to thank Daniel Berdychevsky and Ronald Lepping for providing the solar wind data plot for the Bastille event.

References

Allen, J.H., D.C. Wilkinson, 1992, *Solar-Terrestrial Activity Affecting Systems in Space and on Earth, Solar-Terrestrial Predictions-IV: Proceedings of a Workshop at Ottawa, Canada May 18–22, 1992,* J. Hruska, M.A. Shea, D.F. Smart, G. Heckman (Eds.), 75.

Allen, J.H., H. Sauer, L. Frank, P. Reiff, 1989, Effects of the March 1989 Solar Activity, *EOS* 70–46, 1486–1488.

Aschwanden, M., 2004, Physics of the solar corona – An introduction, 842 pp, Springer-Praxis, Chichester, UK.

Balogh, A., R.G. Marsden, E.J. Smith, 2001, The Heliosphere near solar minimum – The Ulysses perspective, 411 pp, Springer-Praxis, Chichester, UK.

Balogh, A., V. Bothmer (CO-Chairs), N. Cooker, R.J. Forsyth, G. Gloeckler, A. Hewish, M. Hilchenbach, R. Kallenbach, B. Klecker, J.A. Linker, E. Lucek, G. Mann, E. Marsch, A. Posner, I.G. Richardson, J.M. Schmidt, M. Scholer, Y.-M. Wang, and R.F. Wimmer-Schweingruber (Participants), M.R. Aellig, P. Bochsler, S. Hefti and Z. Miki? (Contributing Authors not Participating in the Workshop), 1999, The Solar Origin of Corotating Interaction Regions and Their Formation in the Inner Heliosphere (Report of Working Group 1), Space Sci. Rev., 89, 141–178.

Bartels, J., 1932, Terrestrial-magnetic activity and its relations to solar phenomena, *Terr. Magn. Atmosph. Electr.* 37, 1–52.

Bartels, J., J. Veldkamp, 1949, Geomagnetic and Solar Data – International Data on Magnetic Disturbances, *J. Geophys. Res.* 54, 295–299.

Biermann L., 1951, Kometschweife und solare Korpuskularstrahlung, Z. Astrophys., 29, 274.

Becquerel, H., 1880, Mémoire sur la polarisation atmosphérique et l'influences du magnétisme terrestre sur l'atmosphère, Bern, Dibner.

Bohlin, J.D., N.R. Sheeley Jr., 1978, Extreme Ultraviolet Observations Of Coronal Holes: II. Association of Holes with Solar Magnetic Fields and a Model for their Formation during the Solar Cycle, *Solar Physics* 56, 125 151.

Bothmer, V., 1996, Solar Corona, Solar Wind Structure and Solar Particle Events, Proc. of ESA Workshop on Space Weather Nov. 1998, ESA WPP-155, ISSN 1022–6656, 117.

Bothmer, V., A. Posner, H. Kunow, R. Müller-Mellin, B. Heber, M. Pick, B.J. Thompson, J.-P. Delaboudinière, G.E. Brueckner, R.A. Howard, D.J. Michels, C.St. Cyr, A. Szabo, H.S. Hudson, G. Mann, H.-T. Classen, 1997, Solar energetic particle events and coronal mass ejections: New insights from SOHO, *Proc. 31st ESLAB Symposium*, ESA SP-415, 207.

Bothmer, V., 1999, Magnetic Field Structure and Topology Within CMEs in the Solar Wind, *Solar Wind 9 Conf. Proc.*, AIP, 119–126.

Bothmer, V., 2003, Sources of magnetic helicity over the solar cycle, Proc. ISCS 2003 Symposium, 'Solar Variability as an Input to the Earth's Environment', ESA SP-535, 419.

Bothmer, V., 2004, The Solar and Interplanetary Causes of Space Storms in Solar Cycle 23, IEEE Transactions on Plasma Science, 32, 4.

Bothmer, V. and Schwenn, R., 1994, Eruptive prominences as sources of magnetic clouds in the solar wind, *Space Sci. Rev.*, 70, 215–220.

Bothmer, V., Schwenn, R., 1996, Signatures of fast CMEs in interplanetary space, *Adv. Space Res.*, 17, 319–322.

Bothmer, V., and Schwenn, R., 1998, The Structure and Origin of Magnetic Clouds in the Solar Wind, *Annales Geophysicae*, 1–24.

Bothmer, V., and Rust, D.M., 1997, The field configuration of magnetic clouds and the solar cycle, *AGU Geophys. Monogr.*, 99, 139–146.

Brueckner, G.E., J.-P. Delaboudiniere, R.A. Howard, S.E. Paswaters, O.C. St. Cyr, R. Schwenn, P. Lamy, G.M. Simnett, B. Thompson, D. Wang, 1998, Geomagnetic storms caused by coronal mass ejections (CMEs): March 1996 through June 1997, Geophys. Res. Lett., 25, 3019.

Burlaga, L.F., 1975, Interaction with the earth, Space Sci. Rev., 17, 327.

Burlaga, L.F., K.W. Behannon, L.W. Klein, 1987, Compound streams, magnetic clouds, and major geomagnetic storms, *J. Geophys. Res.* 92, 5725–5734.

Burlaga, L., Fitzenreiter, R., Lepping, R., Ogilvie, K., Szabo, A., Lazarus, A., Steinberg, J., Gloeckler, G., Howard, R., Michels, D., Farrugia, C., Lin, R.P., and Larson, D.E., 1998, A magnetic cloud containing prominence material: January 1997, *J. Geophys. Res.*, 103, 277–285.

Burlaga, L.F., J.H. King, 1979, Intense Interplanetary Magnetic Fields Observed by Geocentric Spacecraft During 1963–1975, J. Geophys. Res., 84, 6633.

Burlaga, L.F., Klein, L., Sheeley, N.R. Jr., Michels, D.J., Howard, R.A., Koomen, M.J., Schwenn, R., Rosenbauer, H., 1982, A magnetic cloud and a coronal mass ejection, *J. Geophys. Res., 9*, 1317–1320.

Burlaga, L., Sittler, E., Mariani, F., and Schwenn, R., 1981, Magnetic loop behind an interplanetary shock: Voyager, Helios and IMP 8 observations, *J. Geophys. Res.*, 86, 6673-6684.

Canfield, R.C., H.S. Hudson, D.E. McKenzie, 1999, Sigmoidal morphology and eruptive solar activity, Geophys. Res. Lett., Vol. 26, 6, 627.

Carrington R.C., 1860, Description of a singular appearance in the Sun on September 1, 1859, Monthly Not. Royal Astron. Soc., 20, 13.

Chapman, S., Ferraro V.A., 1930, A new theory of magnetic storms, Nature, 126.

de Mairan J.J. d'Ortous, 1733, Traite Physique et Historique de l'Aurore Boreale.

Chao, J.K., R.P. Lepping, 1974, A correlative study of ssc's, interplanetary shocks, and solar activity, J. Geophys. res., 79, 13, 1799.

Chen, J., R.A. Santoro, J. Krall, R.A. Howard, R. Duffin, J.D. Moses, G.E. Brueckner, J.A. Darnell, J.T. Burkepile, 2000 Magnetic geometry and dynamics of the fast coronal mass ejection of 1997 September 9, Astrophys. J., 533, 481.

Cremades, H., Bothmer, V., 2004, On the three-dimensional configuration of coronal mass ejections, Astron. & Astrophys., 422, 307.

Cremades, H., V. Bothmer, D. Tripathi, Properties of Structured Coronal Mass Ejections in Solar Cycle 23, in press.

Crooker, N.U., G.L. Siscoe, 1986, The effects of the solar wind on the terrestrial environment, in Physics of the Sun, Vol. III, D. Reidel Publ. Company, 193.

Crooker, N.U., J.T. Gosling (CO-Chairs), V. Bothmer, R.J. Forsyth, P.R. Gazis, A. Hewish, T.S: Horbury, D.S. Intriligator, J.R. Jokipii, J. Kóta, A.J. Lazarus, M.A. Lee, E. Lucek, E. Marsch,

A. Posner, I.G. Richardson, E.C. Roelof, J.M. Schmidt, G.L. Siscoe, B.T. Tsurutani, R.F. Wimmer-Schweingruber (Participants), 1999, CIR Morphology, Turbulence, Discontinuities, and Energetic Particles (Report of Working Group 2), *Space Sci. Rev.*, 89, 179.

Daglis, I.A., Ed., 2004, Effects of Space Weather on Technology Infrastructure, NATO Science Series, II. Mathematics, Physics and Chemistry, 176.

Dal Lago, A.; Vieira, L. E. A.; Echer, E.; Gonzalez, W. D.; de Gonzalez, A. L. C.; Guarnieri, F. L.; Schuch, N. J.; Schwenn, R., 2004, Comparison Between Halo cme Expansion Speeds Observed on the Sun, the Related Shock Transit Speeds to Earth and Corresponding Ejecta Speeds at 1 AU, Solar Physics, 222, 2, 323.

de Mairan, J.J.'Ortous, 1733, Traite Physique et Historique de l'Aurore Boreale.

Dere, K., J. Wang, Y. Yan, Eds., 2004, Coronal and stellar mass ejections, Proc. IAU, Symposium 226, Beijing.

Dessler A.J., Parker E.N., 1959, Hydromagnetic theory of magnetic storms, J. Geophys. Res., 64, 2239.

Fabricius, J., 1611, De Maculis in Sole observatis, et apparente earum cum Sole conversione, Narratio, Witebergae.

Feynman, J., S.F. Martin, 1995, The initiation of coronal mass ejections by newly emerging magnetic flux, J. Geophys. res., 100, 3355

Fitzgerald G.F., 1892, Sunspots and magnetic storms, The Electrician, 30, 48.

Fleck, B., Domino, V., Poland, A., 1995, The SOHO mission, Sol. Phys., 162, 1, Kluwer Academic Publ.

Fleck, B., Svetska, Z., Eds., 1997, The first results from SOHO, Sol. Phys., 170, 1; 172, 2, Kluwer Academic Publ.

Forbes, T.G., 2000, A review on the genesis of coronal mass ejections, J. Geophys. res., 105, 23153.

Gauss C.F., 1839, Allgemeine Theorie des Erdmagnetismus in "Resultate aus den Beobachtungen des magnetischen Vereins im Jahre 1838.

Gibson, S.E., B.C. Low, 2000, Three-dimensional and twisted: An MHD interpretation of on-disk observational characteristics of coronal mass ejections, J. Geophys. Res., 105, A8, 18187.

Gold, T., 1955, Discussion on shock waves and rarefied gases, in Gas Dynamics of Cosmic Clouds, North Holland Publ., Amsterdam, 103.

Gold, T., 1959, J. Geophys. Res., 64, 1219.

Goldstein, E., 1880, Eine neue Form elektrischer Abstossung – A new kind of electrical repulsion, Berlin, Springer.

Gonzalez, W.D., 1990, A unified view of solar wind-magnetosphere coupling functions, Planetary and Space Science, 38, 5, 627.

Gonzalez, W.D., B.T. Tsurutani, 1987, Criteria of interplanetary parameters causing intense magnetic storms (Dst < -100 nT), Planet. Space Sci., 35, 1101.

Gonzalez, W.D., A.L.C. Gonzalez, 1990, Dual-Peak Solar Cycle Distribution of Intense Geomagnetic Storms, Planet. Space Sci., 38, 181.

Goodrich, C.C., J.G. Lyon, M. Wiltberger, R.E. Lopez, K. Papadopoulos, 1998, An overview of the impact of the January 10–11, 1997 magnetic cloud on the magnetosphere via global MHD simulation, Geophys. res. Lett., 25, 14, 2537.

Gopalswamy, N., Lara, A., Yashiro, S., Nunes, S., Howard, R. A., 2003, Coronal mass ejection activity during solar cycle 23, in ESA SP-535: Solar Variability as an Input to the Earth's Environment, 403-414.

Gosling, J.T., Coronal mass ejections and magnetic flux ropes in interplanetary space, 1990, in Physics of Magnetic Flux Ropes, Eds., E.R. Priest, L.C. Lee, C.T. Russell, AGU Geophys. Monogr. 58, 343.

Gosling, J.T., 1993, Coronal mass ejections: The link between solar and geomagnetic activity, Phys. Fluids, B5, 2638.

Gosling, J.T., 1993, The solar flare myth, J. Geophys. Res., 98, A11, 18937.

Gosling, J.T., J.R. Asbridge, S.J. Bame, A.J. Hundhausen, I.B. Strong, 1968, Satellite observations of interplanetary shock waves, J. Geophys. Res., 73, 43.

Gosling, J.T., S.J. Bame, D.J. McComas, J.L. Phillips, 1990, Coronal Mass Ejections And Large Geomagnetic Storms, *Geophys. Res. Lett.* 17, 901–904.

Gosling, J.T., D.J. McComas, 1987, Field line draping about fast coronal mass ejecta: A source of strong out-of-the-ecliptic interplanetary magnetic fields, *Geophys. Res. Lett.* 14, 355–358.

Gosling, J.T., V. Pizzo, 1999, Formation and evolution of corotating interaction regions and their three-dimensional structure, Space Sci. Rev., 89, 1–2, 21.

Gringauz K.I., Bezrukikh V.V., Ozerov V.D., 1961, Results of measurements of the concentration of positive ions in the atmosphere, using ion traps mounted on the Third Soviet Earth Satellite, In Artificial Earth Satellites, ed. L.V. Kurnosova, New York, Plenum Press, 6, 77.

Hale, G.E., 1908, On the probable existence of a magnetic field in sunspots, Astrophys. J., 28, 315.

Hale, G.E., Ellermann, F., Nicholson, S.B., Joy, A.H., 1919, The Magnetic Polarity of Sunspots, *Astrophys. J.*, 49, 153.

Harrison, R.A., 1986, Solar coronal mass ejections and flares, Astron. & Astrophys., 162, 283.

Henke, T., J. Woch, U. Mall, S. Livi, B. Wilken, R. Schwenn, G. Gloeckler, R. von Steiger, R.J. Forsyth, A. Balogh, 1998, Differences in the O+7/O+6 ratio of magnetic cloud and noncloud coronal mass ejections, Geophys. Res. Lett., 25, 3465.

Henke, T., J. Woch, R. Schwenn, U. Mall, G. Gloeckler, R. von Steiger, R.J. Forsyth, A. Balogh, 2001, Ionization state and magnetic topology of coronal mass ejections, J. Geophys. Res., 106, 597.

Hildner, E., Gosling, J. T., MacQueen, R. M., Munro, R. H., Poland, A. I., Ross, C. L., 1976, Frequency of coronal transients and solar activity, Sol. Phys., 48, 127-135.

Hiorter, O.P., 1747, Om Magnetnalens atskillige andreingar, Kongle Swen Wetenskaps Acad Handlingar, 27.

Hoeksema, J.T., 1984, Structure and evolution of the large scale solar and heliospheric magnetic fields, Phd Thesis, Stanford University, CA, USA.

Howard, R.A., A.F. Thernisien, A. Vourlidas, C. Marque, N. Patel, Modeling of CMEs for the STEREO Mission, Proc. Solar Wind 11, in press.

Howard, R. A., et al., 1997, Observations of CMEs from SOHO/LASCO, 10, in Coronal Mass Ejections, Geophysical Monograph 99, 17.

Hundhausen, A.J., C.B. Sawyer, L. House, R.M.E. Illing, W.J. Wagner, 1984, Coronal mass ejections observed during the Solar Maximum Mission: Latitude distribution and rate of occurrence, J. Geophys. Res., 89, 2639.

Isenberg, 1999, American Institute of Physics, 1–56396–865–7/99.

Kivelson, M.G., C.T. Russell, Eds., 1995, Introduction to Space Physics, Cambridge Univ. Press, ISBN 0521457149.

Klassen, A., H. Aurass, G. Mann, et al., 2000, AASS, 141, 357.

Klein, L.W., and Burlaga, L.F., 1982, Interplanetary Magnetic Clouds at 1 AU, *J. Geophys. Res.*, 87, 613–624.

Koomen, M., Howard, R., Hansen, R., Hansen, S., 1974, The Coronal Transient of 16 June 1972, Sol. Phys., 34, 447.

Koskinen, H., E. Tanskanen, R. Pirjola, A. Pulkinnen, C. Dyer, D. Rodgers, P. Cannon, J.-C. Mandeville, D. Boscher, 2000, Space Weather Effects Catalogue, Finish Meteorological Institute, ISSN 0782–6079.

Lang, K., The sun from space, A&A Library, Springer.

Lanzerotti, L.J., C. Breglia, D.W. Maurer, G.K. Johnson, C.G. Maclennan, 1998, Studies of spacecraft charging on a geosynchronous telecommunication satellite, Adv. Space Res., 22, 1, 79.

Lanzerotti, L.J., D.J. Thomson, C.G. Maclennan, Wireless at High Altitudes – Environmental Effects on Space-Based Assets, Bell Labs Technical Journal, 5, Summer 1997.

Lepping, R.P., D.B. Berdichevsky, C.C. Wu, 2003 Sun-Earth electrodynamics: The solar wind connection, Recent Res. Devel. Astrophys., 1, 139.

Lepping, R. P., Berdichevsky, D. B., Burlaga, L. F., Lazarus, A. J., Kasper, J., Desch, M. D., Wu, C.-C., Reames, D. V., Singer, H. J., Smith, C. W., and Ackerson, K. L., 2001, The bastille day magnetic cloud and upstrem shocks: near-earth interplanetary observations, *Solar Physics*, 204, 287-305.

Lepping, R.P., Jones, J.A., and Burlaga, L.F., 1991, Magnetic field structure of interplanetary magnetic clouds at 1 au, *J. Geophys. Res.*, 95, 11,957 – 11,965.

Lepping, R.P., et al., 1995, The Wind magnetic field investigation: The Global Geospace Mission, *Space Sci. Rev.*, 71, 207–229.

Lepping, R. P., Berdichevsky, D. B., Burlaga, L. F., Lazarus, A. J., Kasper, J., Desch, M. D., Wu, C.-C., Reames, D. V., Singer, H. J., Smith, C. W., Ackerson, K. L., 2001, The bastille day magnetic cloud and upstrem shocks: near-earth interplanetary observations, *Solar Physics*, 204, 287.

Lites, B.W., B.C. Low, 1997, Flux emergence and prominences: A new scenario for 3-Dimensional field geometry based on observations with the advanced stokes polarimeter, Sola Phys., 174, 91.

Lodge, O., 1900, Sunspots, magnetic storms, comet tails, atmospheric electricity, and aurorae, Electrician 46, 249.

Low, B.C., 2001, Coronal Mass Ejections, magnetic flux ropes, and solar magnetism, J. Geophys. res., 106, A11, 25141.

Low, B.C., 2001, Coronal mass ejections, magnetic flux ropes, and solar magnetism, *J. Geophys. Res.*, 106, 25141–25163.

Low, B.C., M. Zhang, 2004, Global Magnetic-Field Reversal in the Corona, in AGU Geophys. Monogr., 141, 51.

Lyot, B., 1939, The study of the solar corona and prominences without eclipses, George Darwin Lecture, MNRAS, 99, 580.

Marsden, R.G., 2001, The 3-D heliosphere at solar maximum, reprint from Space Sci. Rev., 97, 1–4, Kluwer Academic Publ.

Marubashi, K., 1986, Structure of the interplanetary magnetic clouds and their solar origins, *Adv. Space Res.*, 6, 335–338.

Marubashi, K., 1997, Interplanetary Magnetic Flux Ropes and Solar Filaments, *AGU Geophys. Monograph*, 99, 147–156.

Marubashi, K., 2002, Interplanetary Magnetic Flux Ropes, J. of the Communications Research Laboratory, 49, 3.

Maunder, E.W., 1904, The great magnetic storms, 1875–1903 and their associations with sunspots, Greenwich Royal Observatory, Royal Astron.Soc., 64, 205.

Mayaud, P.N., 1980, Derivation, Meaning, and Use of Geomagnetic Indices, *AGU Geophysical Monograph* 22, Washington D.C.

McComas, D.J., Bame, S.J., Barker, P., Feldman, W.C., Phillips, J.L., Riley, P., Griffee, J.W.,1998, Solar Wind Electron Proton Alpha Monitor (SWEPAM) for the Advanced Composition Explorer, *Space Sci. Rev.*, 86, 563–612.

McComas, D.J., J.T. Gosling, D. Winterhalter, E.J. Smith, 1988, Interplanetary magnetic field draping about fast coronal mass ejecta in the outer heliosphere, J. Geophys. es., 93, 2519.

McPherron, R.L., 1979, Magnetospheric substorms, Rev. Geophys. Space Phys., 17, 657.

Mulligan, T., C.T. Russell, J.G. Luhmann, 1998, Solar cycle evolution of the structure of magnetic clouds in the inner heliosphere,m Geophys. res. Lett., 25, 2959.

Mulligan, T., C.T. Russell, 2001 Multispaceraft modeling of the flux rope structure of interplanetary coronal mass ejections: Cylindrically symmetric versus nonsymmetric topologies, J. Geophys. Res., 106, 10581.

NASA, 1996, Reference Publication 1390, Spacecraft System Failures and Anomalies Attributed to the Natural Space Environment, K.L. Bedingfield, R.D. Leach, M.B. Alexander, Eds.

NASA, 1996, Foundations of solar particle event risk management, Anser, Arlington, Mission from Planet Earth Study Office, NAGW-4166.

Ness, N.F., 2001, Interplanetary magnetic field dynamics, Space Storms and Space Weather Hazards, Ed. I.A. Daglis, Kluwer Academic Publ., 131.

Neugebauer, M., and C.W. Snyder, 1966, Mariner 2 observations of the solar wind, 1. Average properties, j. Geophys. res., 71.

Odstrcil, D., Linker, J. A., Lionello, R., Mikić, Z., Riley, P., Pizzo, V. J., & Luhmann,J.G., 2002, Merging of coronal and heliospheric numerical two-dimensional MHD models, J. Geophys. Res., 107, A12, SSH 14–1, CiteID 1493, DOI 10.1029/2002JA009334.

Odstrcil, D., 2003, Modeling 3D solar wind structure, *Adv. Space Res.*, 32/4, 497.

Ogilvie, K.W., et al., SWE, 1995, A comprehensive plasma instrument for the Wind spacecraft: The Global Geospace Mission, *Space Sci. Rev.*, 71, 55–57.

Panasenco, O.A., et al., Solar origins of intense geomagnetic stroms in 2002 as seen by the CORONAS-F satellite, Adv. Space Res., xx.

Parker, E.N., 1959, Extension of the solar corona inot interplanetary space, J. Geophys. Res., 64, 1675.

Parks, G.K., 2003, Physics of Space Plasmas: An INtroduction, ISBN 0813341299, Westview Press.

Parker, E.N., 2001, Solar activity and classical physics, Cin. J. Astron. Astrophys., Vol. 1, 2, 99.

Phillips, J.L., J.T. Gosling, D.J. McComas, Coronal mass ejections and geomagnetic storms: Seasonal variations, Proc. Solar-Terrestrial Predictions Workshop, Ottawa, 1992.

Prölss, G., 2004, Physik des erdnahen Weltraums – Eine Einführung, Springer, Berlin.

Reames, D.V., Particle Acceleration by CME-Driven Shock Waves, Highlight Paper, 26th INt. Cosmic Ray Conf., 1999.

Richardson, I.G., E.W. Cliver, H.V. Cane, Sources of geomagnetic storms for solar minimum and maximum conditions during 1972–2000.

Richardson, J.D., Y. Liu, C. Wang, L.F. Burlaga, 2005, ICMEs at very large distances, Adv. Space Res., xxx.

Rostoker, G., 1972, Geomagnetic Indices, *Reviews of Geophysics and Space Physics* 10, 935.

Russell, C.T., 1971, Geophysical Coordinate Transformations, *Cosmic Electrodynamics* 2, 184.

Russell, C.T., R.L. McPherron, 1973, The magnetotail and substorms, *Space Sci. Rev.* 15, 205.

Russell, C.T., R.L. McPherron, 1973, Semiannual variation of geomagnetic activity, J. Geophys. res., 78, 92.

Russell, C.T., T. Mulligan, B.J. Anderson, 2003, Radial variation of magnetic flux ropes: Case studies with ACE and NEAR, Solar Wind Ten: Proc. of the 10th INt. Solar Wind Conf., M. Velli, R. Bruno, F. Malara (Eds.), AIP, 121.

Sabine, E., 1852, On periodic laws discoverable in the mean effects of the larger magnetic disturbances, hilos. Trans. R. Soc. London, 142, 103.

Schlegel, K., 2000, Wenn die Sonne verrückt spielt, Physik in unserer Zeit, 31, 5, 222.

Schreiber, H., 1998, On the periodic variations of geomagnetic activity indices Ap and ap, Ann. Geophys., 16, 510.

Schrijver, C.J., M.L. DeRosa, 2003, Photospheric and heliospheric magnetic fields, Sol. Phys., 212, 165.

Schrijver, C.J., 2001, Catastrophic cooling and high-speed downflow in quiescent solar corobnal loops observed with TRACE, Solar Physics, 198, 325.

Schwabe, S.H., 1844, Solar Observatins During 1843 by Heinrich Schwabe, Astron. Nachrichten, Vol. 21, 233.

Schwenn, R., 1990, Large-Scale Structure of the Interplanetary Medium, in Physics of the inner heliosphere, 1 Large-Scale Phenomena, Springer-Verlag, 99.

Schwenn, R., E. Marsch, Eds., 1990: Physics of the Inner Heliosphere: 1 Large-Scale Phenomena, *Springer*, Berlin/Heidelberg.

Schwenn, R., E. Marsch, Eds., 1991: Physics of the Inner Heliosphere: 2 Particles, Waves and Turbulence, *Springer*, Berlin/Heidelberg.

Secchi, A.P., Le Soleil, Paris – Gauthier-Villars, XVI, 422, 1870.

Sheeley, N.R. Jr., Howard, R.A., Koomen, M.J., Michels, D.J., Schwenn, R., Mühlhäuser, K.-H., Rosenbauer, H., 1985, Coronal mass ejections and interplanetary shocks, *J. Geophys. Res.*, 90, 163–175.

Siebert, M., 1971, Maßzahlen der erdmagnetischen Aktivität, *Handb. Physik* 49/3 (Geophysik 3/3), 206–275, Berlin/Heidelberg/New York.

Smith, C.W., L'Heureux, J., Ness, N.F., Acuna, M.H., Burlaga, L.F., Scheifele, J., 1998, The ACE magnetic field experiment, *Space Sci. Rev.*, 86, 613.

Sonett, C.P., I.J. Abrahms, 1963, The distant geomagnetic field, 3, Disorder and shocks in the magnetopause, J. Geophys. Res., 68, 1233.

St. Cyr, O. C., Burkepile, J. T., Hundhausen, A. J., Lecinski, A. R., 1999, A comparison of ground-based and spacecraft observations of coronal mass ejections from 1980–1989, J. Geophys. Res., 104, 12493.

St. Cyr, O., R.A. Howard, N.R. Sheeley Jr., S.P. Plunkett, D.J. Michels, S.E. Paswaters, M.J. Koomen, G.M. Simnett, B.J. Thompson, J.B. Gurman, R. Schwenn, D.F. Webb, E. Hildner, P.L. Lamy, 2000, Properties of coronal mass ejections: SOHO, LASCO observations from January 1996 to June 1998, J Geophys. Res., 105, 18169.

Seneca, L.A., 4 B.C.-65 A.C., Naturales quaestiones.

Subramanian, P., K.P. Dere, 2001, Source Regions of Coronal Mass Ejections, Astrophys. J., 561, 372.

Tang, F., B.T. Tsurutani, W.D. Gonzalez, S.I. Akasofu, E.J. Smith, 1989, Solar sources of interplanetary southward Bz events responsible for major geomagnetic storms (1978–1979), J. Geophys. res., 94, 3535.

Thompson, B.J., 2001, Moreton waves, in Murdin 2000, xx.

Tsurutani, B.T., 2000, Solar/Interplanetary plasma phenomena causing geomagnetic activity at Earth, Proc. of the Int. School of Physics "Enrico Fermi", Course CXLII, B. Coppi, A. Ferrari and E. Sindoni (Eds.), IOS Press, Amsterdam.

Tsurutani, B.T., 2001, The interplanetary causes of magnetic storms, substorms and geomagnetic quiet, Space Storms and Space Weather Hazards, I.A. Daglis (Ed.), Kluwer Academic Publ., 103.

Tsurutani, B.T., W.D. Gonzalez, F. Tang, Y.T. Lee, 1992, Great Magnetic Storms, Geophys. Res. Lett., 19, 73.

Tsurutani, B.T., W.D. Gonzalez, G.S. Lakhina and S. Alex, The Extreme Magnetic Storm of September 1–2, 1859.

Uchida, Y., et al., Eds., 1994, X-ray solar physics from Yohkoh, Tokyo, University Academy Press.

Veselovsky, I.S., M.I. Panasyuk, S.I. Avdyushin, G.A. Bazilevskaya, A.V. Belov, S.A. Bogachev, V.M. Bogod, A.V. Bogomolov, V. Bothmer, K.A. Boyarchuk, E.V. Vashenyuk, V.I. Vlasov, A.A. Gnezdilov, R.V. Gorgutsa, V.V. Grechnev, Yu.I. Denisov, A. V. Dmitriev, M. Dryer, Yu.I. Yermolaev, E.A. Eroshenko, G.A. Zherebtsov, I.A. Zhitnik, A.N. Zhukov, G.N. Zastenker, L.M. Zelenyi, M.A. Zeldovich, G.S. Ivanov-Kholodnyi, A.P. Ignat'ev, V.N. Ishkov, O.P.

Kolomiytsev, I.A. Krasheninnikov, K. Kudela, B.M. Kuzhevsky, S.V. Kuzin, V.D. Kuznetsov, S.N. Kuznetsov, V.G. Kurt, L.L. Lazutin, L.N. Leshchenko, M.L. Litvak, Yu.I. Logachev, G. Lawrence, A.K. Markeev, V.S. Makhmutov, A.V. Mitrofanov, I.G. Mitrofanov, O.V. Morozov, I.N. Myagkova, A. A. Nusinov, S.N. Oparin, O.A. Panasenco, A.A. Pertsov, A.A. Petrukovich, A.N. Podorol'sky, E.P. Romashets, S.I. Svertilov, P.M. Svidsky, A.K. Svirzhevskaya, N.S. Svirzhevsky, V.A. Slemzin, Z. Smith, I.I. Sobel'man, D. E. Sobolev, Yu.I. Stozhkov, A.V. Suvorova, N.K. Sukhodrev, I.P. Tindo, S.Kh. Tokhchukova, V.V. Fomichev, I.V. Chashey, I.M. Chertok, V.I. Shishov, B.Yu. Yushkov, O.S. Yakovchouk, and V.G. Yanke, Solar and Heliospheric Phenomena in October-November 2003: Causes and Effects, Translated from Kosmicheskie Issledovaniya, Vol.42, No.5, 453–508, 2004; Cosmic Research, Vol.42, No.5, 435–488, 2004.

Veselovsky, I.S., V. Bothmer, P. Cargill, A.V. Dmitriev, K.G. Ivanov, E.R. Romashets, A.N. Zhukov, O.S. Yakovchouk, 2005, Magnetic storm cessation during northward IMF, Adv. Space Res., 36, 2460.

Timothy, A.F., A.S. Krieger, G.S. Vaiana, 1975, Sol. Phys., 42, 135.

Uchida, Y., et al., Eds., 1994, X-ray solar physics from Yohkoh, Tokyo: Univ. Academy Press.

von Humboldt, A., 1808, Magnetische Ungewitter, Annalen der Physik, 29, 25.

Vourlidas, A., Buzasi, D., Howard, R. A., Esfandiari, E., 2002, Mass and energy properties of LASCO CMEs, in ESA SP-506: Solar Variability: From Core to Outer Frontiers, 91-94.

Wang, Y.M., 2000, Astrophys. J., 543, L89.

Wang, Y.M., S.H. Hawley, N.R. Sheeley Jr., 1996, The Magnetic Nature of Coronal Holes, Science, 271, 464.

Wang, Y.M., A.G. Nash, N.R. Sheeley Jr., 1989, Evolution of the sun's polar fields during sunspot cycle 21: Poleward surges and long-term behavior, Astrophys. J., 347, 529.

Wang, Y.M., N.R. Sheeley Jr., 1992, On potential-field models of the olar corona, Astrophys. J., 392, 310.

Wang, Y.M., N.R. Sheeley Jr., 1989, Average properties of bipolar magentic regions during sunspot cycle 21, Solar Phys., 124, 81.

Wang, Y.M., N.R. Sheeley Jr., 1999, Filament eruptions near emerging bipoles, Astrophys. J., 510, L157.

Wang, Y.M., N.R. Sheeley Jr., M.D. Andrews, 2002, Polarity reversal of the solar magnetic field during cycle 23, J. Geophys.Res., 107, A12, 1465, doi: 10.1029/2002JA009463.

Wang, Y.M., N.R. Sheeley, J.H. Walters, G.E. Brueckner, R.A. Howard, D.J. Michels, P.L. Lamy, R. Schwenn, G.M. Simnett, 1998, Origin of streamer material in the outer corona, Astrophys. J., 498, L165.

Webb., D., E.W. Cliver, N.U. Crooker, O.C. St. Cyr, B.J. Thompson, 2000, Relationship of halo coronal mass ejections, magnetic clouds, and magnetic storms, J. Geophys. Res., 105, 7491.

Webb, D.F., N.U. Crooker, S.P. Plunkett, O.C. St. Cyr, 2xxx, The solar sources of geoeffective structures, xx, AGU Monogr. xx, Chapman Conf. on Space Weather, Eds. P. Song, G. Siscoe & H. Singer, xxxx.

Wilson, R.M., and Hildner, E., 1986, On the association of magnetic clouds with disappearing filaments, *J. Geophys. Res.*, 91, 5867–5872.

Yashiro, S., Gopalswamy, N., Michalek, G., St. Cyr, O. C., Plunkett, S. P., Rich, N. B., Howard, R. A., 2004, A catalog of white light coronal mass ejections observed by the SOHO spacecraft, J. Geophys. Res., 7105.

Young, C.A., 1872, The sun and the phenomena of ist atmosphere, New Haven, Conn., C.C. Chatfield & Co.

Yurchyshyn, V.B., Wang, H., Goode, P.R., Deng, Y., 2001, Orientation of the Magnetic Fields in Interplanetary Flux Ropes and Solar Filaments, *Astrophys. J.*, 563, 381.

Zhang G., L.F. Burlaga, 1988, Magnetic clouds, geomagnetic disturbances and cosmic ray decreases, J. Geophys. Res., 93, 2511.

Zhang, J., 2001, On the temporal relationship between coronal mass ejections and flares, Astrophys. J., 559, 452

Zhang, J., K.P. Dere, R.A. Howard, V. Bothmer, 2003, Identification of Solar Sources of Major Geomagnetic Storms between 1996 and 2000, ApJ., 582, 520.

Zhao, X.P., D.F. Webb, Source regions and storm effectiveness of frontside full halo coronal mass ejections, J. Geophys. res., 108, 1234, doi: 10.1029/2002JA009606, 203.

Zhukov, A.N., F. Auchere, 2004, On the nature of EIT waves, EUV dimmings and their link to CMEs, Astron. & Astrophys., 427, 705.

Zhukov, A.N., I.S. Veselovsky, F. Clette, J.-F. Hochedez, A.V. Dmitriev, E.P. Romashets, V. Bothmer, and P. Cargill, 2003, Solar Wind Disturbances and Their Sources in the EUV Solar Corona. In: SOLAR WIND TEN: Proceedings of the Tenth International Solar Wind Conference, Pisa, Italy, 17–21 June 2002. Eds.: Marco Velli, Roberto Bruno, Francesco Malara. AIP Conference Proceedings, Vol. 679, Issue 1, 711.

2 Mercury

Robert G. Strom and Ann L. Sprague

Abstract. Mercury may provide answers to questions regarding the formation and evolution of our Solar System. This article reviews what is known about Mercury from the Mariner 10 flybys of 1974 and 1975 and from thirty years of ground-based telescopic observations with ever improving instrumentation. Many new discoveries, such as possible water ice at Mercury's polar regions, make the anticipation of the arrival of the MESSENGER spacecraft for its first flyby of Mercury in 2008 even more intense.

2.1 Introduction and General Characteristics

Early risers may mistake Mercury for a star in morning or evening twilight. Because Mercury is the closest planet to the Sun and revolves about the Sun in just 88 Earth days, it switches its appearance from east to west in the sky with a slightly variable frequency of about three months. Mercury also has an orbital plane that is tilted with respect to that of the Earth, so sometimes it is either higher or lower than the Sun in declination. The physical parameters and other details regarding Mercury are given in Table 2.1

To date Mariner 10 is the only spacecraft to have explored Mercury. It only imaged about 45% of the surface at a resolution of about 1 km from 10° to 200° W. longitude (Fig. 2.1). Only about 25% of the surface was imaged at low enough sun angles to do terrain analysis. Figures 2.2(a) and (b) and 2.3 show mosaics of the surface of Mercury acquired by Mariner 10 (Davies et al. 1978). Ground-based observers have imaged most of the remaining half of the planet but at considerably lower resolution than Mariner 10. Bright albedo features have been discovered at the locations of bright radar reflection and a large dark region has been discovered that extends from 270° to 300° W. longitude (Fig. 2.4). A montage of ground-based images and the images from Mariner 10 gives the latest and most comprehensive coverage of Mercury's surface to date (Fig. 2.5). This montage includes images from Baumgardner et al. (2000) and Warell (2002).

Like the Moon, Mercury reflects the Sun's light from a highly fractured, comminuted, and glassy regolith that has been gardened by aeons of impacts. The surface is thought to be ~3.9 billion years old (Strom and Neukum 1988, Strom et al. 2005), and, therefore, has been subject to space weathering processes longer than the Moon. Mercury's reflectance function from blue to red is steeper toward the red and of a slightly different shape than that of the Moon (Blewett et al. 1997, Warell 2003). This is because of the complex relationship of impact volatilization, fracturing, brecciation, and space

Table 2.1. Orbital and Physical Data for Mercury

Orbital Data	
Semimajor axis	0.3871 AU (5.79×10^7 km)
Perihelion distance	0.3075 AU (4.60×10^7 km)
Aphelion distance	0.4667 AU (6.98×10^7 km)
Sidereal period	87.97 days
Synodic period	115.88 days
Eccentricity	0.20563
Inclination	7.004°
Mean orbital velocity	47.87 km/s
Rotation period	58.646 days
Physical Data	
Radius	2,439 km
Surface area	7.475×10^7 km^2
Volume	6.077×10^{10} km^3
Mass	3.302×10^{26} g
Mean density	5.44 g/cm^3
Surface gravity	370 cm/s^2
Escape velocity	4.25 km/s
Surface temperatures	90 to 740 K (−183° to 467°C)
Normal albedo (5° phase angle)	0.125
Magnetic dipole moment*	$4.9 \pm 0.2 \times 10^{22}$ gauss cm^3

* If the magnetic field is a currently active dipole and not a remanent field

weathering by cosmic rays and solar wind impact, and composition. Modeling of the reflected continuum between 0.4 and 1.0 micrometer indicates the surface materials are dominated by materials with anorthositic light scattering properties (Warell and Blewett 2004).

2.2 Motion and Temperature

Mercury's motions are very different from any other planet. Its present eccentricity is 0.205 and its present inclination is 7°. However, over periods of a few million years its eccentricity may vary from about 0.1 to 0.28 and its inclination from about 0° to 11°. Because of its large eccentricity Mercury's distance from the Sun varies from 0.3075 AU at perihelion to 0.4667 AU at aphelion, with an average distance of 0.3871 AU. As a consequence, Mercury's orbital velocity averages 47.6 km/s, but varies from 56.6 km/s at perihelion to 38.7 km/s at aphelion. It is, indeed, the speediest planet.

Mercury has the shortest orbital period of 87.969 Earth days, but it rotates very slowly with a period of 58.646 Earth days. Consequently, it has a unique 3:2 resonance between its rotational and orbital periods (Soter and Ulrichs 1967). It makes exactly

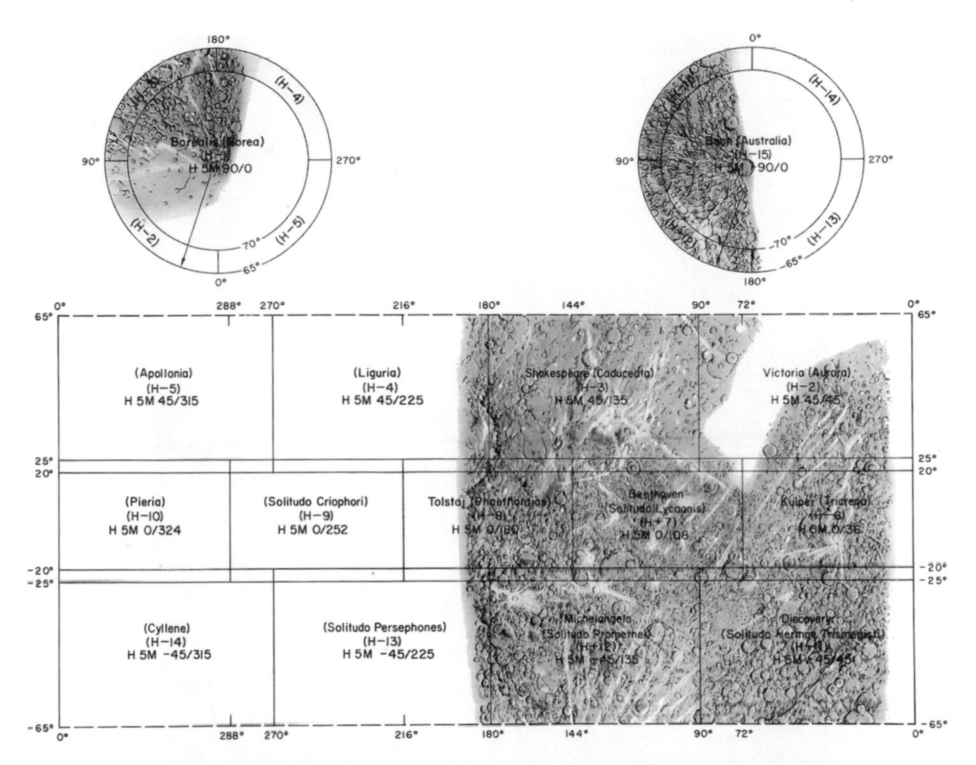

Fig. 2.1. Shaded relief map of Mercury showing the quadrangle names and major features. About 55% of the planet is unknown.

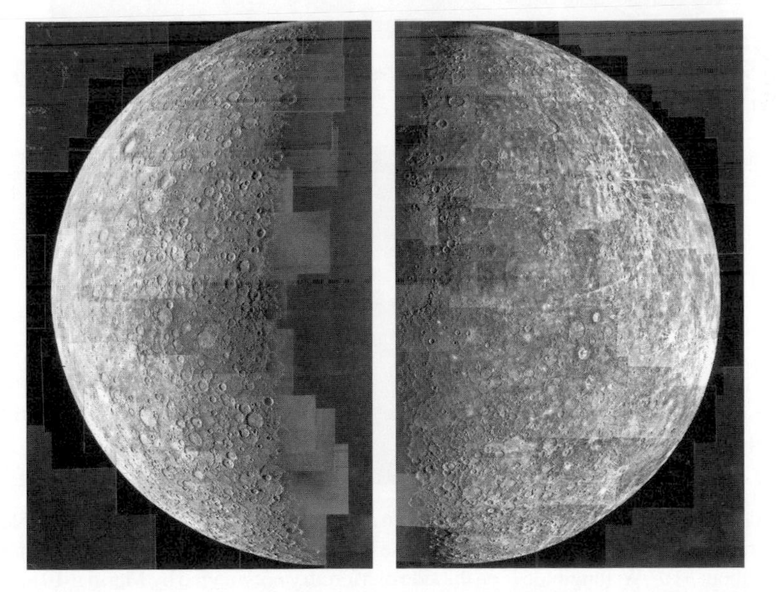

Fig. 2.2. Photomosaics of the incoming (a) and outgoing (b) sides of Mercury as viewed by Mariner 10. Most of the smooth plains are concentrated on the outgoing side, especially around Caloris Basin (Davies et al. 1978).

Fig. 2.3. Mosaic of Mercury's south polar region taken by Mariner 10 on its second encounter (Davies et al. 1978).

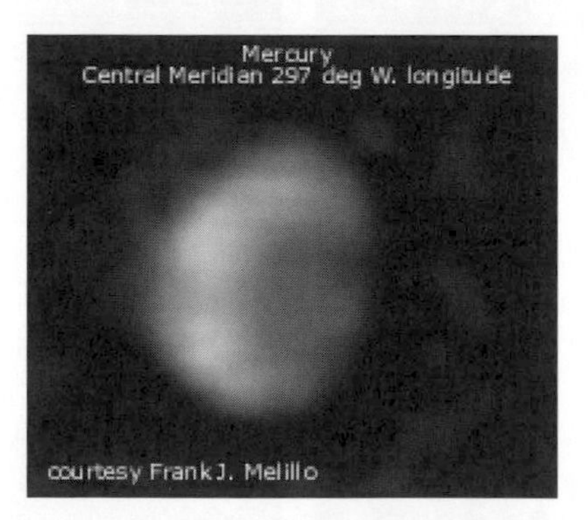

Fig. 2.4. Ground-based image obtained with a 26 cm reflector during a moment of stable Earth's atmosphere from Holtsville, NY, 5 July 2002. The large dark region extending from the terminator at 270° to about 310° W. longitude is on the side of Mercury not imaged by Mariner 10 and appears to surround a brighter region on the terminator just south of the equator. Turbulence in the Earth's atmosphere smears the bright freshly excavated craters and rays to make the indistinct bright regions (courtesy Frank Melillo).

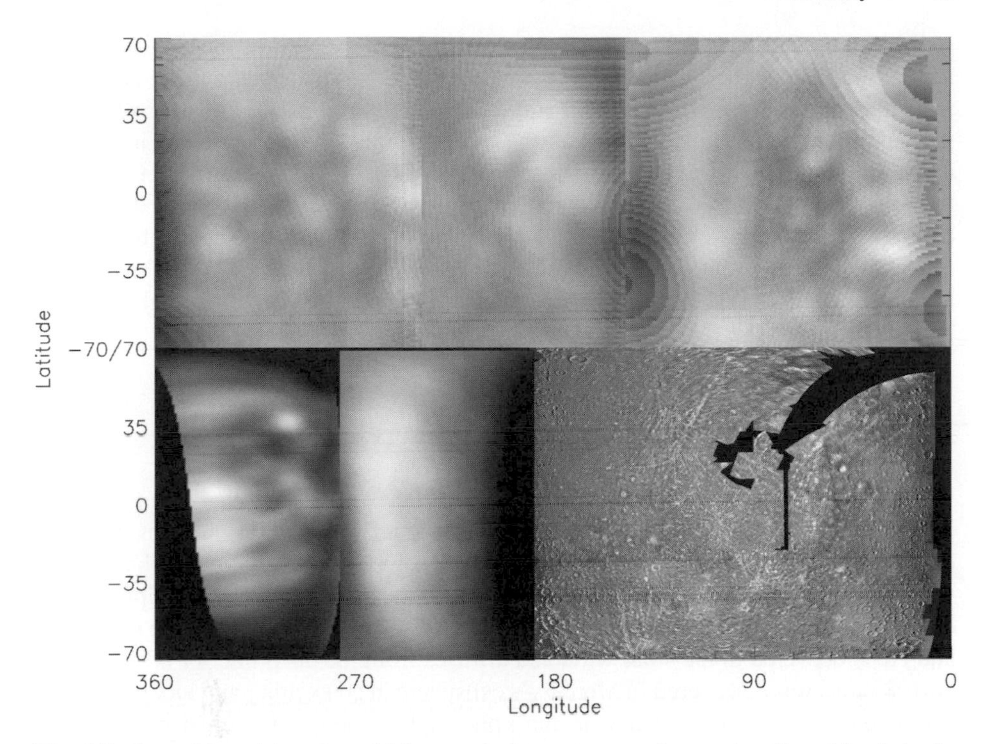

Fig. 2.5. Ground-based imaging of Mercury during moments of exceptionally stable air in the Earth's atmosphere have permitted low resolution, but excellent images of the side of Mercury not imaged by Mariner 10 (200° to 360° W. longitude). The top panel shows all longitudes of Mercury obtained from the Swedish Visible Solar Telescope. The bottom panel shows images (left bottom) Baumgardner et al. (2001) (270° to 360° W longitude), (middle bottom) Steven Massey, Sydney Australia, 52 cm telescope (198° to 270° W longitude), the bright region on the left of this image is the bright limb, the terminator is on the right side, and (right bottom) Mariner 10 mosaic. The gap from 180° to about 225° W. longitude in the bottom panel is empty.

three rotations on its axis for every two orbits around the Sun. As a consequence of this resonance, a Mercury day (sunrise to sunrise) lasts two Mercurian years or 176 Earth days. The obliquity of Mercury is close to 0° and, therefore, it does not experience seasons as do Earth and Mars. Consequently, the polar regions never receive the direct rays of sunlight and are always frigid compared to torrid sunlit equatorial regions.

Another effect of the 3:2 resonance between the rotational and orbital periods is that the same hemisphere always faces the Sun at alternate perihelion passages. This happens because the hemisphere facing the Sun at one perihelion will rotate one and a half times by the next perihelion, placing it directly facing the Sun again. Because the subsolar points of the 0° and 180° longitudes occur at perihelion they are called "hot poles". The subsolar points at the 90° and 270° degree longitudes are called "warm poles'" because they occur at aphelion.

Mercury is the second hottest planet in the Solar System (The surface of Venus is hotter). However, it experiences the greatest range (day to night) in surface temperatures

(650° C) of any planet or satellite in the Solar System because of its close proximity to the Sun, its peculiar 3:2 spin orbit coupling, its long solar day, and its lack of an insulating atmosphere. Its maximum surface temperature is about 740 K or 467° C at perihelion on the equator; hotter than the melting point of zinc. At night just before dawn, the surface temperature plunges to about 90 K or –183°C (Vasavada et al. 1999). Regions of persistently warmer and cooler than model averages have been discovered by cm wavelength observations using the Very Large Array (VLA) and careful thermal modeling. Warm regions correspond roughly to the regions on the "hot poles" (really hot longitudes) of 0° and 180° while persistently cool regions are roughly in high latitudes, but not exclusively so (Mitchell and de Pater 1994).

2.3 Mercury's Surface-bounded Exosphere

Because atoms in Mercury's atmosphere rarely collide with one another the atmosphere is properly termed an exosphere (Hunten et al. 1988). Hydrogen (H), helium (He), and oxygen (O) were identified with the Mariner 10 airglow polychrometer that had 10 wavelength channels to search for light emissions (Broadfoot et al. 1974). Upper limits on the abundance of Ne, Ar, and C were also obtained. The source for H is the solar wind and He is delivered in the solar wind and diffuses from the deep interior (Goldstein et al. 1981). Na was discovered in Mercury's exosphere in an exciting and lucky accident! The amount of light coming from the Sun's photosphere varies with wavelength because atoms in the photosphere absorb sunlight coming from deeper in the Sun. Such absorption lines are called Fraunhofer lines. When sunlight is reflected off the surface of the Moon, or Mercury, or any other solar system body, the Fraunhofer lines appear as distinct regions of less sunlight than the average. The average is called the continuum. While studying a phenomenon called the Ring Effect – the infilling of solar Fraunhofer lines in the reflected continuum from the lunar surface – Potter and Morgan (1985) shifted their view to Mercury and observed significant emission lines high above the continuum at 5890 and 5896 Å. They had discovered the first new species in Mercury's exosphere since Mariner 10 in 1974. This discovery renewed ground-based search efforts to find more atmospheric components, and K and Ca were subsequently discovered in 1986 and 2001 by Potter and Morgan (1986) and Bida et al. (2000) respectively. All of the known species have been discovered by measuring emission of light that is absorbed and emitted at the same wavelength for each particular species. This process is called resonance scattering. The amount of photons emitted by the atoms in the exosphere is measured in units of Rayleighs (R). Sodium emits about 1 million photons per second in a one cm^2 column. This amount is called a mega-Rayleigh (MR).

Table 2.2 shows all known constituents of Mercury's exosphere and their approximate abundances along with some upper limits for others. The abundances of Na, K, and Ca are known to vary by a factor of 10 or more from one measurement to the next or even at different locations above the planet during the same observation period. Probably the abundance of other species also varies but we do not have enough observations to be certain. The Earth's atmosphere has $\sim 2 \times 10^{18}$ molecules square per centimeter. The pressure of Mercury's known exosphere is a few times 10^{-12} bar (b).

Table 2.2. Mercury's Exospheric Species

Constituent	Vertical (zenith) Column Abundance (atoms per cm^2)
Hydrogen (H)	$\sim 5 \times 10^{10}$
Helium (He)	$\sim 2 \times 10^{13}$
Oxygen (O)	$\sim 7 \times 10^{12}$
Sodium (Na)	$\sim 2 \times 10^{12}$
Potassium (K)	$\sim 1 \times 10^{10}$
Calcium (Ca)	$\sim 1 \times 10^{7}$

Because of the extremes of hot and cold on Mercury, and because the exosphere is too thin to retain heat, the atoms of the exosphere can be redistributed into different temperature distributions. Helium extends to 3000–4000 kilometers above the surface, and, thus, was seen against the night side of the planet and well off the sunlit side. Hydrogen was measured to be in both a cool distribution and a hot one on the day side (Broadfoot et al. 1976). For Na, several distributions of atoms have been found with most probable speeds ranging from $\sim 600\,\mathrm{m\,s^{-1}}$ with an equivalent temperature of ~ 500 K up to ~ 1000 K (Killen et al. 1990, Killen et al. 1999). For comparison, the atoms and molecules in Earth's atmosphere have a most probable speed of about $350\,\mathrm{m\,s^{-1}}$ at a temperature of about 250 K. For K, there are no measurements to date which can be used to discriminate multiple speed components in Mercury's exosphere. It is likely however, to also have atoms at many different speeds about the planet. Calcium was observed above and beyond the southern hemisphere of Mercury using a spectrograph and large telescope (Keck I) on Mauna Kea, Hawaii. The emission lines exhibit characteristics of high speed and hot ($\sim$ 12,000°K) atoms. The equivalent most probable speed for Ca is ~ 2.2 km/s (Killen et al. 2005).

Sodium and K emissions are sometimes seen in enhanced bright regions that coincide with bright surface albedo and radar backscatter as seen in Figs. 2.6 and 2.7 (Sprague et al. 1998) or other places such as Caloris basin (Sprague et al. 1990). The amount of sunlight pressure varies considerably as Mercury moves about the Sun because of the distance to the Sun and because the speed of Mercury's motion is changing. As a consequence of radiation pressure, a tail of neutral Na atoms streams behind Mercury in the anti-sunward direction during much of its orbit around the Sun (Fig. 2.8). This phenomenon has been named the Na tail (Potter and Morgan, 2001). Because of pressure from sunlight, some atoms are pushed far from Mercury's surface and beyond the distance where gravitational force is strong enough to keep them bound to Mercury. Atoms are free to escape from the planet when they reach the escape velocity of 4.2 km/s. It is estimated that about 10% of the neutrals in the Na tail exceed the escape velocity. All the neutral atmospheric atoms (H, He, Na, K, O, Ca) will eventually be photoionized and swept away from the planet by electric fields in the interplanetary medium unless they collide with the surface again and are neutralized. Timescales for ionization vary from tens of minutes to Earth days (McGrath et al. 1986, Hunten et al. 1988, Potter and Morgan 1986, Bida et al. 2000).

The sources for the surface-bounded exosphere are meteorite and surface volatilization, diffusion from the regolith and deep interior, sputtering of atoms off the surface by photons, electrons and ions from the surface materials, and delivery from the solar wind

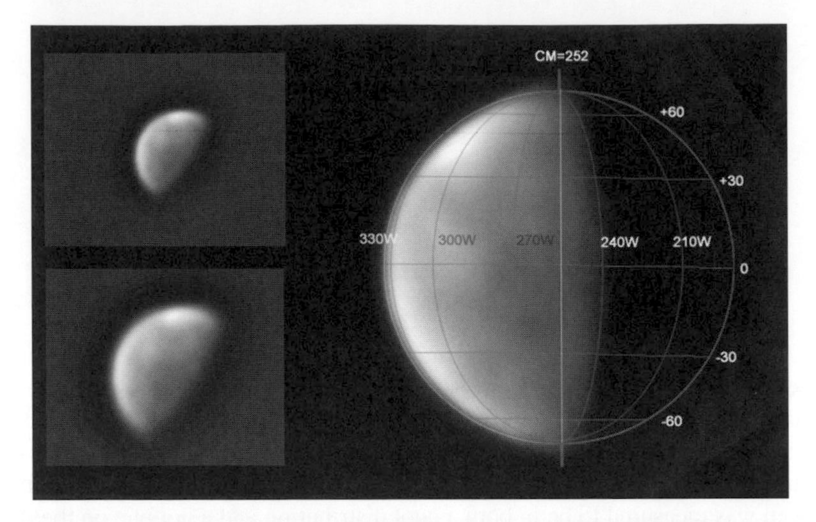

Fig. 2.6. Bright surface features such as the bright region in the north, are likely associated with regions of enhanced sodium emission from neutral sodium atoms in Mercury's surface-bounded atmosphere. The bright region in the northern hemisphere in these images coincides roughly with the radar bright region B. Radar bright region A in the southern hemisphere has not yet rotated into view. (courtesy Steve Massey).

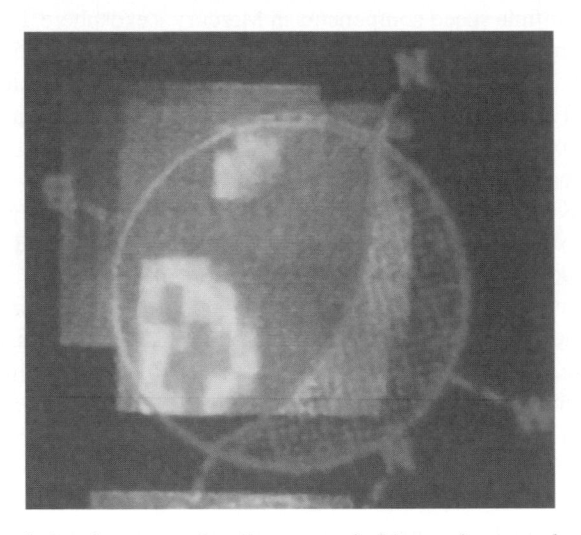

Fig. 2.7. Sodium emission from neutral sodium atoms in Mercury's atmosphere is imaged at the McMath Pierce Solar Telescope at Kitt Peak (Potter and Morgan, 1990). Enhanced emissions are seen at the locations of radar bright regions B (north) and A (south) on 17 February 1989. The source of sodium is likely to be thermal diffusion out of fresh regolith materials of freshly excavated craters at these locations.

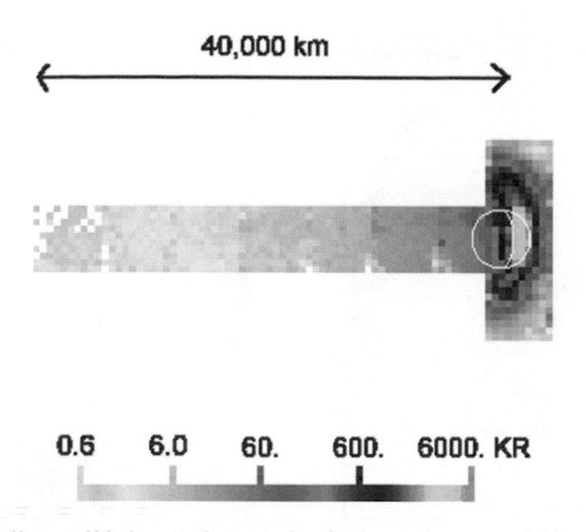

Fig. 2.8. The neutral sodium tail is imaged streaming in the anti-sunward direction from Mercury on 26 May 2001 (Potter et al. 2002). Pressure on Na atoms is at a maximum acceleration of 200 cm/s^2 at true anomaly angles near 64° and 300°. Emission brightness is given in kilo-Rayleighs.

(McGrath et al. 1986, Cheng et al. 1987). The recycling of ions is very dependent on interactions with the magnetic field of Mercury and the electric fields in the solar wind (Ip 1987, Potter and Morgan 1990).

2.4 Polar Deposits

An amazing discovery was made from decimeter radar backscatter measurements of Mercury's polar regions by the Arecibo and Goldstone observatories in 1992. Regions of depolarized coherent backscatter were observed near the polar regions on Mercury (Slade et al. 1992, Harmon and Slade 1992, Butler et al. 1993). On Mars, similar bright depolarized coherent backscatter comes from regions of CO_2 and H_2O ice. Thus, many people think that water ice may be present at Mercury despite its close proximity to the Sun. Locations of very bright depolarized coherent backscatter were found in about 25 craters at latitudes above 72° N (see Figs. 2.9 and 2.10). The south polar radar feature is centered at about 88° S and 150° west longitude, and is largely confined within a crater (Chao Meng-Fu) 150 km in diameter, but a few smaller features occur outside this crater. The area covered by these deposits (both north and south) is estimated to be 30,000 km^2. This would be equivalent to 4×10^{16} to 8×10^{17} grams of ice, or 40–800 km^3 for a 2–20 meter thick deposit. Each meter thickness of ice would be equivalent to about 10^{13} kilograms of ice.

These deposits are located in the permanently shadowed regions of fresh, undegraded craters (Harmon et al. 1994, 2001). Because Mercury has no or very little tilt to its polar axis, sunlight cannot penetrate to the bottom of deep craters with high crater walls within the latitude regions where the bright radar backscatter was observed (Paige et al. 1992). Temperatures in the polar regions should be <135 K (Vasavada et al. 1999). Crater studies show that some of the craters with high coherent backscatter have shallower depths than expected from lunar counterparts (Vilas et al. 2005). This may be indicative of infilling

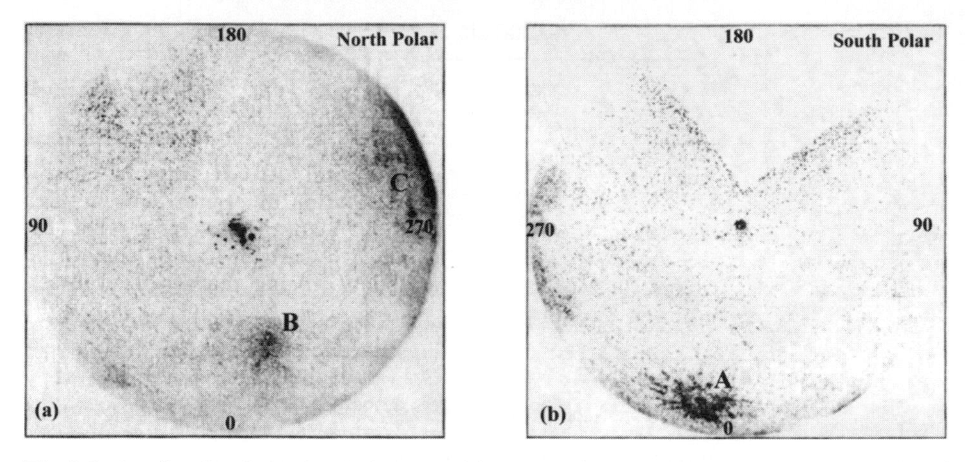

Fig. 2.9. Arecibo depolarized radar images of the (a) northern, and (b) southern hemispheres of Mercury, in polar projection. The radar features A, B, and C are indicated on the figure. The polar deposits can also be seen near the poles (from Harmon, 1997).

by the material causing the depolarized backscatter. Computations show that water ice, even if covered with a thin cover of dust, must be relatively young to survive the thermal radiation of Mercury and the high energy radiation of the Sun (Butler 1997, Moses et al. 1999). It is possible a recent comet or water-rich asteroid impact deposited the water in the deepest craters at high latitudes and the buried ice remains beneath a layer of regolith (Crider et al. 2005). Two large radar bright impact craters are apparently recent and could be the source of the water. The bright spot in Fig. 2.7 is one of these craters. It is possible they are recent comet or water-rich asteroid impacts that delivered water to Mercury. Following impact the water vapor could have migrated to high latitudes and become sequestered in the permanently shadowed regions of deep craters.

Other materials probably have similar radar backscattering properties required for the observed radar signals. Sulfur (Sprague et al. 1995) and pyroclastic silica glass (Starukina 2001) have been suggested for the infilling material. A one meter thick layer of water ice is stable for one billion years at a temperature of −161°C while sulfur is stable at a considerably higher temperature of −55°C. It is easier to preserve sulfur deposits at high latitudes. However, the region surrounding permanently shadowed craters is less than −55°C, there are no radar reflective deposits there thus detracting from the case for sulfur as the source of the high radar backscatter material.

2.5 Interior and Magnetic Field

Mercury has a unique internal structure. Its mean density of 5.44 g/cm^3 is only slightly less than Earth's (5.52 g/cm^3), and larger than Venus' (5.25 g/cm^3). However, Earth's large internal pressures result in an uncompressed density of only 4.4 g/cm^3 compared to Mercury's uncompressed density of 5.3 g/cm^3. Mercury, therefore, contains a much larger fraction of iron than any other planet or satellite in the Solar System (Solomon

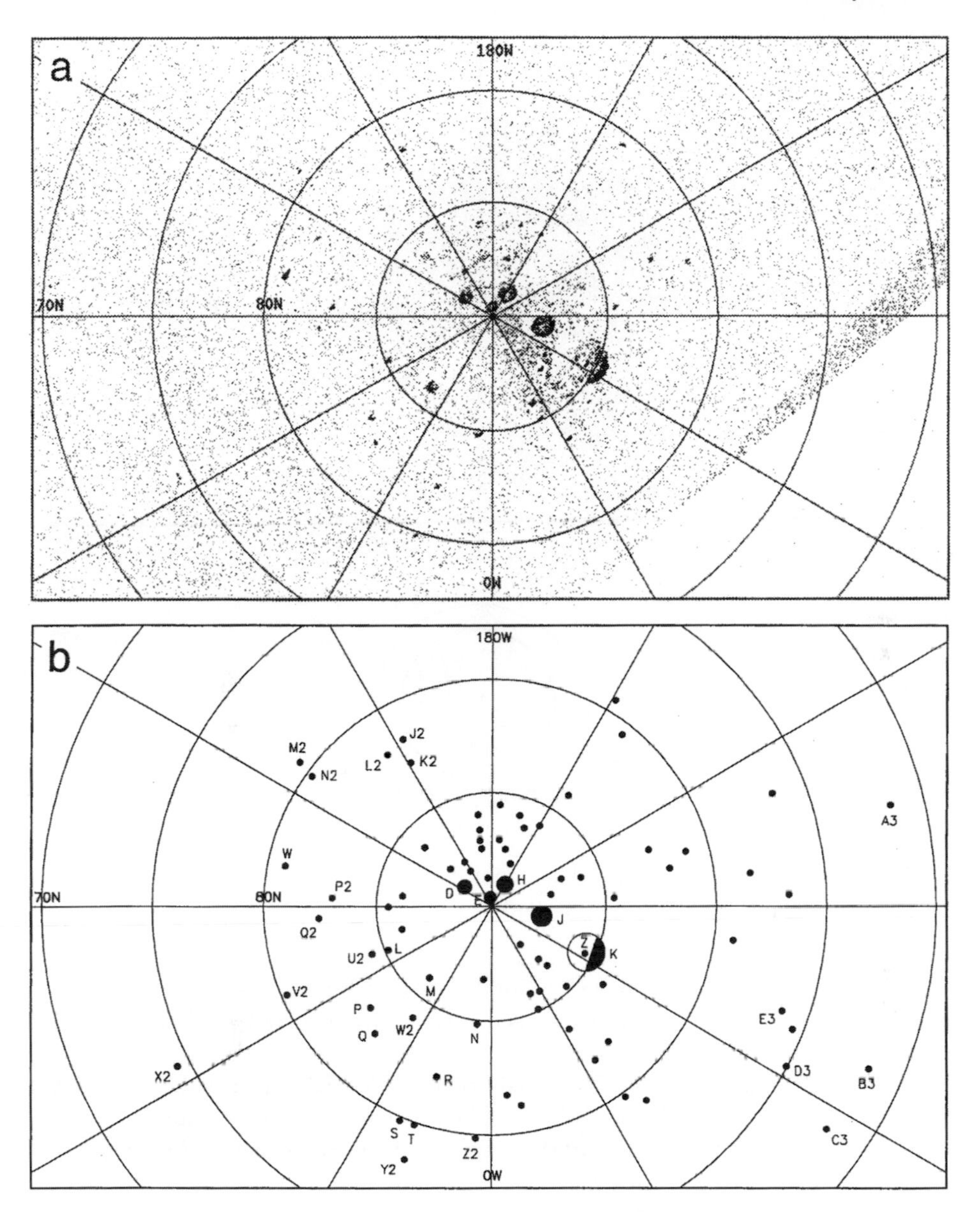

Fig. 2.10. Detailed high-resolution radar image showing the radar-bright deposits in the north polar regions, and with a superposed map of their locations and designations (from Harmon et al. 2001).

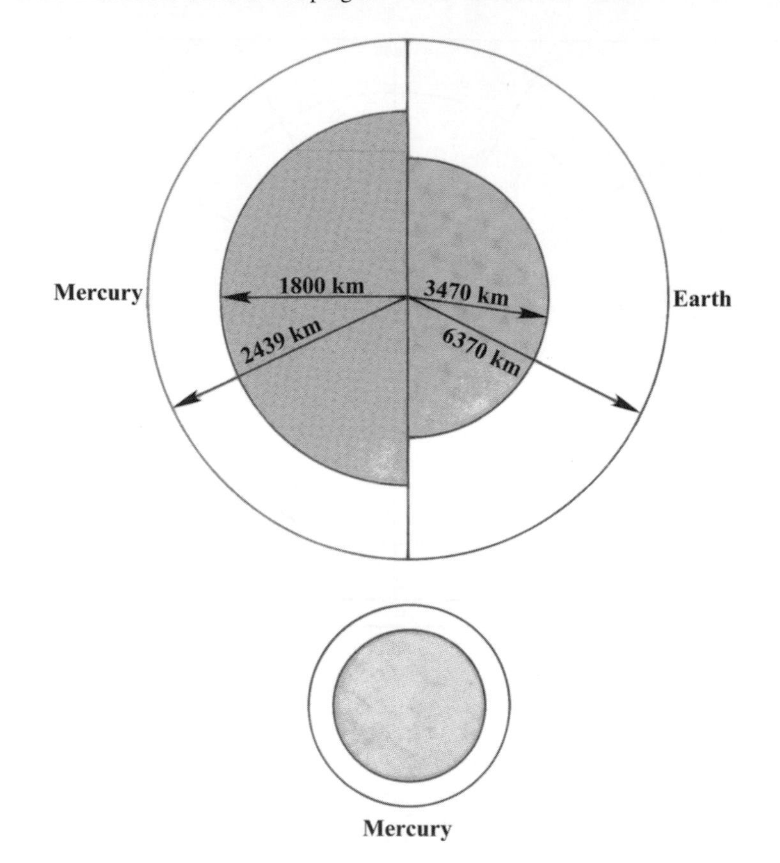

Fig. 2.11. Mercury's interior compared and normalized to the Earth's. The bottom diagram is Mercury's interior to the same scale as the Earth.

1977), indicating an iron core about 75% of the planet diameter, or some 42% of its volume (Fig. 2.11). Thus, its silicate mantle and crust is only about 600 km thick. (Schubert et al. 1988).

Aside from Earth, Mercury is the only other terrestrial planet with a significant magnetic field. Based on Mariner 10's short (only 30 minutes) observation of the field, investigators constructed a picture of the magnetic field environment at Mercury based on analogy with that of Earth's magnetic field and particle environment (Ness et al. 1974, Ness et al. 1975). Mercury probably lacks the ionosphere and trapped radiation zones of Earth's magnetosphere, so many of these comparisons are inappropriate (Ip 1987).

The measured magnetic field is strong enough to hold off the solar wind to form a bow shock (Hood and Schubert 1979). As the spacecraft approached the planet it measured a sudden increase in the field strength that represented the bow shock, and other signals indicating the entrance to and exit from a magnetopause surrounding a magnetospheric cavity that is about 20 times smaller than the Earth's. The magnetopause sub-solar distance is estimated to be about 1.35 ± 0.2 Mercury radii, and the bow shock distance

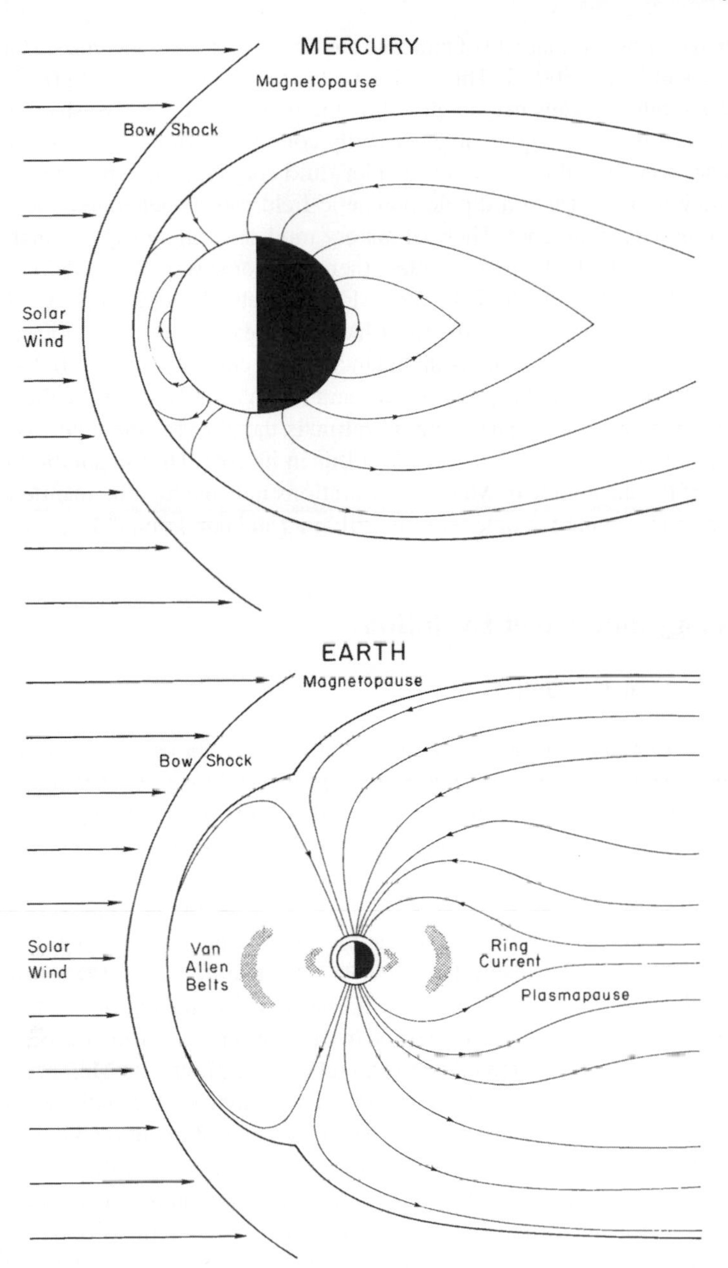

Fig. 2.12. Comparison of the Earth's and Mercury's magnetic fields. Mercury's field is much smaller than Earth's but it structure is similar.

is about 1.9 ± 0.2 Mercury radii. The polarity of the field is the same as Earth's. The magnetic strength increased as the spacecraft approached the planet. The interplanetary field is about 25 nT in the vicinity of Mercury, but it increased to 100 nT at closest approach to Mercury. If that rate of increase continued to the surface, the surface strength would be about 200 to 500 nT. This is about 1% of the Earth's strength (Fig. 2.12).

Although other models may be possible, the maintenance of terrestrial planet magnetic fields is thought to require an electrically conducting fluid outer core surrounding a solid inner core. The thickness of the outer fluid core is unknown at present, but theoretical studies suggest that a dipole magnetic field can be generated and maintained even in a thin outer fluid core. Thermal history models strongly suggest that Mercury's core would have solidified long ago unless there was some way of maintaining high core temperatures throughout geologic history. Most theoretical studies consider the addition of a light alloying element to be the most likely cause of a currently molten outer core (Schubert et al. 1988). For a sulfur abundance in the core of less than 0.2%, the entire core should be solidified at the present time, and for an abundance of 7% the core should be entirely fluid at present. Therefore, if sulfur is the alloying element, then Mercury probably contains between 0.2% and 7% sulfur in its core. High-resolution radar measurements of the magnitude of Mercury's librations indicate that the mantle is detached from the core indicating the outer core is still in a fluid condition (Margot et al. 2004).

2.6 Geology and Planet Evolution

2.6.1 Geologic surface units

The 45% of Mercury's surface viewed by Mariner 10 can be divided into 1) impact craters and basins, and their ejecta deposits, 2) hilly and lineated terrain, 3) intercrater plains, and 4) smooth plains. There may be other units on the unexplored part of the planet. Earth-based images show some bright and dark features on the side not imaged by Mariner 10 (Strom and Sprague 2003).

The heavily cratered uplands record the period of late heavy meteoroid bombardment that ended about 3.8 billion years ago on the Moon, and presumably at about the same time on Mercury. This period of late heavy bombardment occurred throughout the inner Solar System and is also recorded by the heavily cratered regions on the Moon and Mars. Mercury craters, however, have some marked differences from those on the Moon. These differences can be explained by the larger surface gravity on Mercury ($3.70\,m/s^2$) compared to the Moon ($1.62\,m/s^2$). For a given rim diameter the radial extent of Mercurian continuous ejecta is uniformly smaller by a factor of about 0.65 than that for the Moon. Furthermore, the maximum density of secondary impact craters occurs closer to the crater rim than for similar-sized lunar craters: the maximum density occurs at about 1.5 crater radii from the rim of Mercurian primaries, while on the Moon the maximum density occurs at about 2–2.5 crater radii. Another consequence of the larger surface gravity is that the transition diameter from simple (bowl-shaped) to complex (central peak and terraces) craters occurs at a smaller diameter on Mercury (10 km) than the Moon (20 km).

Twenty-two multi-ring basins have been recognized on the part of Mercury viewed by Mariner 10. The 1300 km diameter Caloris impact basin is the largest well-preserved impact structure (Fig. 2.13), although the much more degraded Borealis basin is larger (1530 km). The floor structure of the Caloris basin consists of closely spaced ridges and

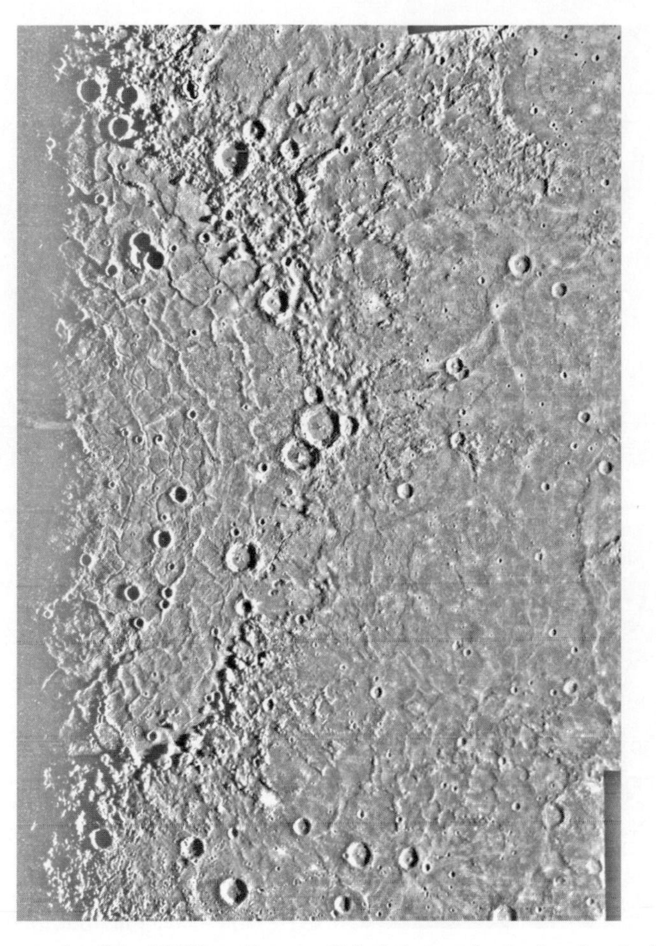

Fig. 2.13. Photomosaic of the 1300 km diameter Caloris impact basin showing the highly ridged and fractured nature of its floor.

troughs arranged in both a concentric and radial pattern. The ridges are probably due to contraction, while the troughs are probably extensional graben that post-date the ridges. The fractures get progressively deeper and wider toward the center of the basin. This pattern may have been caused by subsidence and subsequent uplift of the basin floor.

On the other side of Mercury directly opposite the Caloris basin (the antipodal point of Caloris) is the unusual hilly and lineated terrain that disrupts pre-existing landforms. The hills are 5 to 10 km wide and about 0.1 to 1.8 km high. Linear depressions that are probably extensional fault troughs form a roughly orthogonal pattern. Crater counting and the relationship of the fault patterns to neighboring surface units suggest that the age of this terrain is the same as the Caloris basin. The hilly and lineated terrain is thought to be the result of shock waves generated by the Caloris impact and focused at the antipodal region (Fig. 2.14). The large size of the disrupted terrain may be the result of enhanced shock wave focusing due to the large iron core.

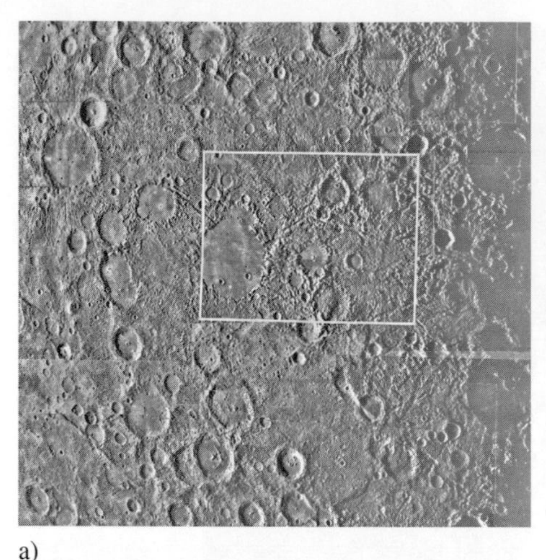

a)

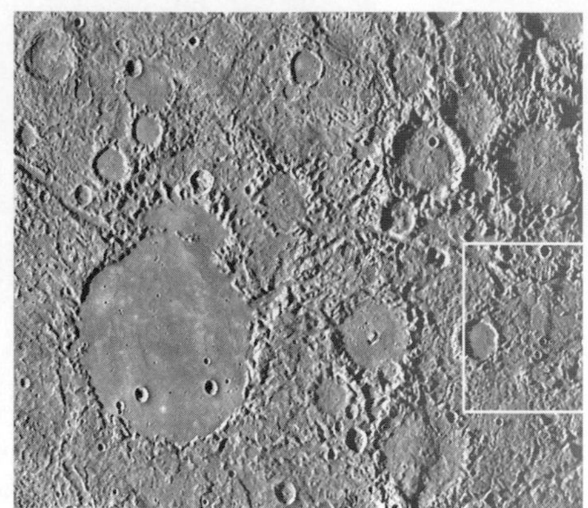

b)

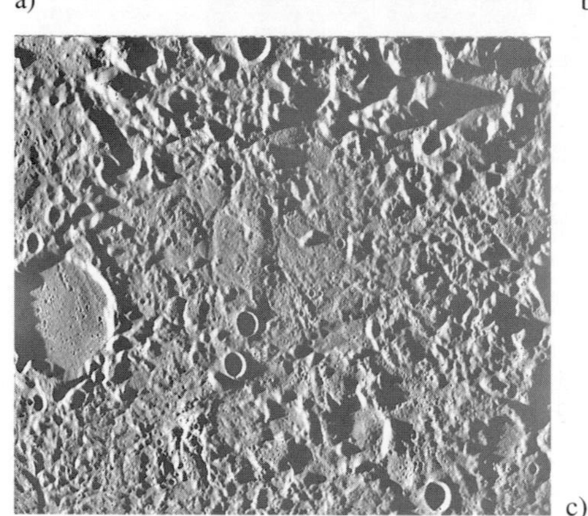

c)

Fig. 2.14. (a) A portion of the hilly and lineated terrain antipodal to the Caloris impact basin. The image is 543 km across. (b) Detail of the hilly and lineated terrain. The largest crater in (b) is 31 km in diameter.

Mercury has two plains units that have been interpreted as either impact basin ejecta or as lava plains. The older intercrater plains are the most extensive terrain on Mercury (Fig. 2.15). They both partially fill and are superimposed by craters in the heavily cratered uplands. Intercrater plains were emplaced over a range of ages contemporaneous with the period of late heavy bombardment. Although no landforms diagnostic of volcanic activity have been discovered, there are also no obvious source basins to provide ballistically emplaced ejecta. The global distribution of intercrater plains and the lack of source basins for ejecta deposits are indirect evidence for a volcanic origin. Additional evidence for a volcanic origin is recent Mariner 10 enhanced color images (Robinson and Lucey 1999) showing color boundaries that coincide with geologic unit boundaries of some intercrater plains. If intercrater plains are volcanic then they are probably lava flows erupted from fissures early in Mercurian history. Intercrater plains are probably $\geq$ 3.9 billions years old.

Almost 40% of the total area imaged by Mariner 10 is covered by the younger smooth plains. About 90% of these plains are associated with large impact basins. They also

Fig. 2.15. View of the intercrater plains surrounding clusters of craters in the Mercurian highlands. Several lobate scarps (thrust faults) can also be seen.

fill smaller basins and large craters. The largest occurrence of smooth plains fill and surround the Caloris basin (Fig. 2.13), and occupy a large circular area in the north polar region that is probably an old impact basin (Borealis basin). They are similar in morphology and mode of occurrence to the lunar maria. Craters within the Borealis, Goethe, Tolstoj and other basins have been flooded by smooth plains (Fig. 2.16). This indicates the plains are younger than the basins they occupy. This is supported by the fact that the density of craters superimposed on the smooth plains that surround the Caloris basin is substantially less than all major basins including Caloris. The smooth plains' youth relative to the basins they occupy, their great areal extent, and other stratigraphic relationships suggest they are volcanic deposits erupted relatively late in Mercurian history.

Mariner 10 enhanced color images (Robinson and Lucey 1999) show the boundary of smooth plains within the Tolstoj basin is also a color boundary, further strengthening the volcanic interpretation for the smooth plains (Robinson and Taylor 2001). Based on the shape and density of the size/frequency distribution of superimposed craters, the smooth plains probably formed near the end of late heavy bombardment. They may have an average age of about 3.8 billion years as indicated by crater densities. If so, they are, in general, older than the lava deposits that constitute the lunar maria.

Three large radar-bright anomalies (Fig. 2.9) have been identified on the side of Mercury not imaged by Mariner 10. They are designated as A (345° W longitude, −32° latitude), B (345° W longitude, 58° longitude), and C (240° W longitude, 0° latitude). Both features A and B are relatively fresh impact craters with radar-bright ejecta blankets and rays similar to the radar image of Kuiper crater. Feature A is about 140 km in diameter with an extensive ray system and a rough radar-bright floor, consistent with a fresh impact crater. Feature B is smaller ($\sim$ 70 kilometers) with radar-bright rays and

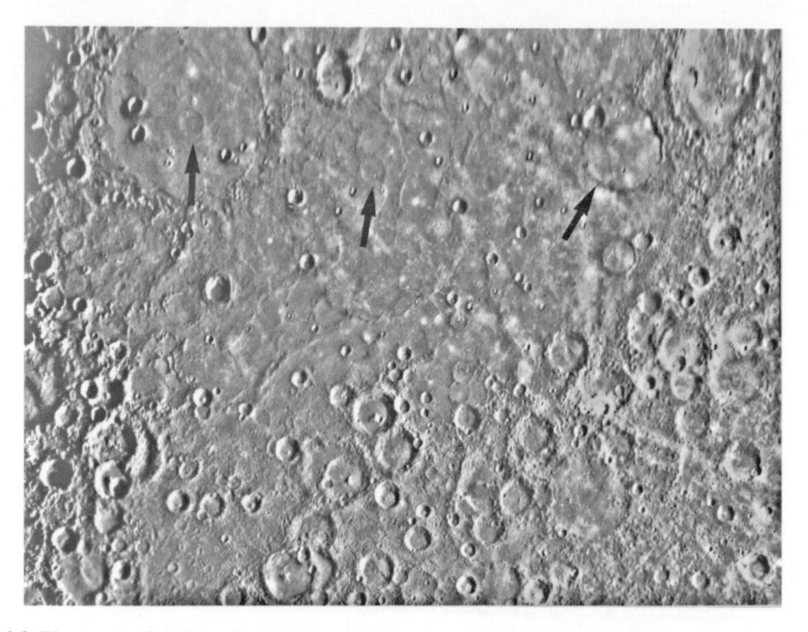

Fig. 2.16. Photomosaic of the Borealis basin showing numerous craters (arrowed) that have been flooded by smooth plains. The largest crater is the Goethe basin 340 km in diameter.

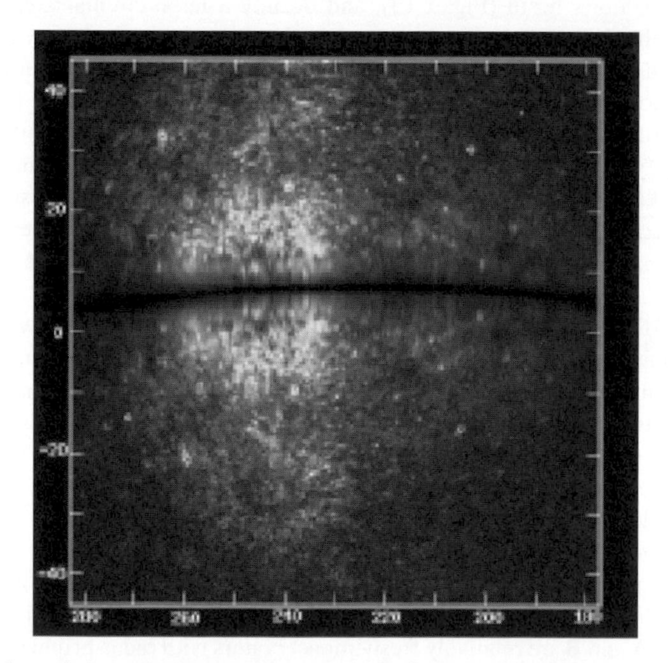

Fig. 2.17. High resolution radar image of Feature B. The upper half is a mirror image of the bottom half (Harmon and Campbell 2001).

a radar-dark floor. Unlike Feature A, the radar-dark floor indicates it is smooth at the 12.6 cm wavelength of the image. Figure 2.17 is an excellent Earth-based image of a portion of the surface not imaged by Mariner 10 (Massey, 2005, Fig. 5). One outstanding feature is the bright spot at 50° N latitude and 245° longitude. This spot is radar feature B. It is bright in visible light, as would be expected from a fresh impact crater. It is also a source of sodium atoms that shows up as bright resonant scattering in the exosphere.

Feature C consists of a large circular region about 1000 kilometers in diameter consisting of small radar-bright spots covering a region to the west and south of Caloris basin. The region is centered at about 0° latitude and extends from 220° to 280° W. longitude (Fig. 2.9). It has been interpreted as a dense cluster of small ring features, probably a cluster of impacts. It exhibits different radar properties than features A and B (Harmon 2002). In Fig. 2.4 there is a dark circular feature in the terminator that could be a lava filled impact basin. The true geologic nature of this feature will have to await further exploration.

2.6.2 Surface composition

There is some knowledge of the chemical makeup of Mercury's surface. Ground-based spectroscopic observations have indicated that there may be basalts and anorthosites on Mercury's surface (Sprague et al. 1994). Basalts and anorthosites have very different origins. Basalt is formed by a petrogenic process called partial melting. As the temperature of a parent rock is increased, the low-melting point fraction melts first. It has a characteristic composition of low-temperature melts and rises to erupt on the surface. Examples from the Moon are both the mare and KREEP (potassium, rare earth elements, phosphorous). Anorthosite, on the other hand, is formed by a process called fractional crystallization. In this process, when an igneous melt begins to crystallize, the denser crystals sink through the residual liquid and the lesser density crystals float to the top. Therefore, layers accumulate in which one mineral is greatly concentrated. The lunar highlands anorthositic rocks are characterized by a super-abundance of one mineral, plagioclase feldspar ($CaAl_2Si_2O_8$). This composition is far removed from low-melting point liquids and was formed by fractional crystallization.

About 40% of Mercury's surface has been measured spectroscopically in the near- and mid-infrared. From these observations we know that Mercury's surface composition is heterogeneous. Regions near Homer and the Murasaki Crater Complex appear feldspathic and appear to be more Na-rich than the lunar anorthosites (Sprague and Roush 1998). Bulk compositions are of intermediate silica content. Some mixed compositions of more basic silica content are present in the regions from 68°–160° longitude but are not at all locations measured (Cooper et al. 2001). One candidate for the rocks at these locations is low-iron basalt – mixtures of feldspar, pyroxene, and minor olivine of intermediate (52–57 weight % SiO_2) or basic (46–51 weight %) SiO_2. This interpretation is very generalized because the areal extent of the spatial footprint is no smaller than 200 km by 200 km for the very best spatially resolved observations, and as much as 1000 km by 1000 km for the least spatially resolved region. A strong 5 micrometer emission feature in a spectrum from 45°–85° longitude closely resembles that of laboratory clino-pyroxene powders (Sprague et al. 2002). The best fit is to diopside, a low-FeO clino-pyroxene.

Table 2.3. Comparison of FeO Abundance in the Solar System

Terrestrial Body	Typical Weight % FeO
Mercury	1–3
Venus	8–10
Earth	5–28
Moon	4–27
Mars	10–40
Asteroids	4–25

Another important fact is that Mercury's surface is very low in oxidized iron (McCord and Clark 1976, Vilas 1986, Vilas et al. 1988., Warell and Blewett 2004). The absorption bands of reflected light from Mercury's surface show very shallow, if any, absorption caused by an electronic transition in an iron cation (Fe^{2+}) that is when Fe is bound to O in a lattice of a silicate. Recently, the first unambiguous detection of the 1 μm absorption band characteristic of iron-bearing olivine was found at the north and south polar regions centered on 247° W. longitude (Warell et al. 2006). Both absorptions are very shallow and indicate only a trace of oxidized iron (Fig. 2.18). A comparison of the abundance of FeO inferred from many measurements of all the terrestrial planets, the Moon, and some asteroids is given in Table 2.3.

Based on these observations, petrologists of the lunar surface think that lunar anorthosites (Robinson and Taylor 2001), perhaps more sodium-rich like labradorite ($(Na,Ca)Si_3O_8$) provide the best analogue for the composition of much of Mercury (Sprague and Roush 1998). An alternative explanation for the apparent match to plagioclase feldspar at so many locations is that the spectrum comes from the glassy soil on Mercury's surface that is very mature after aeons of meteoritic bombardment. Scientists have shown that if lunar soils are very mature, much of the FeO is removed from the glasses and they have the spectral appearance of feldspar in laboratory spectral measurements (Nobel and Pieters 2001, Sasaki and Kurahashi 2004).

There are also regions on Mercury's surface that give evidence of other compositions. Mariner 10 imaging has revealed information about Mercury's surface, the Rudaki (3° S, 56° W longitude) and Tolstoj (16° S, 164° W longitude) smooth plains. At both locations there are embaying boundaries indicative of infilling lava flows and with scattering properties similar to pyroclastics and glasses on the Moon (Robinson and Taylor 2001). On the Moon lava flows are basalt. On Mercury there is no evidence for the Fe-bearing basalts so common on the Moon. However there is ample evidence for fluid lava flows (Strom and Sprague 2003). Such fluid lava flows could be low-iron basalts or other types of lava that retain their fluid nature. At other locations high alkali basalts and rocks formed from other late stage magmas may be present (Jeanloz et al. 1995).

2.6.3 Tectonic framework

Mercury has a unique tectonic framework. It consists of a system of contractional thrust faults called lobate scarps (Fig. 2.19). Individual scarps vary in length from about 20 km to over 500 km, and have heights from a few 100 m to about 2 km. They have a random

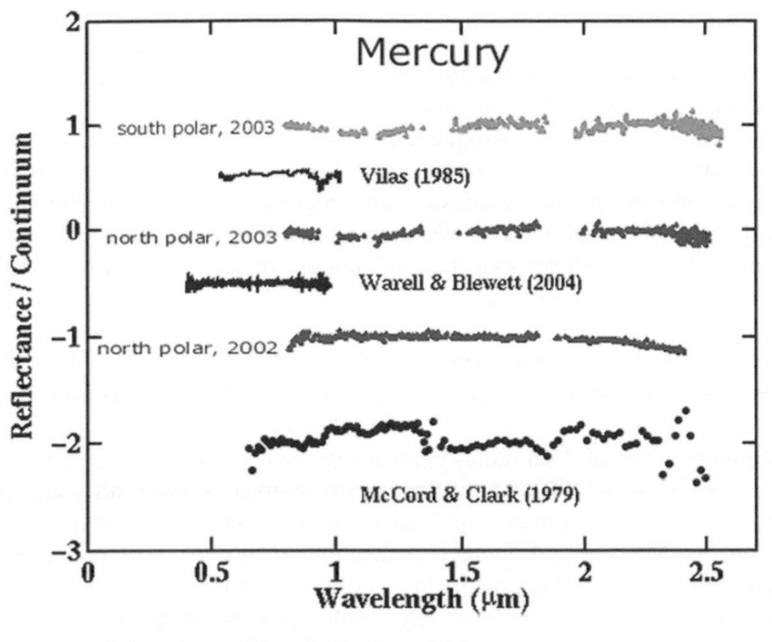

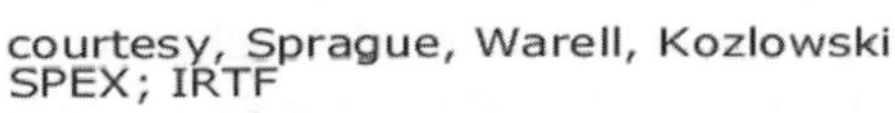

Fig. 2.18. Spectra obtained at the IRTF using SpeX (grey data) from three different locations on Mercury's surface are shown after removal of solar reflectance and thermal emission components and division by a linear fit to the continuum at 0.77 and 1.6 μm. All spectra are normalized at 1 μm for easy comparison to one another and to spectra previously published by others. The data covering most of the 0.8 to 2.5 μm spectral region are the first data to unambiguously show the FeO absorption band from 0.8 to 1.1 μm The Mercury (2002 N) spectrum from ~ 250° W longitude at high north latitudes show no evidence for the FeO absorption band. Shallow bands centered at 1.1 μm are present in both the north and south latitude spectra at 107°–194° and 120°–194° W longitude, respectively. (Figure from Warell et al. 2006).

spatial and azimuthal distribution over at least half of the planet and presumably are globally distributed. This indicates that, at least in its latest history, the entire planet was subjected to global contractional stresses (Melosh and McKinnon 1988, Thomas et al. 1988). The only occurrences of features indicative of extensional stresses are localized on the floor of the Caloris basin and at its antipode. Both of these occurrences are the direct or indirect result of the Caloris impact. No lobate scarps have been embayed by intercrater plains on the region viewed by Mariner 10, and they cut across fresh as well as degraded craters. Few craters are superimposed on the scarps. Therefore, the system of thrust faults appears to post-date the formation of intercrater plains and formed relatively late in Mercurian history. This tectonic framework was probably caused by crustal shortening resulting from a decrease in the planet radius due to cooling of the planet. The amount of radius decrease is estimated to have been between 0.5 and 2 km.

2.6.4 Thermal history

All thermal history models of planets depend on compositional assumptions, such as the abundance of uranium, thorium and potassium in the planet. Since our knowledge of the composition of Mercury is so poor, these models can only provide a general idea of the thermal history for certain starting assumptions. Nevertheless, they are useful in providing insights into possible modes and consequences of thermal evolution. Starting from initially molten conditions for Mercury, thermal history models with from 0.2 to 5% sulfur in the core indicate that the total amount of planetary radius decrease due to cooling is from about 6 to 10 km depending on the amount of sulfur (Fig. 2.20(a)). About 6 km of this contraction is solely due to mantle cooling during about the first 700 million years before the start of inner core formation. The amount of radius decrease due to inner core formation alone is about 1 km for 5% sulfur and about 4 km for 0.2% sulfur.

If the upper value of 2 km radius decrease inferred from the thrust faults was due solely to cooling and solidification of the inner core, then the core sulfur abundance is probably 2 to 3% and the present fluid outer core is about 500 or 600 km thick. If the lower value of 0.5 km radius decrease is correct then there must be more than 5% sulfur in the core and the present fluid outer core would be over 1000 km thick.

If the smooth and intercrater plains are volcanic flows, then they had to have some way to easily reach the surface to form such extensive deposits. Early lithospheric compressive stresses would make it difficult for lavas to reach the surface , but the lithosphere may have been relatively thin at this time ($<$50 km). Large impacts would be expected to strongly fracture it, possibly providing egress for lavas to reach the surface and bury compressive structures.

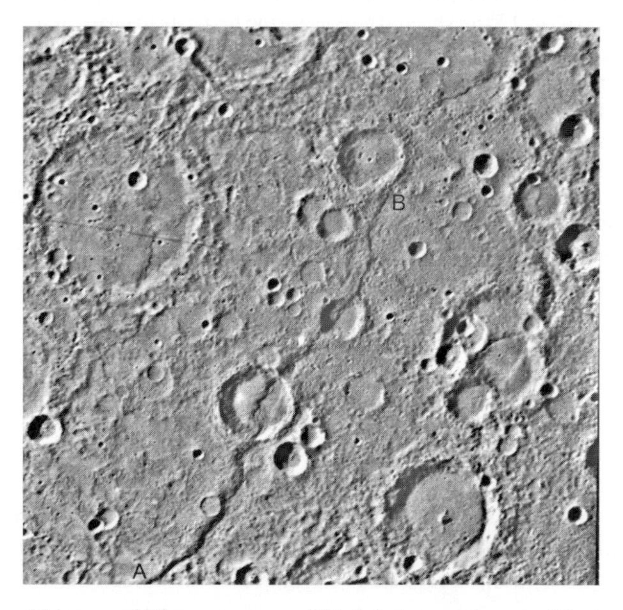

Fig. 2.19. Mariner 10 image of Discovery scarp. This lobate scarp is one of the largest thrust faults imaged by Mariner 10 (see text for description).

2.6.5 Geologic history

Mercury's earliest history is very uncertain. If a portion of the mantle was stripped away, as invoked by most scenarios to explain its high mean density, then Mercury's earliest recorded surface history began after core formation, and a possible mantle-stripping event (see Sect. 2.7). The earliest events are the formation of intercrater plains ($\sim 3.9 \times 10^9$ years ago) during the period of late heavy bombardment. These plains may have been erupted through fractures caused by large impacts in a thin lithosphere. Near the end of late heavy bombardment the Caloris basin was formed by a large impact that caused the hilly and lineated terrain from shock waves focused at the antipodal region. Further eruption of lava within and surrounding the Caloris and other large basins formed the smooth plains about 3.8×10^9 years ago. The system of thrust faults formed after the intercrater plains, but how soon after is not known. If the observed thrust faults resulted only from core cooling, then they may have begun after smooth plains formation, and resulted in a decrease in Mercury's radius (Fig. 2.20(b)). As the core continued to cool

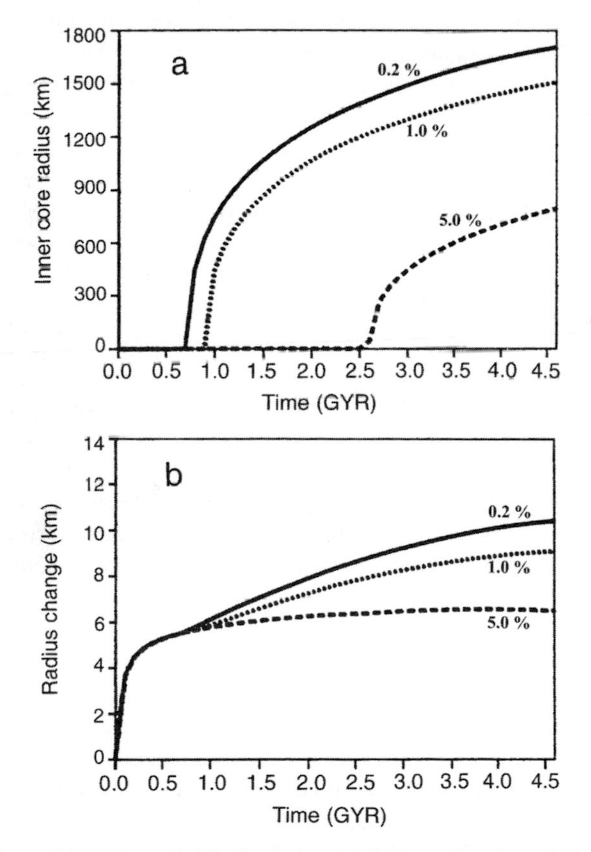

Fig. 2.20. (a) A thermal history modal for inner core radius as a function of time for three values of initial core sulfur content. The lines show the ranges in sulfur content from 0.2% to >5%, and the solid, dotted and dashed lines are for sulfur contents of 0.2%, 1%, and 5% respectively. (b) Decrease in Mercury's radius due to mantle cooling and inner core growth for three values of initial core sulfur content as in (a). (Schubert et al. 1988).

and the lithosphere thickened contractive stresses closed off the magma sources and volcanism ceased near the end of late heavy bombardment (Solomon 1977). All of Mercury's volcanic events probably took place very early in its history, perhaps the first 700 to 800 million years (Schubert et al. 1988). Since that time only occasional impacts of comets and asteroids have occurred. Today the planet may still be contracting as the present fluid outer core continues to cool.

2.7 Origin

Probably the central question about Mercury is how it acquired such a large fraction of iron compared to the other terrestrial planets. For Mercury's present position, chemical equilibrium condensation models in that part of the solar nebula cannot account for the large fraction of iron that must be present to explain its high density (Lewis 1988). Revised models that take into account material supplied from feeding zones in more distant regions of the inner Solar System can only account for a mean uncompressed density of about 4.2 g/cm^3, rather than the observed 5.3 g/cm^3. Also, at Mercury's present distance the cosmochemical models predict almost no sulfur (100 parts per trillion FeS). However, considerably more sulfur is apparently required to account for the presently molten outer core. Other volatile elements and compounds, such as water, should also be severely depleted ($H < 1$ ppb).

Three hypotheses have been offered to explain the discrepancy between the predicted and observed iron abundance. One is called selective accretion where the enrichment of iron results from the mechanical and dynamical accretion processes in the innermost part of the Solar System (Weidenschilling 1978). The other two (post-accretion vaporization (Cameron 1985) and giant impact (Wetherill, 1988; Benz et al. 1988)) involve the removal of a large fraction of the silicate mantle from a once larger proto-Mercury. In the selective accretion model, the differential response of iron and silicates to impact fragmentation and aerodynamic sorting leads to iron enrichment owing to the higher gas density and shorter dynamical time scales in the innermost part of the solar nebula. In this model the removal process for silicates from Mercury's present position is more effective than for iron, leading to iron enrichment. The post-accretion vaporization hypothesis proposes that intense bombardment by solar electromagnetic and corpuscular radiation in the earliest phases of the Sun's evolution vaporized and drove off much of the silicate fraction of Mercury leaving the core intact. In the giant impact hypothesis, a planet-sized object impacts Mercury and essentially blasts away much of the planet's silicate mantle leaving the core largely intact.

It may be possible to decide between these models from the chemical composition of Mercury's silicate mantle. For the selective accretion model, Mercury's silicate portion should contain about 3.6 to 4.5% alumina, about 1 percent alkali oxides (Na and K) and between 0.5 and 6% FeO. Post-accretion vaporization should lead to very severe depletion of alkali oxides (0%) and FeO ($<0.1\%$), and extreme enrichment of refractory oxides (40%) (Fegley and Cameron 1987). If a giant impact stripped away the crust and upper mantle late in accretion then alkali oxides may be depleted (0.01 to 0.1%), with refractory oxides between about 0.1 to 1% and FeO between 0.5 and 6%. To date knowledge of Mercury's silicate composition is extremely poor, but near and mid-infrared spectroscopic measurements favor low FeO and alkali bearing feldspars. If the tenuous atmosphere of sodium and potassium is being outgassed from crustal materials, then the post-accretion vaporization model may be unlikely. Deciding between the other two models is not possible with our current state of ignorance about the silicate composition of the crust and mantle. Since the selective accretion hypothesis requires Mercury to have formed near its present position, then sulfur should be nearly absent, unless the solar nebula temperatures in this region were considerably lower than predicted by the chemical equilibrium condensation model.

The giant impact hypothesis is supported by three-dimensional computer simulations of terrestrial planet formation for several starting conditions (Wetherill, 1988). Since these simulations are by nature stochastic, a range of outcomes is possible. They suggest, however, that significant fractions of the terrestrial planets may have accreted from material formed in widely separated parts of the inner Solar System. The simulations indicate that during its accretion Mercury may have experienced large excursions in its semimajor axis. These semimajor axis excursions may have ranged from as much as 0.4 to 1.4 AU due to energetic impacts during accretion (Fig. 2.21). Consequently, Mercury could have accumulated material originally formed over the entire terrestrial planet range of heliocentric distances. About half of Mercury's mass could have accumulated at distances between about 0.8 and 1.2 AU (Fig. 2.22). If so, then Mercury may have acquired its sulfur from material that formed in regions of the solar nebula where sulfur was stable. Plausible models estimate FeS contents of 0.1 to 3%. However, the most extreme models of accretional mixing result in homogenizing the entire terrestrial planet region, contrary to the observed large systematic density differences.

The simulations also indicate that by-products of terrestrial planet formation are planet-sized objects up to 3 times the mass of Mars that become perturbed into eccentric

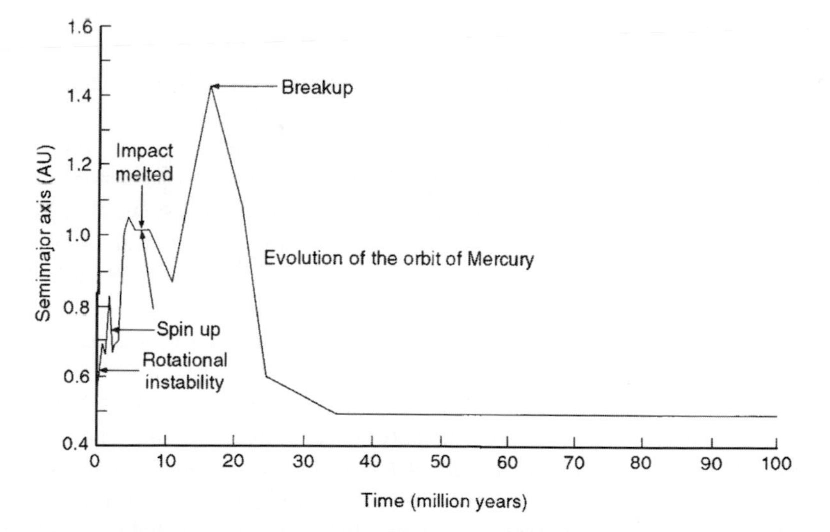

Fig. 2.21. Results of a computer simulation of terrestrial planet evolution showing the change of "Mercury's" semimajor axis during its accretion. In this case "Mercury's" semimajor axis spans the entire terrestrial planet region (0.5 to 1.4 AU) during the planet's growth. (Wetherill, 1988).

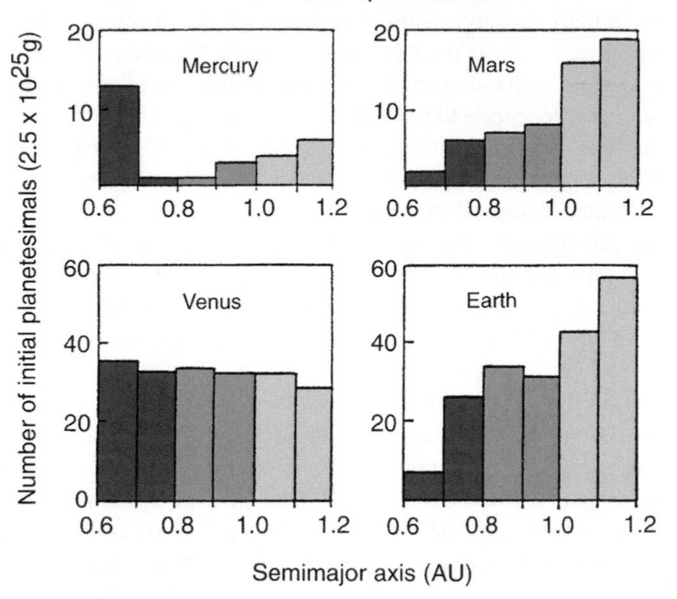

Fig. 2.22. Results of a computer simulation of terrestrial planet evolution showing the region (semimajor axis) from which the terrestrial planets acquired their mass. In this simulation, "Mercury" acquires about half its mass from regions between 0.8 and 1.2 AU. (Wetherill, 1988).

orbits (mean e $\sim$ 0.15 or larger) and eventually collide with the terrestrial planets during their final stages of growth. The final growth and giant impacts occur within the first 50 million years of Solar System history. Such large impacts may have resulted in certain unusual characteristics of the terrestrial planets, such as the slow retrograde rotation of Venus, the origin of the Moon (discussed by Harrison Schmitt in Chapter 4 of this volume), the martian crustal dichotomy, and Mercury's large iron core.

In computer simulations where proto-Mercury was 2.25 times the present mass of Mercury with an uncompressed density of about 4 g/cm^3, nearly central collisions of large projectiles with iron cores impacting at 20 km/s, or non-central collisions at 35 km/s resulted in a large silicate loss and little iron loss (Benz et al., 1988). In the former case, although a large portion of Mercury's iron core is lost, an equally large part of the impactor's iron core is retained resulting in about the original core size. At Mercury's **present** distance from the Sun, the ejected material re-accretes back onto Mercury if the fragment sizes of the ejected material are greater than a few centimeters. However, if the ejected material is in the vapor phase or fine-grained ($\leq$ 1 cm), then it will be drawn into the Sun by the Poynting-Robertson effect in a time shorter than the expected collision time with Mercury (about 106 years). The proportion of fine-grained to large-grained material ejected from such an impact is uncertain. Therefore, it is not known if a large impact at Mercury's present distance could exclude enough mantle material to account for its large iron core. However, the disruption event need not have occurred at Mercury's present distance from the Sun. It could have occurred at a much greater distance, e.g., >0.8 AU (Fig. 2.21). In this case the ejected mantle material would be mostly swept up by the larger terrestrial planets, particularly Earth and Venus.

2.8 The MESSENGER Mission

A spacecraft called MESSENGER is now on its way to orbit Mercury (Santo et al. 2001). The name MESSENGER is an acronym for **ME**rcury, **S**urface, **S**pace **EN**vironment, **GE**ochemisty, and **R**anging. To the ancients, Mercury was the messenger of the gods. On August 3, 2004 the MESSENGER spacecraft was launched from Cape Canaveral, Florida to explore Mercury for the first time in over 30 year. After the Earth flyby that took place in August 2005, it will make two flybys of Venus (Oct. 2006 and June 2007) and three flybys of Mercury (Jan. and Oct. 2008, and Sept. 2009) before it is inserted into Mercury orbit in March 2011. The reason it takes 7 years to put the spacecraft in orbit around Mercury is because it must make 6 planetary encounters to slow the spacecraft enough to put it in orbit with a conventional retro-rocket.

There are seven main objectives of this NASA mission, all of which are important to understanding the origin and evolution of Mercury and the inner planets (Solomon et al. 2001). One is to determine the nature of the polar deposits including their composition. Another objective is to determine the properties of Mercury's core including its diameter and the thickness of its outer fluid core. This is accomplished by accurately measuring Mercury's libration amplitude from the laser altimeter and radio science experiments. A third objective is to determine variations in the structure of the lithosphere and whether or not convection is currently taking place. A fourth objective is to determine the nature of the magnetic field and confirm whether or not it is a dipole. There are several instruments (Goldsten, et al. 2005, Schlemm et al. 2005) to study the chemical and mineralogical composition of the crust that should place constraints on Mercury's origin, and hopefully decide between the three competing hypotheses. Also these data will be extremely useful to decipher Mercury's geology. The geologic evolution of Mercury will be addressed by the dual camera system that will image the entire surface at high resolution and at a variety of wavelengths. Finally, the exosphere will be studied to determine its composition (McClintock and Lankton 2005) and how it interacts with the magnetosphere and surface. There are eight science experiments on board the spacecraft (Gold et al. 2001) as listed in Table 2.4.

MESSENGER will be placed in an elliptical orbit with a 200 km periapse altitude located at about 60° north latitude. The orbit has a 12-hour period when data will be collected and read out. The spacecraft will also collect valuable data on its three flybys of Mercury prior to orbit insertion. MESSENGER should provide the data necessary to answer most of the questions raised in this chapter.

Table 2.4. Scientific Payload on MESSENGER Spacecraft

Instrument	**Observation**
Dual Imaging System (1.5° and 10.5° field of view)	Surface mapping in stereo (10 color filters)
Gamma Ray and Neutron Spectrometer	Surface composition (O,Si,Fe,H,K)
X-Ray Spectrometer (1 to 10 KeV)	Surface composition (Mg,Al,Fe,Si,S,Ca,Ti)
Atmospheric and Surface Spectrometer	Surface and Exosphere composition
Magnetometer	Magnetic field
Laser Altimeter	Topography of northern hemisphere
Energetic Particles and Plasma Spectrometer	Energetic particles and plasma
Radio Science (X-band transponder)	Gravity field and physical libration

References

Baumgardner, J., Mendillo, M. and Wilson, J.K., 2000. A digital high definition imaging system for spectral studies of extended planetary atmospheres: 1. Initial result in white light showing features on the hemisphere of Mercury unimaged by Mariner 10. *Astron. J.* **119**, 2458–2464.

Benz, W., W.L. Slattery, and A.G.W. Cameron, 1988. Collisional stripping of Mercury's mantle. *Icarus* **74**, 516–528.

Bida, T.A., R.M. Killen, and T.H. Morgan, 2000. Discovery of calcium in Mercury's atmosphere. *Nature* **404**, 159–161.

Blewett, D.T., P.G. Lucey, B.R. Hawke, G.G. Ling, and M.S. Robinson, 1997. A Comparison of Mercurian Reflectance and Spectral Quantities with Those of the Moon. *Icarus* **129**, 217–231.

Broadfoot, A.L., Kumar, S., Belton, M. and McElroy, M.B., 1974. Mercury's atmosphere from Mariner 10: Preliminary results. *Science* **185**, 166–169. Broadfoot, A.L., D.E. Shemansky, and S. Kumar, 1976. Mariner 10: Mercury atmosphere, *Geophys. Res. Lett.* **3**, 577–580.

Butler, B., Muhleman, D. and Slade, M., 1993. Mercury: Full-disk radar images and the detection and stability of ice at the north pole. *J. Geophys. Res.* **98**, 15,003–15,023.

Butler, B.J. 1997. The migration of volatiles on the surfaces of Mercury and the Moon. *J. Geophys. Res.* **102** (E8), 19,283–19,291.

Cheng, A.F., Johnson, R.E., Krimigis, S.M. and Lanzerotti, L.J., 1987. Magnetosphere, Exosphere, and Surface of Mercury. *Icarus* **71**, 430–440.

Cameron, A. G. W., 1985. The partial volatilization of Mercury. *Icarus* **64**, 285–294.

Cooper, B., A. Potter, R. Killen, and T. Morgan, 2001. Mid-Infrared Spectra of Mercury. *J. Geophysical Res.* **106** (E12), 32,803–32,814.

Crider, D. H. and R. M. Killen, 2005. Burial rate of Mercury's polar volatile deposits. *Geophys. Res. Lett.* **32** (L12201), 10.1029/2005GL022689.

Davies, M.E., S.E. Dwornik, D.E. Gault, and R.G. Strom, 1978. *NASA SP-423, Atlas of Mercury*, Scientific and Technical Information Office, National Aeronautics and Space Administration, US Government Printing Office, Washington, D.C.

Fegley, B., Jr. and A.G.W. Cameron, 1987. A vaporization model for iron/silicate fractionation in the Mercury protoplanet. *Earth Planet. Sci. Lett.* **82**, 207–222.

Gold, R. E., et al., 2001. The MESSENGER mission to Mercury: scientific payload. *Planet. Space Sci.* **49**, 1467–1479.

Goldsten, J.O., et al., 2005. The MESSENGER Gamma-Ray/Neutron Spectrometer. *Space Sci. Rev.* submitted.

Goldstein, B.D., Suess, S.T. and Walker, R.J., 1981. Mercury: Magnetospheric processes and the atmospheric supply and loss rate. *J. Geophys. Res.* **86**, 5485–5499.

Harmon, J.K. and Slade, M.A., 1992. Radar mapping of Mercury: Full-disk images and polar anomalies. *Science* **258**, 640–642.

Harmon, J.K., M.A. Slade, R.A. Velez, A. Crespo, M.J. Dryer, and J.M. Johnson, 1994. Radar mapping of Mercury's polar anomalies. *Nature* **369**, 213–215.

Harmon, J.K., P.J. Perillat, and M.A. Slade, 2001. High-Resolution radar imaging of Mercury's north pole. *Icarus* **149**, 1–15.

Harmon, J.K., and D.B. Campbell, 2002. Mercury Radar Imaging at Arecibo in 2001. Lunar and Planetary Science Conference 33, 11–15 March 2002, Houston, Texas, Abstract #1858.

Hood, L.L. and Schubert, G., 1979. Inhibition of solar wind impingement on Mercury by planetary induction currents. *J. Geophys. Res.* **84**, 2641–2647.

Hunten, D.M., T.H. Morgan, and D. Shemansky, 1988. The Mercury atmosphere. In *Mercury*, eds. F. Vilas, C.R. Chapman, and M.S. Matthews, Univ. of Arizona Press, Tucson, Arizona, pp. 562–612.

Ip, W.H., 1987. Dynamics of electrons and heavy ions in Mercury's magnetosphere. *Icarus* **71**, 441–447.

Jeanloz, R., D.L. Mitchell, A.L. Sprague, and I. de Pater, 1995. Evidence for a basalt-free surface on Mercury and implications for internal heat. *Science* **268**, 1455–1457.

Killen, R.M., T.H. Morgan, and A.E. Potter, 1990. Spatial distribution of sodium vapor in the atmosphere of Mercury. *Icarus* **85**, 145–167.

Killen, R.M., A.E. Potter, A. Fitzsimmons, and T.H. Morgan, 1999. Sodium D2 line profiles: clues to the temperature structure of Mercury's exosphere. *Plan. Space Sci.*, **47**, 1449–1458.

Killen, R.M., T.A. Bida, and T.H. Morgan, 2005. The calcium exosphere of Mercury, *Icarus* **173**, 300–311.

Lewis, J. S., 1988. Origin and Composition of Mercury. In *Mercury*, eds. F. Vilas, C.R. Chapman, and M.S. Matthews, Univ. of Arizona Press, Tucson, Arizona, pp. 651–667.

Margot, J.L., S.J. Peale, R.F. Jurgens, M.A. Slade, and I.V. Holin, 2004. Earth-based radar measurements of Mercury's longitude librations. In *35th COSPAR Scientific Assembly*, 18–25 July 2004, Paris, France, p.3693.

McClintock, W. and M.R. Lankton, 2005. The Mercury Atmospheric and Surface Composition Spectrometer for the MESSENGER Mission. *Space Sci. Rev.* submitted.

McCord, T.B., and R.N. Clark, 1979. The Mercury soil: Presence of Fe2+. *J. Geophys. Res.* **84**, 7664–7668.

McGrath, M.A., Johnson, R.E. and Lanzerotti, L.J., 1986. Sputtering of sodium on the planet Mercury. *Nature* **323**, 696–696.

Melosh, H.J., and W.B. McKinnon, 1988. The tectonics of Mercury. In *Mercury*, eds. F. Vilas, C.R. Chapman, and M.S. Matthews, Univ. of Arizona Press, Tucson, Arizona, pp. 374–400.

Mitchell, D. and I. de Pater 1994. Microwave imaging of Mercury's thermal emission at wavelengths from 0.3 to 20.5 cm. *Icarus* **110**, 2–32.

Moses, J. I., K. Rawlins, K. Zahnle, and L. Dones, 1999. External Sources of Water for Mercury's Putative Ice Deposits. *Icarus* **137**, 197–221.

Ness, N.F., Behannon, K.W., Lepping, R.P., Whang, Y.C. and Schatten., K.H., 1974. Observations at Mercury encounter by the plasma science experiment on Mariner 10. *Science* **185**, 159–170.

Ness, N.F., K.W. Behannon, R.P. Lepping, and Y.C. Whang, 1975. Magnetic field of Mercury confirmed. *Nature* **255**, 204–205.

Nobel, S.K., and C.M. Pieters, 2001. Space Weathering in the Mercurian Environment. In *Mercury: Space Environment, Surface, and Interior*. Proceedings of a workshop held at The Field Museum, 4–5 October, 2001, Chicago, Illinois. *LPI Contribution No. 1097*, Lunar and Planetary Science Institute, Houston, Texas, pp. 68–69.

Paige, D.A., S.E. Wood, and A.R. Vasavada, 1992. The thermal stability of water ice at the poles of Mercury. *Science* **258**, 643–646.

Potter, A.E. and Morgan, T.H., 1985. Discovery of sodium in the atmosphere of Mercury. *Science* **229**, 651–653.

Potter, A.E. and Morgan, T.H., 1986. Potassium in the atmosphere of Mercury. *Icarus* **67**, 336–340.

Potter, A.E., and T.H. Morgan, 1990. Evidence for magnetospheric effects on the sodium atmosphere of Mercury. *Science* **248**, 835–838.

Potter, A.E., R.M. Killen, and T.H. Morgan, 2002. The sodium tail of Mercury. *Meteoritics & Planetary Science* **37** (9), 1165–1172.

Robinson, M. S., and P. G. Lucey, 1997. Recalibrated Mariner 10 color mosaics: implications for mercurian volcanism. *Science* **275**, 197–200.

Robinson, M.S., and G.J. Taylor, 2001. Ferrous oxide in Mercury's crust and mantle. *Met. Planet. Sci*. **36**, 841–847.

Santo, A. G. et al., 2001. The MESSENGER mission to Mercury: spacecraft and mission design. *Planet. Space Sci*. **49**, 1481–1500.

Sasaki, S., and E. Kurahashi, 2004. Space weathering on Mercury. *Adv. Space Res*. **33**, 2152–2155.

Schlemm, C.E., et al., 2005. The X-Ray Spectrometer (XRS) on the MESSENGER Spacecraft. *Space Sci. Rev.* (submitted).

Schubert, G., M.N. Ross, D.J. Stevenson, and T. Spohn, 1988. Mercury's thermal history and the generation of its magnetic field, In *Mercury*, eds. F. Vilas, C.R. Chapman, and M.S. Matthews, Univ. of Arizona Press, Tucson, Arizona, pp. 429–460.

Slade, M., B. Butler, and D. Muhleman, 1992. Mercury radar imaging: Evidence for polar ice. *Science* **258**, 635–640.

Solomon, S.C., 1977. The relationship between crustal tectonics and internal evolution in the moon and Mercury. *Phys. Earth Planet. Int*. **15**, 135–145.

Solomon, S.C. et al., 2001. The MESSENGER mission to Mercury: scientific objectives and implementation. *Planet. Space Sci*. **49**, 1445–1465.

Soter, S.L., and J. Ulrichs, 1967. Radiation and Heating of the Planet Mercury. *Nature* **214**, 1315.

Sprague, A.L., and T.L. Roush, 1998. Comparison of Laboratory Emission spectra with Mercury Telescopic Data. *Icarus* **133**, 174–183.

Sprague, A.L., R.W.H. Kozlowski, and D.M. Hunten, 1990. Caloris Basin: An enhanced source for potassium in Mercury's atmosphere. *Science* **249**, 1140–1143.

Sprague, A.L., R.W.H. Kozlowski, F.C. Witteborn, D.P. Cruikshank, and D.H. Wooden, 1994. Mercury: Evidence for anorthosite and basalt from mid-infrared (7.5–13.5 μm) spectroscopy. *Icarus* **109**, 156–167.

Sprague, A.L., D.M. Hunten, and K. Lodders, 1995. Sulfur at Mercury, elemental at the poles and sulfides in the regolith. *Icarus* **118**, 211–215.

Sprague, A.L., W.J. Schmitt, and R.E. Hill, 1998. Mercury: Sodium Atmospheric Enhancements, Radar Bright Spots, and Visible Surface Features. *Icarus* **135**, 60–68.

Sprague, A.L., J.P. Emery, K.L. Donaldson, R.W. Russell, D.K. Lynch, and A.L. Mazuk, 2002. Mercury: Mid-infrared (3–13.5 micrometer) observations show heterogeneous composition, presence of intermediate and basic soil types, and pyroxene. *Met. Planet. Sci*. **37**, 1255–1268.

Starukhina, L.V., 2001. Water detection on atmosphereless celestial bodies: alternative explanations of the observations. *J. Geophys. Res*. **106** (E7), 14701–14710.

Strom, R.G., R. Malhotra, T. Ito, F. Yoshida, and D. Kring, 2005. The origin of planetary impactors in the inner solar system. *Science* **309**, 1847–1850.

Strom, R.G., and A.L. Sprague, 2003. *Exploring Mercury: The Iron Planet*, 216 pp. Springer-Praxis, Chichester, UK.

Strom, R.G., and G. Neukum, 1988. The cratering record on Mercury and the origin of impacting objects. In *Mercury*, eds. F. Vilas, C.R. Chapman, and M.S. Matthews, Univ. of Arizona Press, Tucson, Arizona, pp. 336–373.

Thomas, P.G., P. Masson, and L. Fleitout, 1988. Tectonic history of Mercury. In *Mercury*, eds. F. Vilas, C.R. Chapman, and M.S. Matthews, Univ. of Arizona Press, Tucson, Arizona, pp. 401–428.

Vasavada, A.R., D.A. Paige, and S.E. Wood, 1999. Near-surface temperatures on Mercury and the Moon and the stability of polar ice deposits. *Icarus* **141**, 179–193.

Vilas, F., 1985. Mercury – Absence of crystalline Fe(2+) in the regolith. *Icarus* **64**, 133–138.

Vilas, F., 1988. Surface composition of Mercury from reflectance spectrophotometry. In *Mercury*, eds. F. Vilas, C.R. Chapman, and M.S. Matthews, Univ. of Arizona Press, Tucson, Arizona, pp. 59–76.

Vilas, F., M.A. Leake, and W.W. Mendell, 1984. The dependence of reflectance spectra of Mercury on surface terrain. *Icarus* **59**, 60–68.

Vilas, F., P.S. Cobian, N.G. Barlow, and S.M. Lederer, 2005. How much material do the radar-bright craters at the mercurian poles contain? *Planet. Space Sci.* **53**, 1496–1500.

Warell, J., 2002. Properties of the Hermean Regolith: II. Disk-Resolved Multicolor Photometry and Color Variations of the "Unknown" Hemisphere. *Icarus* **156**, 303–317.

Warell, J., 2003. Properties of the Hermean regolith: III. Disk-resolved vis-NIR reflectance spectra and implications for the abundance of iron. *Icarus* **161**, 199–222.

Warell, J. and D.T. Blewett, 2004. Properties of the Hermean regolith: V. New optical reflectance spectra, comparison with lunar anorthosites, and mineralogical modelling. *Icarus* **168**, 257–276.

Warell, J., A.L. Sprague, J.P. Emery, R.W.H. Kozlowski, and A. Long, 2006. The 0.7–5.3 micrometer IR spectra of Mercury and the Moon: Evidence for high-Ca clinopyroxene on Mercury. *Icarus*, in press.

Weidenschilling, S.J., 1978. Iron/Silicate Fractionation and the Origin of Mercury. *Icarus* **35**, 99–111.

Wetherill, G.W., 1988. Accumulation of Mercury from Planetesimals. In *Mercury*, eds. F. Vilas, C.R. Chapman, and M.S. Matthews, Univ. of Arizona Press, Tucson, Arizona, pp. 670–691.

3 The Atmosphere of Venus: Current Knowledge and Future Investigations

Dmitri V. Titov, Hakan Svedhem and Fred W. Taylor

Abstract. As the Earth's nearest planetary neighbour, Venus has been studied by ground-based observers for centuries, and has been visited by more than 20 spacecraft. However, in the last decade and a half Venus research has been relatively neglected, despite the fact that a great many major questions about its atmosphere, surface and interior remain unanswered. Several of these questions relate to the unique position of Venus as the Earth's near twin, in terms of size, density and proximity to the Sun, which led early astronomers to expect an Earth-like environment on the planet, possibly one fit for human habitation, and perhaps even the seat of indigenous life. The picture which has emerged from missions to the planet is quite different, raising questions about the evolution and stability of terrestrial planet environments that are both intriguing and possibly of practical relevance to global change problems on the Earth. This chapter reviews the scientific issues, and goes on to describe two new missions to Venus, the European Venus Express and the Japanese Planet-C orbiters, which will take place in the next few years to address some of them in depth. Other questions will remain unanswered, and further missions will be required, including landing on Venus and sample return.

3.1 Introduction

Venus has always fascinated even casual observers, as it is as one of the brightest objects in the sky, and following the invention of the telescope it soon became an attractive target for astronomers. Venus was the very first celestial object to which Earthlings sent spacecraft at the dawn of the space era (Colin, 1983; Titov et al., 2002). The first phase of spacecraft investigations of Venus, from 1960 to 90, included a set of orbiters, descent probes, and balloons delivered to Venus by the Soviet Union in the framework of the extensive Venera and Vega programmes, the US Mariner 2 and 10 flyby spacecraft, the orbiter and multiprobe Pioneer Venus missions, the dedicated radar orbiter Magellan. More recently, the Galileo and Cassini spacecraft observed Venus during encounters made *en route* to their ultimate targets in the outer Solar System.

Ground-based observations, which have always significantly contributed to studies of Venus, revealed that our nearest planetary neighbour is in many respects the twin of the Earth, leading to expectations that conditions would be similar on both planets. However, the first spacecraft discovered an entirely different and exotic world hidden behind a thick cloud veil.

Today, despite exploration by more than 20 spacecraft, the "Morning Star" still keeps its mysteries. All these studies gave us a basic knowledge of the conditions on the planet, but generated a string of fundamental questions concerning the atmospheric composition, chemistry, structure, dynamics, surface-atmosphere interactions, atmospheric and geological evolution, and the plasma environment. It is time to proceed from the discovery phase to a thorough investigation and deep understanding of what lies behind Venus' complex chemical, dynamical, and geological phenomena. The most interesting questions are how the two neighbouring planets that apparently began their life in similar initial conditions could have evolved so differently, and how sensitive the present conditions on the terrestrial planets are to the initial state and differences in evolutionary and climate forming processes. Answers to these will help us to understand the destiny of our own planet.

This chapter begins with reviewing our present knowledge of the conditions as well as our current understanding of the processes on Venus. Next, it describes the European Venus Express and Japanese Planet-C missions to the planet, discusses the expected results as well as science issues that will likely remain unsolved, and gives scientific rationale and technological requirements for future studies of Venus. Finally, we conclude with a discussion of the place of Venus studies in context of comparative planetology and climatology.

3.2 Current Knowledge and Outstanding Puzzles

3.2.1 Temperature structure

Existing observations of the lower atmosphere (0–60 km), hidden below the clouds, are largely limited to in situ measurements acquired by 16 descent probes, mostly at near-equatorial latitudes (Sieff, 1983; Sieff et al., 1985). The probes showed that the temperature structure below 30 km is characterized by a lapse rate of between 8 and 9 K/km, and high surface temperatures of 735 ± 6 K resulting from an extremely powerful greenhouse effect (Fig. 3.1). The temperature profile in the troposphere is almost independent of latitude and local solar time, as a result of the very high thermal inertia of the massive atmosphere. The lapse rate is close to the adiabatic value indicating that convective equilibrium dominates in this region. However, precise measurements of small deviations of the lapse rate from the adiabatic are important for understanding vertical stability and momentum transport, and have been made only in a few locations so far.

The Venusian mesosphere (60–100 km) is easily accessible from space and was studied remotely by the Pioneer Venus infrared radiometer and Venera-15 Fourier spectrometer (Taylor et al., 1980; Zasova et al., 1999). They found that the mesospheric temperature structure shows significant latitude variations, with temperatures increasing from equator to pole (Fig. 3.2). This trend is the opposite of expectations from radiative-convective equilibrium, and indicates a significant role of atmospheric dynamics in the poleward branch of the Hadley cell. However, the physical mechanisms responsible for the observed behavior remain poorly understood, partly because the limited duration of the early observations resulted in incomplete latitude and local time coverage.

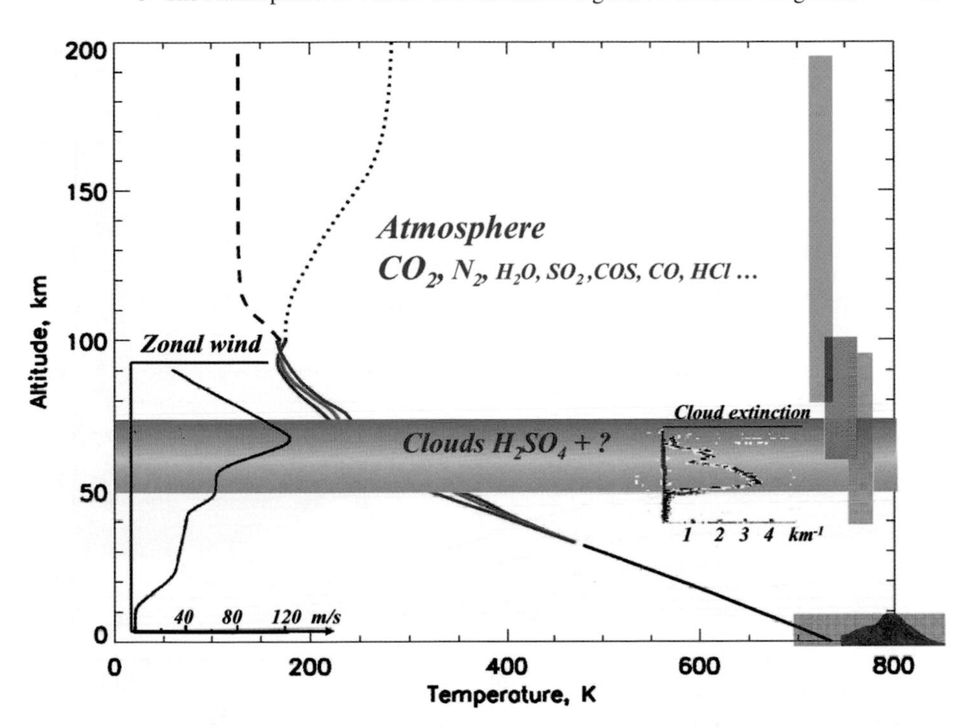

Fig. 3.1. Temperature profiles in the middle atmosphere of Venus (Seiff et al., 1985): red — low latitudes, green — 60° N, blue — 85° N. Profiles in the upper atmosphere (Keating et al., 1985): dots – day side, dashes – night side. Subplots show vertical profiles of zonal wind (lower left) (Gierasch et al., 1997) and cloud extinction (lower right) (Ragent et al., 1985). Colour rectangles on the right hand side show altitude ranges sounded by the Venus Express experiments: SPICAV stellar occultation (magenta), PFS and VIRTS thermal sounding (blue), VeRa radio occultation (yellow) and thermal imaging of the mountain slopes (pink).

Temperatures in the thermosphere (100–200 km) are controlled by the balance between solar UV radiation and molecular conductivity. Our knowledge of this region is based on the Pioneer Venus remote observations of UV airglow and in situ measurements of neutral densities (Keating et al., 1985; Fox and Bougher, 1991). These observations show relatively low ($\leq$300 K) dayside temperatures above 140 km, despite Venus' proximity to the Sun, and a sharp collapse to very cold night side temperatures across the terminator (Fig. 3.1). The existence of such a "cryosphere" is explained by the strong radiative cooling provided by CO_2 molecules, but is in stark contrast with what was expected and has no counterpart anywhere else in the Solar System.

Our knowledge of the structure of Venus' thermosphere remains incomplete because (i) in-situ measurements provided limited spatial sampling, (ii) airglow features have poor vertical resolution, typically probing 15–20 km of vertical range, (iii) most of the data were collected in the equatorial regions. The Magellan drag measurements provided only limited and tantalizing information regarding the high latitude structure. Thus, a global characterization of the density and temperature structure of the Venus thermosphere especially in high latitudes is still needed.

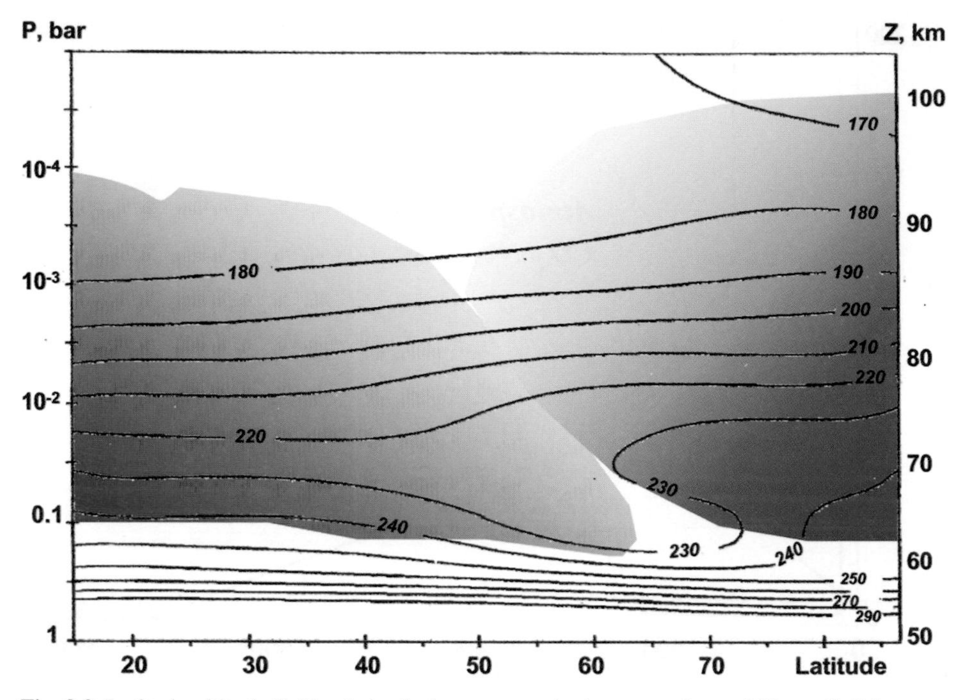

Fig. 3.2. Latitude-altitude fields of physical parameters in the mesosphere of Venus. Solid curves are isolines of atmospheric temperature derived from the Venera-15 remote sounding experiment (Lellouch et al., 1997). Pink and blue shaded areas mark the regions with net radiative heating and cooling, with maximum values up to ~10 K/day (Crisp and Titov, 1997).

3.2.2 Composition and chemistry

The main components of the Venus atmosphere are CO_2 (96.5%) and N_2 (3.5%). Sulphur bearing gases, carbon and chlorine compounds, and water vapour are present in the atmosphere in amounts from few to few hundred parts per million (ppm) (Esposito et al., 1997; de Bergh et al., 2006). Figure 3.3 shows the abundances of the main trace gases and their vertical profiles as measured by the earlier missions to Venus. Strong altitude and latitude variations imply that the minor constituents are involved in a number of chemical cycles and dynamical processes. Carbon monoxide is very abundant in the upper atmosphere due to the dissociation of CO_2 by solar ultraviolet radiation. The observed increase of CO amount from equator to pole in the lower atmosphere (Taylor et al., 1997) was tentatively attributed to the advective transport of this gas in the descending branch of the Hadley cell at high latitudes. The upper troposphere (60–70 km) is a "photochemical factory" where the reactions between CO_2, SO_2, H_2O and chlorine compounds induced by the solar UV radiation eventually lead to the formation of sulphuric acid aerosols at the top of the clouds (Krasnopolsky and Pollack, 1994; Imamura and Hashimoto, 2001). These processes result in an abrupt decrease of sulphur dioxide and water vapour with altitude at the cloud tops (Fig. 3.3), where both gases demonstrate strong variability. The spatial variations of H_2O abundance reach a factor of 4 (Ignatiev et al., 1999), while comparisons of the Pioneer Venus observations showed

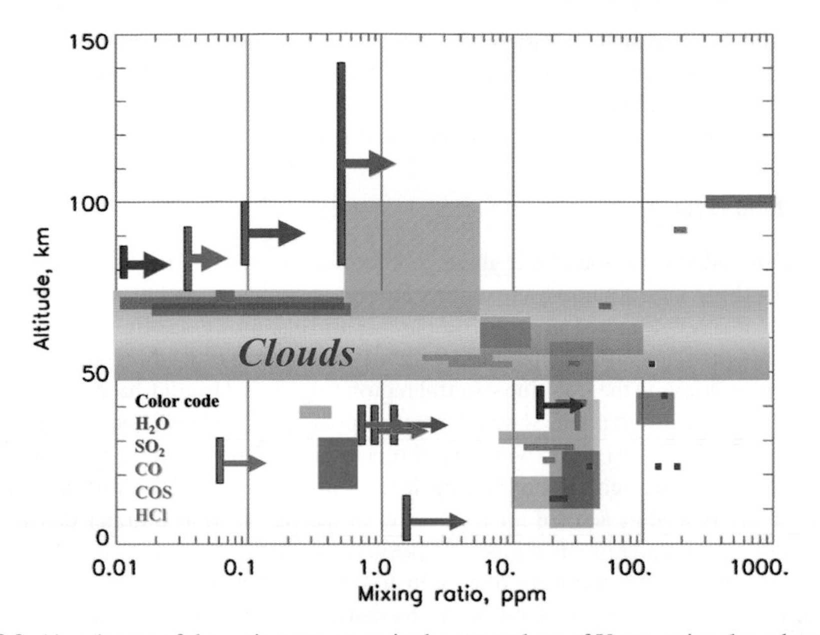

Fig. 3.3. Abundances of the main trace gases in the atmosphere of Venus, using the colour code shown in the lower left corner. The rectangles are results from earlier missions (Von Zahn and Moroz (1985); De Bergh et al., (2006)). The bars with arrows show the limits of detectability by the Venus Express experiments.

the existence of strong peak of vapour abundance in the early afternoon (Schofield et al., 1982). These facts probably indicate strong local convective activity in the upper cloud region close to the sub-solar point. Various observations of the sulphur dioxide at the Venus cloud tops have indicated about a factor of 10 decrease of its abundance in the period from 1978 to 1990 (Esposito et al., 1997). These observations were attributed to a strong outburst of volcanic activity in late 1970s that caused a global increase of SO_2 mixing ration at the cloud tops, with a subsequent decline due to photochemistry. There is a general problem with SO_2 present in the Venus atmosphere in amount of $\sim$150 ppm since it is then in thermal disequilibrium with the carbonates expected to be on the Venus surface.

The chemistry of the lower atmosphere is dominated by the thermal decomposition of sulphuric acid right below the clouds, and includes thermochemical cycles that involve sulphur and carbon species (SO_2, CO, COS) and water vapour. The fast increase in the carbonyl sulfide (COS) abundance below the clouds, seen in Fig. 3.3, results from the decomposition of sulphuric acid molecules. Surface minerals can also play significant role in buffering the abundance of certain other gases, such as HCl and HF, in the lower atmosphere (Fegley et al., 1997). The surprisingly low amount of oxygen found in the Venus atmosphere may be also attributable to binding of this gas by the minerals.

While earlier missions to Venus provided general description of the atmospheric composition, our understanding of the atmospheric chemistry and surface –atmosphere interactions is far from complete. An important task which remains is to investigate the latitude and local solar time variability of trace gases, particularly in the atmosphere below the clouds, where there are only very limited observations by previous Venus missions. The atmospheric composition is also important for understanding the radiative balance of the planet and the details of the greenhouse effect.

3.2.3 Cloud layer

Venus is shrouded by a cloud layer about 22 km thick located in altitude between 70 km and 48 km (Fig. 3.1). Although the clouds are not very dense, and the visibility inside the clouds is a few hundred meters, comparable to fogs on the Earth, its total opacity varies between 20 and 40. The clouds are almost featureless in visible light but display prominent markings in the UV-blue spectral region (Fig. 3.4). Ground-based polarimetric and spectroscopic observations showed that the upper cloud, at least, consists of micron-sized droplets of 75% H_2SO_4, which is photochemically produced at the cloud tops. Measurements on the Venera and Pioneer descent probes sounded the vertical structure of the clouds as well as aerosol microphysical properties such as number density, size distribution, shape and refractive index (Esposito et al., 1983; Krasnopolsky, 1989). One of the surprises was the discovery of a multi-modal size distribution that could indicate several distinct processes involved in the formation of the cloud system. The optical properties of the micron-size particles, for which the number density is $\sim 50\,\mathrm{cm}^{-3}$, were found to be consistent with those of sulphuric acid. The composition and nature of the rest of the particle population is still a mystery, although the sub-micron mode

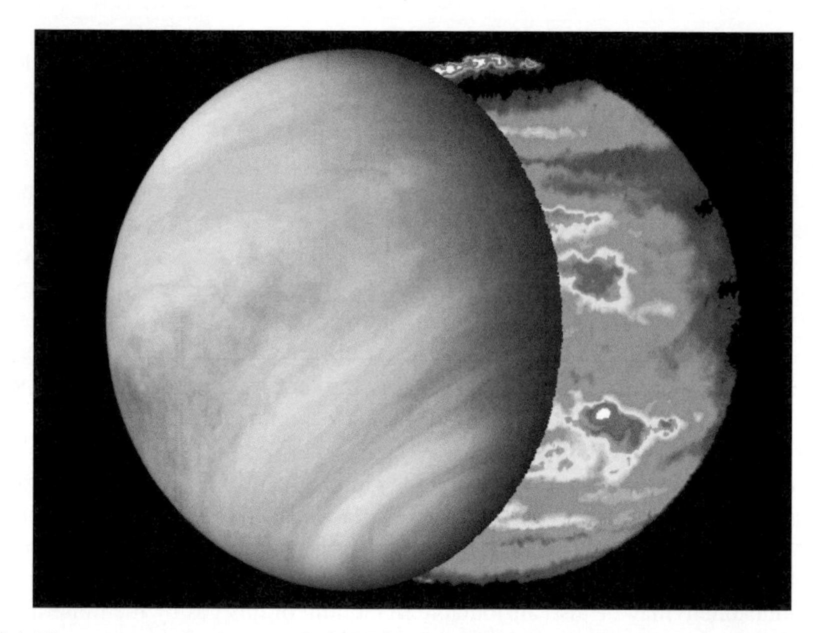

Fig. 3.4. Venus as seen from space. Left: UV image of the day side captured by the Mariner 10 spacecraft. Right: night side image taken by NIMS/Galileo in the 2.3 μm transparency "window". False colours are used to mark the intensity of the radiation escaping from the lower atmosphere, which depends on the cloud opacity. (courtesy of NASA).

with number density exceeding $1000\,cm^{-3}$ presumably consists of the yellowish species whose vertical and spatial variations produce the well-known UV contrasts on the disc of Venus (Fig. 3.4). The large particles with diameter >3 microns and number density $\sim 10\,cm^{-3}$ are very probably solid crystals and thus cannot be composed of sulphuric acid.

In situ analyses of the chemical composition of the aerosols on Venus descent probes have led to controversial results so far. Sulphuric acid was definitely detected, but chlorine and phosphorous compounds were found to also constitute a significant portion of the bulk cloud mass. The total mass loading of the clouds varies from 1 to $10\,mg/m^3$.

The discovery of the near-IR "transparency windows" in the spectrum of Venus (Allen and Crawford, 1984), through which thermal radiation from the hot lower atmosphere can leak into space, gave a new and very powerful tool to investigate variations of the cloud total opacity. These observations revealed significant spatial variations of the cloud structure (Fig. 3.4). Venus was thought to be a planet without weather, with the deep atmosphere resembling the Earth's ocean in this regard. The observations showed that this is not the case and there is significant level of dynamical variability in the clouds. Future progress in understanding the Venus cloud system requires long-term observations of cloud structure and morphology, as well as detailed in-situ analysis of the aerosol composition at different locations and altitudes.

3.2.4 Radiative effects in the Venus atmosphere

Solar radiation is the main driving force for many processes on all of the planets. In particular, the distribution of energy sources and sinks, defined by the balance between the incoming solar and escaping thermal radiation, determines temperature conditions in planetary atmospheres as well as their dynamical state.

Venus is the brightest among the planets. Its spherical albedo exceeds 75% in the visible spectral range (Moroz et al., 1985) due to the ubiquitous clouds that scatter back the incoming solar radiation. The high albedo means that Venus absorbs less solar energy than the Earth does, despite being closer to the Sun. About half of the solar flux absorbed by Venus is deposited within the upper cloud layer (58–65 km) due to the presence of the unknown UV absorber (Ekonomov et al., 1984; Bertaux et al., 1996), while only a few percent of the solar energy that falls on the top of the atmosphere reaches the surface. Solar heating of the mesosphere and its cooling to space result in net heating of the mesosphere at the cloud tops at low latitudes and cooling at high latitudes (Fig. 3.2). The thermal wind field correlates with the distribution of energy sources and sinks in the mesosphere, indicating the role of radiative forcing. Another manifestation of the coupling between radiative balance and atmospheric dynamics can be found in the latitude distribution of the radiative balance of the planet. Both on the Earth and on Venus, the amount of solar energy received by the planet drops towards the poles. The emitted thermal energy has a similar latitudinal trend for the Earth, but on Venus the thermal emission to space remains virtually constant with latitude, pointing to meridional energy transport (Schofield and Taylor, 1982).

High surface temperature of about 735 K results from the powerful greenhouse effect created by the presence of sulphuric acid clouds and certain gases (CO_2, H_2O, SO_2) in the atmosphere (Pollack et al., 1980). Although less than 10% of the incident solar radiation

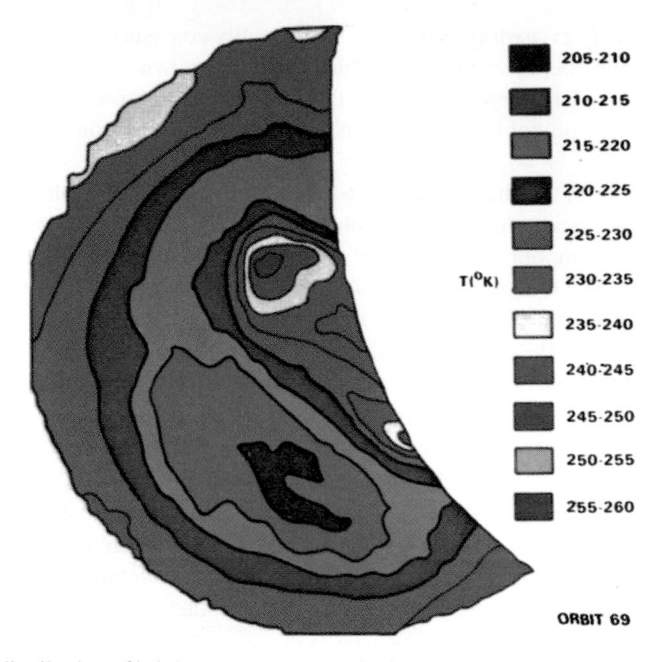

Fig. 3.5. The distribution of brightness temperature in the polar "dipole", observed in the Northern hemisphere by the Pioneer-Venus OIR experiment (Taylor et al., 1980).

penetrates through the atmosphere and heats the surface, the presence of these species in the atmosphere prevents thermal radiation from escaping to space and cooling the surface and lower atmosphere. The result is about 450 K difference between the surface temperature and that of the cloud tops, an absolute record among the terrestrial planets (Fig. 3.1).

Carbon dioxide provides the largest contribution to the greenhouse effect on Venus. Removal of this gas from the atmosphere would result in the temperature decreasing by $\sim$400 K, with smaller contributions by water vapour ($\sim$200 K) and the sulphuric acid clouds ($\sim$100 K). The contributions from the various agents are not additive, since the the infrared absorption bands which produce the greenhouse effect strongly overlap. The recently-discovered near-IR spectral "windows" are "holes" between the strong bands through which thermal radiation can leak to space. The efficiency of the leakage is highly variable and depends on the cloud opacity and the minor species abundances in the lower atmosphere (Crisp and Titov, 1997). The effect of cloud opacity variations on the outgoing thermal flux is clearly seen in Fig. 3.4.

The greenhouse effect is also in action on Mars and Earth. On Mars its efficiency is low ($\sim$7 K) due to the very thin atmosphere. On the Earth the greenhouse effect heats the surface by about 33 K, which is of crucial importance for the maintenance of habitable conditions on our planet.

3.2.5 Atmospheric dynamics

The Venusian atmosphere has two global dynamical regimes: *retrograde zonal super-rotation* in the troposphere and mesosphere (Gierasch et al., 1997) and *solar-antisolar circulation* across the terminator in the thermosphere (Bougher et al., 1997). Tracking of the UV markings, descent probes, and Vega balloon trajectories all showed that the lower atmosphere has zonal retrograde velocities decreasing from $\sim$100 m/s at the cloud tops to almost 0 at the surface (Fig. 3.8). Above the cloud tops the cyclostrophic assumption, which assumes a balance between the pressure gradient and centrifugal force of acting on a zonally rotating air parcel, reduces the Navier-Stokes equations to the thermal wind equation with proportionality of the vertical shear of zonal wind to the latitudinal derivative of atmospheric temperature. The mesospheric temperature field retrieved from the Pioneer Venus and Venera-15 thermal radiance measurements, in which the main feature is a gradual increase of temperature towards the poles, yields a wind field that declines with altitude above the cloud tops. This, however, does not explain the existence of a significant zonal wind component in the lower thermosphere.

In addition to zonal super-rotation of the upper troposphere there appears to be a slower overturning of the atmosphere from equator to pole with meridional velocities of $\leq$10 m/s and giant vortices at each pole recycling the air downwards. Only the North polar vortex has been observed in any detail (Fig. 3.9). It has a double 'eye' surrounded by a collar of much colder air, the difference in brightness temperature between the two being $\sim$50 K at a wavelength of 12.5 µm. The 'dipole' rotates with a period of 2.8 (Earth) days. The dynamics of the vortex was derived from a relatively short period of Pioneer Venus observations – 72 days – and with $\sim$100 km resolution mapping only every four days. Attempts to model the structure and dynamical behavior of the vortex have shown that such limited observations are quite inadequate for the task. Until then, the giant Venusian polar vortices remain one of the great mysteries of the solar system.

A further attribute of Venus' atmospheric dynamics, which also defies explanation, has been revealed in the post-Pioneer and Venera era following the discovery of bright near-infrared markings on the night side of the planet in the near-infrared windows (Allen and Crawford, 1984). The best view of these was obtained by the NIMS on the Galileo spacecraft *en route* to Jupiter in February 1990 (Carlson et al., 1993) (Fig. 3.4). This shows that the clouds on Venus are highly inhomogeneous, not only vertically but also horizontally. The origin of the contrasts must be dynamical, and is probably due to variable condensation of cloud material in large-scale cumulus-type dynamics in the deep atmosphere of Venus. What drives this we can only guess; however, it may be a manifestation of vertically-propagating waves, a possible carrier of the momentum which we observe in the middle atmosphere as the zonal super-rotation.

All attempts to model the zonal super-rotation have had limited success so far, indicating that the basic mechanisms of the phenomenon are unclear. There is an even more basic problem, and that is that we have so few observations of the deep atmosphere that we do not know what to model. The only real solution is to make new observations to gather a more basic description of the lower atmosphere and then to think about the problem again.

What is most puzzling about the regime represented by this scenario is how the atmosphere is accelerated to such high speeds on a slowly-rotating planet. Associated questions include (1) whether the meridional circulation is one large 'Hadley' cell extending from the upper atmosphere to the surface, or a stack of such cells, or something else altogether; (2) how the polar vortices couple the two main components of the global circulation and why they have such a complex shape and behaviour; and (3) what the observed (and observable) distributions of the minor constituents in Venus' atmosphere, including the clouds, are telling us about the motions.

The Hadley circulation is expected to be efficient on Venus in transporting warm air polewards and cooler air equatorwards, and is not subject to the same instabilities that restrict this kind of circulation on the relatively-rapidly rotating Earth. The observed movements of the cloud markings are consistent with such a regime, but whether it actually exists on Venus is not yet proven. Discontinuities in the vertical temperature gradient, observed by the Venera, Pioneer Venus, and Vega entry probes at characteristic altitudes, have been interpreted as possible evidence for a stack of Hadley cells on top of each other, rotating in alternate directions.

To understand what is really happening will require new and systematic observations from orbit. Galileo flew by Venus so quickly that only two full maps of the planet's nighttime northern hemisphere were possible, hardly an adequate sample for calculating wind velocities and directions.

The occurrence of lightning is probably related to the extent and vigour of convection in the Venusian clouds. The sulphuric acid particles that compose the clouds, in the upper layers at least, can accumulate charge and offer the potential of lightning. Reports of optical flashes have been made based on terrestrial telescopic sightings and orbital data. Very low frequency waves generated in the atmosphere were detected by the Venera landers and the Pioneer Venus orbiter. Radio frequency spherics were observed during the Galileo flyby of Venus but no lightning was detected in an optical search from the Cassini flyby. Thus the nature and occurrence, and even the existence, of lightning on Venus remains a topic of much debate (Grebowsky et al., 1997). Resolving this with new data is important both for understanding how the atmospheres of the terrestrial planets become electrified and discharged and to determine how lightning affects the chemistry of the Venusian atmosphere.

3.2.6 Plasma environment and solar wind-atmosphere interaction

Venus has no internally generated magnetic field. This leads to important differences between Venus and Earth in their plasma environments and their interaction with the solar wind, atmospheric escape, and energy deposition processes. Since the upper atmosphere of Venus is not protected by a magnetic field from direct interaction with the solar wind, a large portion of the exosphere resides in the shocked solar wind flow, and photo-ionisation, charge exchange and electron impact processes effectively remove ionized exospheric components. The tailward convection of the plasma mantle, situated between the shocked solar wind flow and the ionosphere, leads to another type of atmospheric loss in which the ions gyrating around the magnetic field embedded in plasma re-enter the atmosphere and cause sputtering (Luhmann et al., 1997). Yet another loss mechanism occurs when the solar wind interacts with the top of the ionosphere to form a complex array of plasma clouds, tail rails, filaments and ionospheric holes on the night side through which a substantial amount of material leaves the planet (Fig. 3.6). These escape

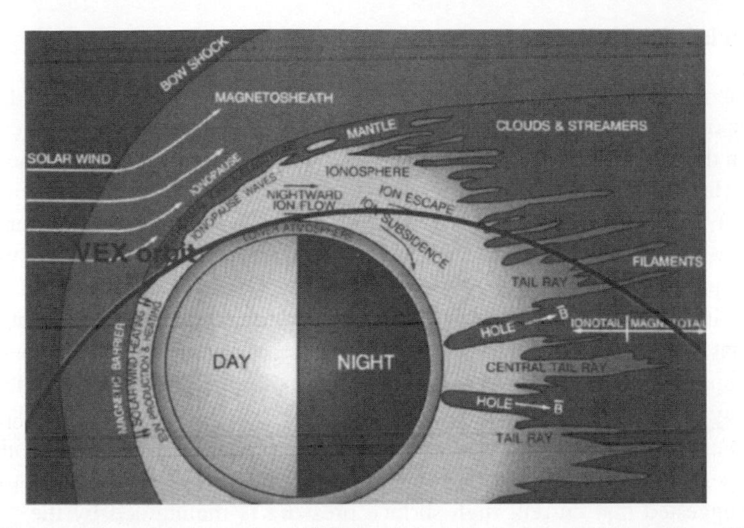

Fig. 3.6. The structure of the ionosphere of Venus induced by solar wind interactions (Brace and Kliore, 1991). The black curve shows the Venus Express orbit.

mechanisms, induced by the solar wind, are the dominant ones for the loss of heavy atmospheric gases such as oxygen because the planetary gravitational force inhibits the Jeans escape even for non-thermal components.

The Venera and Pioneer Venus orbiters found that the current induced by the solar wind electric field forms a magnetic barrier that deflects the most of the solar wind flow around the planet and leads to the formation of a bow shock (Cravens et al., 1997). The bow shock compresses the ionosphere on the dayside causing rapid anti-sunward convection and tail rails on the nightside. However, the short lifetime of the Venera-9 and 10 orbiters, and insufficient temporal and mass resolution in the Pioneer Venus plasma instrument, did not allow a study of the mass and energy exchange between the solar wind and the upper atmosphere of Venus in sufficient detail. Any study of atmospheric escape by in situ measurements is inherently limited by the spatial coverage possible with a single spacecraft and, can be interpreted only statistically. Imaging of energetic neutrals, ions and electrons would provide deeper insight into the plasma properties and help to quantify atmospheric escape.

While the plasma dynamics in near-Venus space is governed by the interplanetary magnetic field, the plasma transport in the ionosphere is determined by the local field due to local currents in the ionospheric plasma, forming domains such as the magnetosheath, magnetic barrier, ionopause, and magnetotail. In turn the geometry of these structures constrains the possible plasma escape channels, so that it is difficult to interpret the in situ plasma measurements and associated ionospheric structures without the help of magnetic field measurements. The vertical distribution of species in the exosphere and the plasma environment near Venus is also very important for understanding the evolution of terrestrial atmospheres and understanding the Earth's environment during epochs of weak magnetic field.

3.2.7 Surface and its interactions with atmosphere

The radar images of the surface delivered by the Pioneer Venus, Venera 15/16, and Magellan orbiters surprisingly revealed that Venus is among the most geologically active planets in the Solar System. Volcanism and tectonics have strongly altered the Venusian surface (Hansen et al., 1997; Crumpler et al., 1997) forming highly deformed plateau (Tesserae) and extensive lowlands (Planitiae) -- vast volcanic plains covering about 80% of the surface. Relatively rare and uniformly distributed impact craters suggest that global resurfacing of Venus happened about 500 My ago (Basilevsky et al., 1997). If this is so, it corresponds to a mechanism which is unique on the terrestrial planets, and means that 90% of Venus' geological history is lost. Many of the volcanic and tectonic features appear to be formed very recently, suggesting that the internal activity of the planet may be ongoing.

The Venus surface and interior seem to play an important role in buffering current conditions on the planet and in the evolution of climate. More than 50 years ago Urey (1953) suggested that current high surface pressure is maintained by the "carbonate buffer" – carbonate-silicate reactions on the surface (Fegley et al., 1997). However, the existence of carbonates on the Venus surface would result in conversion of the atmospheric SO_2 into sulphates within geologically short time of $\sim$2 My. Currently observed large abundance of SO_2 could therefore indicate ongoing volcanic activity that replaces sulphur dioxide removed by reaction with the minerals. Since there is no experimental evidence for the existence of carbonates on the surface, Hashimoto and Abe (2005) considered an alternative "pyrite buffer" that is in equilibrium with the observed amount of atmospheric SO_2. However the stability of pyrite under Venus conditions is also an open issue. The surface could also be a buffer for atmospheric gases like HCl and HF as well as taking up oxygen released in the photodissociation of water vapour and carbon dioxide. All these reactions between the surface and the atmosphere have little observational confirmation so far, due to the lack of data about the composition of the lower atmosphere and surface.

3.2.8 Evolution of the atmosphere and climate

Venus most likely received similar amounts of volatiles from the proto-planetary nebula as the Earth. After the planets were formed, various processes affected the volatile inventories, many of them poorly understood. The primitive atmospheres seem to have been almost completely lost from the terrestrial planets by hydrodynamic escape induced by the strong UV flux of the young Sun. Today's atmospheres were the formed by the net effect of outgassing from the interior, escape from the top of the atmosphere, and chemical interaction with the surface. Irregular events such as giant impacts could also alter the composition by eroding the atmosphere, while comets could add large amounts of volatiles.

Traces of these processes are recorded in the abundance patterns of noble gases and their isotopes. Measurements during the last several decades (Fig. 3.7) find two remarkable features of the rare gases abundances. The first is that the abundance patterns on Earth and Mars are explicable by massive hydrodynamic escape of the early atmospheres with subsequent replenishing of the lost volatiles by outgassing from the interiors (Pepin,

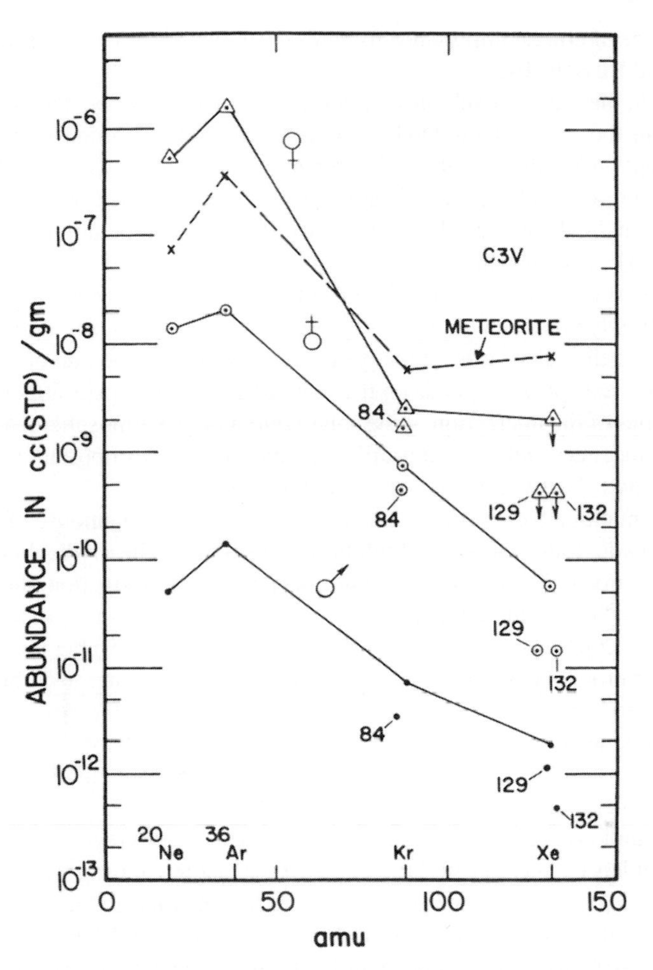

Fig. 3.7. Abundances of the noble gases in the atmospheres of the terrestrial planets and in meteorites (class 3V carbonaceous chondrites) per gram of the planet mass (Donahue and Pollack, 1983).

1991). Another approach assumes a significant role for delivery of volatiles to the inner Solar System by comets (Owen and Bar-Nun, 1995). Analysis of the noble gas compositions in the SNC meteorites from Mars, Earth mantle-derived rocks, and laboratory measurements of trapped rare gases in ices at $\sim$50 K, led to the conclusions that, firstly, the Martian atmosphere was severely eroded by impacts in the early stages, and, secondly, that both Earth and Mars received their atmospheres from two major sources: outgassing of their interiors and cometary supply. The pattern on Venus strongly differs, with greater relative amounts of neon and argon, placing the planet closer in this respect to meteorites and the Sun than the other terrestrial planets. The implication is that either Venus' atmosphere was formed almost completely of outgassed volatiles, or, if there was an external contribution, it should be a comet formed at $<$30 K in order to keep the noble gas abundances close to solar. New measurements of the xenon isotopes on

Venus could be of crucial importance for understanding of the origin of the atmosphere (Donahue and Russell, 1997).

At present, there is relatively little water in any form on Venus. The history of water is recorded in the value of the D/H ratio, and deuterium is found to be ~150 times more abundant on Venus than on Earth (Donahue et al., 1997). This enrichment can be explained by preferential escape of H atoms from the upper atmosphere, so the present water content and D/H ratio can be interpreted either as the signature of a lost primordial ocean, a steady state in which water is continuously supplied to the atmosphere by comets or volcanism, or a combination of both. The lifetime of the atmospheric water on Venus is highly uncertain but is probably less than 1 Myr, so that a primordial ocean is unlikely to be the sole source of the present water. The escape rate of H, D and other atoms from the exosphere depends on their abundance in the upper atmosphere and on the peculiarities of the interaction of the solar wind with the atmosphere. Measurements of atomic abundances and vertical profiles will be extremely important for quantifying the loss rates and clarifying the history of water on Venus.

The present climate on Venus is controlled by a strong greenhouse effect produced by carbon dioxide, water vapour, and sulphuric acid clouds. These greenhouse agents are influenced by processes like escape, cloud formation and dissipation, and interactions with the surface, as considered in models of the evolution of climate on Venus (Bullock and Grinspoon, 2001). The time scale of H escape from the atmosphere is about 170 My. A much faster process is destruction of atmospheric SO_2 by surface carbonates, which has characteristic time of ~20 My. The model shows that in the absence of a supply of sulphur dioxide from the interior, the sulphuric acid clouds of Venus would disappear in a relatively short time, giving way to thin, high water clouds with a resulting decrease in planetary albedo that has the surface temperature reaching 850–900 K. The existence of sulphuric acid clouds in the present epoch implies that a mechanism of SO_2 replenishment has been effective during at least the last few tens of My.

The geological record has been interpreted as showing that, about 500 My ago, Venus underwent global resurfacing as a result of an outflow of fresh lavas lasting a relatively short time (10–100 My). This event should have also produced massive outgassing and the injection of great amounts of SO_2 and H_2O into the atmosphere. The Bullock and Grinspoon model shows that, in the absence of additional sources, the planet should have lost both gases by the present. This model however assumes the dominance of carbonates on the surface, which in their model would make the climate of present-day Venus unstable. The alternative model proposed by Hashimoto and Abe in which the surface interaction is dominated by pyrites can overcome this problem and equilibrate the surface with the currently observed atmospheric composition. Data on the actual composition of the surface of Venus, and the extent to which it couples with the atmosphere, is obviously a pressing issue.

3.3 Venus Express and Future Missions to Venus

Despite early expectations, which saw Venus as another Earth and therefore probably another harbour of life, our neighbour was found to be an exotic world in which well-known processes are observed in extreme and bizarre forms. The first phase of Venus spacecraft exploration in 1960–90 brought many discoveries that turned out to be fundamental puzzles for the future (Moroz, 2002; Taylor, 2006). The data from ground-based observations and previous space missions is very limited in space and time coverage, and, prior to the discovery of the near infrared spectral windows, lacked the capability to sound the lower atmosphere of Venus remotely and study the phenomena hidden behind the thick cloud deck from orbit.

While the long-term exploration of Venus will require in situ measurements from probes and balloons, leading to sample return, so many key questions about Venus remain unanswered that even a relatively simple orbiter mission to the planet can lead to a rich harvest of high quality scientific results. After more than a decade of neglect, two remote sensing missions are now participating in a return to the "forgotten planet". This section describes the main investigations planned by the European Venus Express and Japanese Planet-C projects, and their expected scientific results, as well as possible strategy for future Venus exploration after they have completed their missions.

3.3.1 Venus Express -- the first European mission to Venus

Venus Express is the first European mission to Venus (Svedhem et al., 2006). Its main goal is to provide a remote sensing investigation of the global atmosphere and the plasma environment around Venus. The mission will also address some important aspects of the surface physics. The spacecraft (Fig. 3.8) was launched on November 9, 2005 by a Russian Souz-Fregat launcher from the Baykonur cosmodrome in Kazakhstan. In April 2006 Venus Express will reach the planet, there to be inserted in a polar orbit with pericentre and apocentre altitudes of 250 and 66,000 km and a revolution period of 24 hours. Science instruments will collect data both in the vicinity of the planet and from a distance of some ten planetary radii, thus combining global context and detailed close-up views (Titov et al., 2006). The observations will begin in Spring 2006 and continue for 500 days of nominal mission with possible extension for another 500 days.

The mission is based on the re-use of the Mars Express spacecraft and seven scientific instruments, five of which were inherited from the Mars Express and Rosetta projects. Although the spacecraft was not specifically designed for Venus, the versatile bus and powerful payload makes it very well suited to the task of making a breakthrough in the study of Venus.

The payload core consists of a suite of spectro-imaging remote sensing instruments.

The imaging spectrometer VIRTIS will map Venus in a broad spectral range from the UV to the thermal IR (0.25–5 μm) with moderate spectral resolution (R∼200). In addition it will provide high-resolution ($R \sim 1200$) spectra for the 2–5 μm range (Drossart et al., 2006). It has an instantaneous field of view of 0.25 mrad that corresponds to a spatial resolution ranging from hundreds of meters at pericentre to ∼15 km at apocentre. The high sensitivity of the instrument makes it ideally suited for sounding the composition of the lower atmosphere, total cloud opacity, and surface thermal mapping by measuring weak night-side emissions in the transparency "windows", while its mapping capabilities will be used to study atmospheric dynamics by tracking cloud features in the UV and IR ranges. VIRTIS will monitor the temperature and aerosol structure of the mesosphere between 60 and 90 km by measuring thermal infrared emission in the 4–5 μm range.

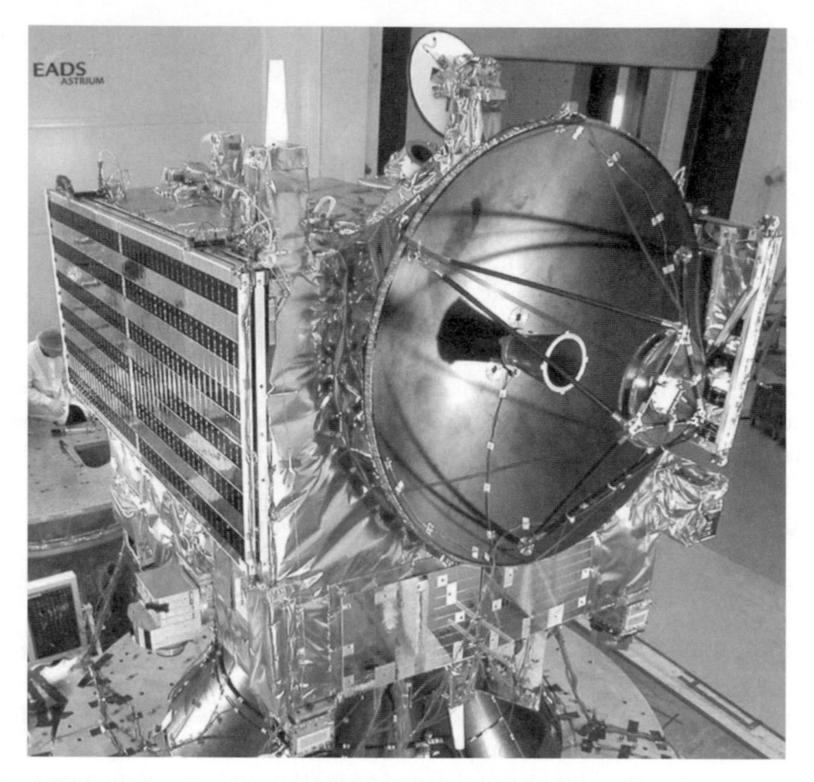

Fig. 3.8. The Venus Express spacecraft during environmental testing (courtesy of ESA).

The Planetary Fourier Spectrometer (PFS) is an infrared spectrometer optimised for atmospheric studies (Formisano et al., 2006). The instrument covers the spectral range from 0.9 to 45 μm with a spectral resolution of $\sim 1\,\mathrm{cm}^{-1}$ and has a field of view of ~2 degrees that corresponds to a spatial resolution of about 10 km at pericentre. The main science objective of PFS is to study temperature, aerosol structure and composition in the Venusian mesosphere (60–100 km). Observations of the night side emissions with high spectral resolution will monitor trace gases in the lower atmosphere and total cloud opacity.

The SPICAV instrument combines three spectrometers for the UV and near-IR range to study the vertical structure and composition of the mesosphere (60–100 km) and lower thermosphere (100–200 km) in solar and stellar occultation, limb and nadir geometries (Bertaux et al., 2006). Occultation observations provide high sensitivity to the abundance of minor species (SO_2, COS, CO, HCL, HF, H_2O, and HDO) and give the experiment the potential to search for new trace gases, in particular hydrocarbons (CH_4, C_2H_2), nitrogen oxides (NO, N_2O), and chlorine bearing compounds (CH_3Cl, ClO_2). These observations will significantly enhance our knowledge of mesospheric composition and chemistry and will contribute to the problem of atmospheric evolution, in particular to the question of water escape.

The Venus Monitoring Camera (VMC) is a wide-angle camera for observations of the atmosphere and the surface in four narrow-band filters centred at 0.365, 0.513, 0.965, and 1.01 μm (Markiewicz et al., 2006). The overall field of view is 17 degrees with 0.75 mrad/pixel, corresponding to a spatial resolution ranging from 0.2 km at pericentre to 50 km at apocentre. The main goal of the VMC is to investigate the cloud morphology and atmospheric dynamics by tracking the cloud features in the UV and near-IR ranges. Observations on the night side will be used for thermal mapping of the surface and to study any spatial variations in water vapour in the lower atmosphere.

The Venus Express Radio Science Experiment (VeRa) uses radio signals emitted by the spacecraft radio system in the X- and S- bands (3.5 cm and 13 cm) to sound the structure of the neutral atmosphere and ionosphere with a vertical resolution of a few hundred metres (Haeusler et al., 2006). The experiment will also investigate the gravity field and surface properties of Venus, and the solar corona. An ultra-stable oscillator provides a high quality onboard reference frequency source for the spacecraft transponder. The same measurements will sound the abundance of H_2SO_4 vapor below the clouds (40–50 km), looking for variability that can be seen as a tracer for atmospheric motions.

Two experiments onboard Venus Express are focused on the analysis of the plasma environment and the interaction of the solar wind with the atmosphere. The ASPERA instrument comprises four sensors: two detectors of energetic neutral atoms (ENA), plus electron and ion spectrometers (Barabash et al., 2006). It will measure the composition and fluxes of neutrals, ions and electrons to address how the interplanetary plasma and electromagnetic fields affect the Venus atmosphere and identify the main escape processes. These investigations will greatly extend those on previous missions to Venus to help understand similarities and differences in the solar wind interaction with the other terrestrial planets, Earth and Mars.

MAG is a magnetometer with two fluxgate sensors to measure the magnitude and direction of the magnetic field around Venus (Zhang et al., 2006). It will map the magnetic field in the magnetosheath, magnetic barrier, ionosphere, and magnetotail with high sensitivity and temporal resolution and characterize the boundaries between various plasma regions. These observations will provide important auxiliary data for the plasma measurements by ASPERA. The MAG experiment will also contribute to the search for lightning by measuring the strength and occurrence of electromagnetic waves associated with any atmospheric electrical discharges.

3.3.2 Planet-C — the Japanese meteorological mission to Venus

Planet-C, also known as Venus Climate Orbiter, is another remote sensing mission to Venus that is currently being designed and built by the Japanese Space Agency JAXA (Imamura et al., 2006). The main goal is to investigate the dynamics and meteorology of the Venusian atmosphere from orbit. The payload consists of five cameras to map Venus in several spectral bands from the UV to the thermal IR, thus sounding different levels in the atmosphere. The spacecraft will be inserted in an equatorial elongated orbit with a period of 30 hours, to allow the satellite to be in co-rotation with the zonal super-rotation at the cloud tops for a major portion of the revolution for cloud feature tracking and monitoring of the evolution of the cloud morphology. The Planet-C observations will strongly complement the Venus Express investigations in the field of atmospheric and cloud dynamics.

3.3.3 Results expected from Venus Express and Planet-C

The Venus Express and Planet-C missions are expected to significantly improve our knowledge about the atmosphere of Venus in the next decade by providing global observations of the atmospheric structure, composition, dynamics, and plasma properties with complete coverage in latitude and local solar time (Fig. 3.1). The upper atmosphere (80–180 km) will be studied with high vertical resolution using solar and stellar occultations (Bertaux et al., 2006), while spectroscopic observations in the 4.3 µm and 15 µm CO_2 bands will provide temperature sounding of the mesosphere (60–100 km) with a spatial resolution of a few tens of kilometers (Formisano et al., 2006; Drossart et al., 2006). Deeper sounding down to 40 km will be carried out by the radio-occultation experiment with a vertical resolution of a few hundred meters (Haeusler et al., 2006). Mapping the high elevation regions in sub-micron spectral "windows" on the night side will determine the surface temperature as a function of altitude. Meadows and Crisp (1996) used similar ground based observations to constrain the thermal profile and lapse rate in the lower 10 kilometers assuming the near-surface air temperature is equal to that of the surface. This will investigate the static stability and constrain the dynamics and turbulence in this region.

An important task for Venus Express is to investigate the variability of the trace gases with latitude and local solar time, particularly in the atmosphere below the clouds using the near IR transparency "windows" where few previous measurements exist. Specific objectives include abundance measurements of H_2O, SO_2, COS, CO, H_2O, HCl, and HF and their horizontal and vertical variations, to improve our understanding of the chemistry, dynamics, and radiative balance of the lower atmosphere, and to search for localized volcanic activity. The abundance of SO_2, SO, H_2O, HCl, and CO between 80 km and the cloud tops will give clues on the processes of the photochemical production of sulphuric acid and constrain models of the cloud formation and evolution. The CO profile will be measured from the cloud tops up to ~120 km covering the region of photochemical formation of this trace gas. Figure 3.3 shows the abundance levels and altitude ranges to be covered by the Venus Express spectroscopic measurements.

The remote sensing instruments on board Venus Express will also sound the structure, composition, dynamics, and variability of the cloud layer. The spatial and vertical distribution of the unknown UV absorber will be investigated by the imaging instruments (Markiewicz et al., 2006; Drossart et al., 2006), and correlated with the abundance of SO_2 and other trace gases in an attempt to unveil the nature of this mysterious species. The vertical structure and microphysical properties of the cloud and hazes above it will be studied in stellar and solar occultation (Bertaux et al., 2006), and by limb observations and nadir sounding in the thermal infrared (Formisano et al., 2006).

The open problems in Venus atmospheric dynamics can be addressed by detailed observations of winds at various altitudes. Venus Express will quantify the velocities at ~70 km by tracking the UV markings and at ~50 km by observing the motions of near-IR cloud features on the night side (Drossart et al., 2006; Markiewicz et al., 2006). Temperature sounding will allow the thermal wind field in the mesosphere to be derived (Formisano et al., 2006), to validate the cyclostrophic balance assumption and its limitations. Observation of O_2, NO, O, and H airglow observations will trace the global circulation in the thermosphere between 110 and 160 km. Similar method of using the minor species as tracers of atmospheric motions can be applied to the lower atmosphere (Taylor et al., 1997). Thus, the Venus Express observations will eventually yield a 3-D time-variable picture of the global atmospheric dynamics. Local phenomena like convection and waves at the cloud tops will be studied by the imaging instruments with spatial resolution of few hundreds of meters. The Venus Express investigations will be followed and complemented by the Planet-C meteorological observations from equatorial orbit.

Venus Express will measure the outgoing radiative fluxes over a broad spectral range, which, when combined with temporarily- and latitudinally-resolved cloud and constituent mapping will give an insight into the radiative and dynamical heat transport, and the role of various species in the planetary heat balance and greenhouse effect. The mission will investigate the plasma environment by in-situ measurements of the energetic neutral atoms, ions, electrons, and magnetic field (Barabash et al., 2006; Zhang et al., 2006) and combine these with high-resolution spectroscopic observations of CO_2 and H_2O and their isotopes from the cloud tops up to $\sim$200 km (Bertaux et al., 2006) to provide quantitative characterization of the escape processes at Venus. The vertical structure of the ionosphere will be also studied in the radio-occultation experiment (Haeusler et al., 2006).

Venus Express will provide a long-term set of electromagnetic and optical observations to determine the strength of the flashes and their rates of occurrence, and hence to investigate the nature of lightning on Venus.

Despite the fact that the Venus Express payload does not include specific instruments for surface studies, the mission will address this subject in several ways (Titov et al., 2006). The bi-static radar experiment (Haeusler et al., 2006), which consists of sending radio signals to the planet and receiving them on Earth, will focus on Aphrodite Terra, Beta and Atla Regio, and Maxwell Montes – the highlands that showed anomalously high reflectivity in the Magellan radar images. The VeRa gravity field investigation will study the mass distribution in Atalanta Planitia. The optical instruments will perform thermal mapping of the surface in the 1 μm "window" on the night side to measure the temperature and estimate emissivity of the surface. Indirect indications of surface-atmosphere interactions will be obtained from spectroscopic observations of the lower atmosphere composition.

Venus might be the site of significant seismic activity, with quakes generating waves in the solid planet that can propagate though the atmosphere. The resulting pressure and temperature perturbations at the top of the atmosphere above epicenter could be detected by mapping non-LTE fluorescent or LTE thermal emissions (Drossart et al., 2006).

3.3.4 Future Venus exploration

The Venus Express and Planet-C missions will provide global and comprehensive observations of atmospheric and plasma processes, but still have limitations and will not be able to solve all of the outstanding problems. The following measurement objectives are not expected to be attained by Venus Express and Planet-C and will be targets for future missions to Venus:

1) The isotopic composition, especially that of noble gases, which provides information on the origin and evolution of Venus and its atmosphere.
2) The chemical composition below the clouds with more detail than is possible using remote sensing, in order to fully characterize the chemical cycles involving clouds, surface and atmospheric gases.
3) The surface composition and mineralogy at several locations representing the main types of Venus landforms.
4) A search for seismic activity and seismology on the surface, and measurements at multiple locations to sound the interiors.
5) In situ investigation of the atmospheric dynamics, for instance by tracking the drift of floating balloons.
6) The composition and microphysics of the cloud layer at different altitudes and locations, by direct sampling.

These measurements and associated investigations must be made within the atmosphere and on the surface. The natural next step after Venus Express and Planet-C could be a mobile geochemical/geophysical mission using an orbiter, entry probes, and balloons to fully exploit the available *in situ* techniques and provide a detailed picture of the molecular and isotopic composition of the deep atmosphere and the surface (Crisp et al., 2003; Chassefiere et al., 2006). The experience gained in the earlier Soviet and American Venus programmes has already been expanded by the Venus entry probe and sample return studies recently carried out by ESA. Missions for in-situ exploration of Venus can be scheduled for launch in 2015–2020 after the results of Venus Express and Planet-C are analyzed and assimilated.

3.4 Venus' Place in Comparative Planetology and Climatology

Venus research plays an extremely important role in the comparative planetology and climatology of the terrestrial planets, and may be very important for understanding the current conditions and their evolution and stability on the Earth. Venus provides a natural laboratory to study various phenomena such as atmospheric dynamics on a slowly rotating planet and atmospheric chemistry at very high temperatures, with possibly strong catalytic and buffering effects due to the surface. Venus gives a striking example of a powerful greenhouse effect that makes us worry about the limits of the Earth's climate stability.

The general question behind our interest in Venus is why these two planets evolved in such a different way despite similar initial conditions. What is the role of each of Venus' peculiarities, like the absence of an internal magnetic field and plate tectonics, in the overall evolution of the planet, its atmosphere, and climate? What are the feedback mechanisms coupling the surface and the atmosphere? Does Venus represent the past or the future of the Earth?

Knowing what harsh conditions exist on Venus, it seems hardly possible to expect that life once originated or evolved on the planet. However there have been suggestions that, if life had evolved before the planet arrived at its current inhabitable state, it might have found shelter at the cloud level where the temperatures are quite comfortable by terrestrial standards. While this seems unlikely, Venus still gives an important example for the discussion of the habitability of terrestrial planets and its limitations that we need to understand.

The recent progress in the ongoing search for extrasolar planets will very likely soon lead to the discovery of terrestrial planets around other stars. Will those planets be Earth- or Venus-like? What kind of climate-markers or even bio-markers should be used in order to distinguish between those two types? How broad is the habitability zone around other stars? Venus Express is opening up a field of investigation that will be with us for many decades to come before definitive answers are found.

References

Allen, D., and J. Crawford, 1984. Discovery of cloud structure on the dark side of Venus. *Nature* **307**, p. 222.

Barabash et al., 2006. ASPERA experiment for Venus Express, *Planet. Space Sci.*, in press.

Basilevsky, A.T., J.W. Head, G.G. Schaber, and R.G. Strom, 1997. The resurfacing history of Venus. In *Venus II*, eds. S.W. Bougher, D.M. Hunten, and R.J. Phillips, Univ. of Arizona Press, Tucson, Arizona, pp. 1047–1084.

De Bergh, C., V. Moroz, F.W. Taylor, D. Crisp, B. Bezard, L.V. Zasova, 2006. The composition of the atmosphere of Venus below 100 km altitude: an overview. *Planet. Space Sci.*, in press.

Bertaux, J.-L., T. Widemann, A. Hauchecorne, V.I. Moroz, A.P. Ekonomov, 1996. Vega-1 and Vega-2 entry probes: an investigation of local UV absorption (220–400 nm) in the atmosphere of Venus (SO_2, aerosols, cloud structure). *J. Geophys. Res.* **101**, 12709–12745.

Bertaux et al., 2006. SPICAV for Venus Express, *Planet. Space Sci.*, in press.

Bougher, S.W., M.J. Alexander, and H.G. Mayr, 1997. Upper atmosphere dynamics: global circulation and gravity waves. In *Venus II*, eds. S.W. Bougher, D.M. Hunten, and R.J. Phillips, Univ. of Arizona Press, Tucson, Arizona, pp. 259–291.

Brace, L.H. and A.J. Kliore, 1991. The structure of the Venus ionosphere. *Space Sci. Rev.* **55**, 81–163.

Bullock, M.A. and D.H. Grinspoon, 2001. The recent evolution of climate on Venus. *Icarus* **150**, 19–31.

Carlson, R.W., L.W. Kamp, K.H. Baines, J.B. Pollack, D.H. Grinspoon, Th. Encrenaz, P. Drossart, and F.W. Taylor, 1993. Variations in Venus cloud particle properties: a new view of the Venus' cloud morphology as observed by Galileo near-infrared mapping spectrometer. *Planet. Space Sci.* **41**, 477–485.

Chassefiere, E. et al., 2006. Proceedings of the Venus Entry Probe Workshop, February 19–20, 2006, ESTEC, The Netherlands.

Colin, L., 1983. Basic facts about Venus. In *Venus*, eds. D.M. Hunten, L. Colin, T.M. Donahue, and V.I. Moroz, Univ. of Arizona Press, Tucson, Arizona, pp. 10–26.

Crisp, D. and D.V. Titov, 1997. The thermal balance of the Venus atmosphere. In *Venus II*, eds. S.W. Bougher, D.M. Hunten, and R.J. Phillips, Univ. of Arizona Press, Tucson, Arizona, pp. 353–384.

Crisp, D. et al., 2003. Divergent evolution among the Earth-like planets: The case for Venus exploration. The future of Solar System Exploration, 2003–2013, ASP Conference series.

Crumpler, L.S., J.C. Aubele, D.A. Senske, S.T. Keddie, K.P. Magee, and J.W. Head, 1997. Volcanoes and centres of volcanism on Venus. In *Venus II*, eds. S.W. Bougher, D.M. Hunten, and R.J. Phillips, Univ. of Arizona Press, Tucson, Arizona, pp. 697–756.

Donahue, T.M. and J.B. Pollack, 1983. Origin and evolution of the atmosphere of Venus. In *Venus*, eds. D.M. Hunten, L. Colin, T.M. Donahue, and V.I. Moroz, Univ. of Arizona Press, Tucson, Arizona, pp. 1003–1036.

Donahue, T.M. and C.T. Russell, 1997. The Venus atmosphere and ionosphere and their interaction with the solar wind: an overview. In *Venus II*, eds. S.W. Bougher, D.M. Hunten, and R.J. Phillips, Univ. of Arizona Press, Tucson, Arizona, pp. 3–31.

Donahue, T. M., Grinspoon, D. H., Hartle, R. E., and Hodges R. R. Jr., 1997. Ion neutral escape of hydrogen and deuterium: evolution of water. In *Venus II*, eds. S.W. Bougher, D.M. Hunten, and R.J. Phillips, Univ. of Arizona Press, Tucson, Arizona, pp. 385–414.

Drossart P., et al., 2006. VIRTIS for Venus Express, *Planet. Space Sci.*, in press.

Ekonomov, A.P., V.I. Moroz, B.E. Moshkin, V.I. Gnedykh, Yu.M. Golovin, and A.V. Grigoriev, 1984. Scattered UV solar radiation within the clouds of Venus. *Nature* **307**, 345–346.

Esposito, L.W., R.G. Knollenberg, M.Ya. Marov, O.B. Toon, and R.P. Turco, 1983. The clouds and hazes of Venus. In *Venus*, eds. D.M. Hunten, L. Colin, T.M. Donahue, and V.I. Moroz, Univ. of Arizona Press, Tucson, Arizona, pp. 484–564.

Esposito L.W., J-L. Bertaux, V.A. Krasnopolsky, V.I. Moroz, and L.V. Zasova, 1997. Chemistry of lower atmosphere and clouds. In *Venus II*, eds. S.W. Bougher, D.M. Hunten, and R.J. Phillips, Univ. of Arizona Press, Tucson, Arizona, pp. 415–458.

Fegley, B. Jr, G. Klinghoefer, K. Lodders, and T. Widemann, 1997. Geochemistry of surface-atmosphere interactions on Venus. In *Venus II*, eds. S.W. Bougher, D.M. Hunten, and R.J. Phillips, Univ. of Arizona Press, Tucson, Arizona, pp. 591–636.

Fox, J. L., and S. W. Bougher, 1991. Structure, luminosity, and dynamics of the Venus thermosphere. *Space Sci. Rev.* **55**, 357.

Formisano V. et al., 2006. PFS for Venus Express, *Planet. Space Sci.*, in press.

Gierasch P.J., R.M. Goody, R.E. Young, D. Crisp, C. Edwards, R. Kahn, D. McCleese, D. Rider, A. Del Genio, R. Greeley, A. Hou, C.B. Leovy, and M. Newman, 1997. The general circulation of the Venus atmosphere: an assessment. In *Venus II*, eds. S.W. Bougher, D.M. Hunten, and R.J. Phillips, Univ. of Arizona Press, Tucson, Arizona, pp. 459–500.

Grebowsky, J.M., R.J. Strangeway, and D.M. Hunten, 1997. Evidence for Venus lightning. In *Venus II*, eds. S.W. Bougher, D.M. Hunten, and R.J. Phillips, Univ. of Arizona Press, Tucson, Arizona, pp. 125–157.

Haeusler B. et al., 2006. VeRa experiment for Venus Express, *Planet. Space Sci.*, in press.

Hansen, V.L., J.J. Willis, and W.B. Banerdt, 1997. Tectonic overview and synthesis. In *Venus II*, eds. S.W. Bougher, D.M. Hunten, and R.J. Phillips, Univ. of Arizona Press, Tucson, Arizona, pp. 797–844.

Hashimoto, G.L. and Y. Abe, 2005. Climate control on Venus: comparison of carbonate and pyrite models. *Planet. Space Sci.* **53**, 839–848.

Imamura T. and G. Hashimoto, 2001. Microphysics of Venusian clouds in rising tropical air. *J. Atmos. Sci.* **58**, 3597–3612.

Imamura,T., M. Nakamura, M. Ueno, N. Iwagami, T. Satoh, S. Watanabe, M. Taguchi, Y. Takahashi, M. Suzuki, T. Abe, G.L. Hashimoto, T. Sakanoi, S. Okano, Y. Kasaba, J. Yoshida, M. Yamada, N. Ishii, T. Yamada, and K.-I. Oyama, 2006. Planet-C: Venus climate orbiter mission of Japan, *Planet. Space Sci.*, in press.

Ignatiev, N.I., V.I. Moroz, L.V. Zasova, and I.V. Khatuntsev, 1999. Water vapour in the middle atmosphere of Venus: an improved treatment of the Venera 15 IR spectra. *Planet. Space Sci.* **47**, 1061–1075.

Keating G.M., J.L. Bertaux, S.W. Bougher, T.E. Cravens, R.E. Dickinson, A.E. Hedin, V.A. Krasnopolsky, A.F. Nagy, J.Y. Nicholson III, L.J. Paxton, and U. von Zahn, 1985. Models of Venus upper atmosphere: structure and composition. *Adv. Space Res.* **5**, 117–171.

Koukouli, M.E., P.G.J. Irwin, and F.W. Taylor, 2005. Water vapour abundance in Venus' middle atmosphere from Pioneer Venus OIR and Venera-15 FTS measurements. *Icarus* **173**, 84–99.

Krasnopolsky, V.A., 1989. Vega mission results and chemical composition of Venusian clouds. *Icarus* **80**, 202–210.

Krasnopolsky, V.A. and J.B. Pollack, 1994. H_2O-H_2SO_4 system in Venus' clouds and OCS, CO, and H_2SO_4 profiles in Venus' troposphere. *Icarus* **109**, 58–78.

Luhmann, J.G., S.H Brecht, J.R. Spreiter, S.S. Stahara, R.S. Steinolfson, and A.F. Nagy, 1997. Global models of the solar wind interaction with Venus. In *Venus II*, eds. S.W. Bougher, D.M. Hunten, and R.J. Phillips, Univ. of Arizona Press, Tucson, Arizona, pp. 33–59.

Markiewicz, W.J., D.V. Titov, N. Ignatiev, H.U. Keller, D. Crisp, S.S. Limaye, R. Jaumann, R. Moissl, N. Thomas, L. Esposito, S. Watanabe, B. Fiethe, T. Behnke, I. Szemerey, H. Michalik, H. Perplies, M. Wedemeier, I. Sebastian, W. Boogaerts, C. Dierker, B. Osterloh, W. Bker, M. Koch, H. Michaelis, D. Belyaev, A. Dannenberg and M. Tschimmel, 2006. Venus monitoring camera for Venus Express, *Planet. Space Sci.*, in press.

Meadows, V. and D. Crisp, 1996. Ground-based near-infrared observations of the Venus night side: the thermal structure and water abundance near the surface. *J. Geophys. Res.* **101**, 4595–4622.

Moroz, V.I, A.P. Ekonomov, B.E. Moshkin, H.E. Revercomb, L.A. Sromovsky, J.T. Schofield, D. Spaenkuch, F.W. Taylor, and M.G. Tomasko, 1985. Solar and thermal radiation in the Venus atmosphere. *Adv. Space Res.* **5**, 197–232.

Moroz, V.I., 2002. Studies of the atmosphere of Venus by means of spacecraft: solved and unsolved problems. *Adv. Space Res.* **29**, 215–225.

Owen, T. and A. Bar-Nun, 1995. Comets, impacts, and atmospheres. *Icarus* **116**, 215–226.

Pepin, R.O., 1991. On the origin and early evolution of terrestrial planet atmospheres and meteoritic volatiles. *Icarus* **92**, 2–79.

Pollack, J.B., O.B. Toon, and R. Boese, 1980. Greenhouse models of Venus' high surface temperature, as constrained by Pioneer Venus measurements. *J. Geophys. Res.* **85**, 8223–8231.

Schofield, J.T. and F.W. Taylor, 1982. Net global thermal emission from the Venus atmosphere. *Icarus* **52**, 245.

Schofield, J.T., Taylor, F.W. and McCleese, D.J. 1982. The global distribution of water vapour in the middle atmosphere of Venus. *Icarus* **52,** 263–278.

Sieff, A., 1983. Temperature structure of the Venus atmosphere. In *Venus*, eds. D.M. Hunten, L. Colin, T.M. Donahue, and V.I. Moroz, Univ. of Arizona Press, Tucson, Arizona, pp. 215–279.

Sieff, A., J.T. Schofield, A.J. Kliore, F.W. Taylor, S.S. Limaye, H.E. Revercomb, L.A. Sromovsky, V.V. Kerzhanovich, V.I. Moroz, and M.Ya. Marov, 1985. Models of the structure of the atmosphere of Venus from the surface to 100 km altitude. *Adv. Space Res.* **5**, 3–58.

Svedhem H., D.V. Titov, D. McCoy, J.-P. Lebreton, S. Barabash, J.-L. Bertaux, P. Drossart, V. Formisano , B. Häusler , O. Korablev, W.J. Markiewicz , D. Nevejans, M. Pätzold, G. Piccioni, T.L. Zhang, F.W. Taylor, E. Lellouch, D. Koschny, O. Witasse, M. Warhaut, A. Accommazzo, J. Rodriguez-Canabal, J. Fabrega, T. Schirmann, A. Clochet, M. Coradini, 2006. Venus Express – the first European mission to Venus. *Planet. Space Sci.*, in press.

Taylor, F.W., Beer, R., Chahine, M.T., Diner, D.J., Elson, L.S., Haskins, R. D. , McCleese, D.J., Martonchik, J.V., Reichley, P.E., Bradley, S.P., Delderfield, J., Schofield, J.T., Farmer, C.B., Froidevaux, L., Leung, J., Coffey, M.T., and Gille, J.C., 1980. Structure and meteorology of the middle atmosphere of Venus: infrared remote sounding from the Pioneer Orbiter. *J. Geophys. Res.* **85,** 7963–8006.

Taylor, F.W., D. Crisp, and B. Bezard, 1997. Near-infrared sounding of the lower atmosphere of Venus. In *Venus II*, eds. S.W. Bougher, D.M. Hunten, and R.J. Phillips, Univ. of Arizona Press, Tucson, Arizona, pp. 325–351.

Taylor, F.W., 2006. Venus before Venus Express. *Planet. Space Sci.*, in press.

Titov, D.V., K.H. Baines, A.T. Basilevsky, E. Chassefiere, G. Chin, D. Crisp, L.W. Esposito, J.-P. Lebreton, E. Lellouch, V.I. Moroz, A.F. Nagy, T.C. Owen, K.-I. Oyama, C.T. Russell, F.W. Taylor, R.E. Young., 2002. Missions to Venus. Proc. ESLAB 36 Symposium, *Earth-like planets and moons*, ESA SP-514.

Titov, D.V., H. Svedhem , D. Koschny, R. Hoofs, S. Barabash, J.-L. Bertaux, P. Drossart, V. Formisano, B. Häusler, O. Korablev, W.J. Markiewicz, D. Nevejans, M. Pätzold, G. Piccioni, T.L. Zhang, D. Merritt, O. Witasse, J. Zender, A. Accommazzo, M Sweeney, D. Trillard, M. Janvier, and A. Clochet, 2006. Venus Express science planning. *Planet. Space Sci.* in press.

Zasova, L.V., I.A. Khatountsev, V.I. Moroz, and N.I. Ignatiev, 1999. Structure of the Venus middle atmosphere: Venera-15 Fourier spectrometry data revisited. *Adv. Space Res.* **23**, 1559–1568.

Zhang, T.L., W. Baumjohann, M. Delva, H.-U. Auster, A. Balogh, C. T. Russell, S. Barabash, M. Balikhin, G. Berghofer, H.K. Biernat, H. Lammer, H. Lichtenegger, W. Magnes, R. Nakamura, T. Penz, K. Schwingenschuh, Z. Vörös, W. Zambelli, K.-H. Fornacon, K.-H. Glassmeier, I. Richter, C. Carr, K. Kudela, J. K. Shi, H. Zhao, U. Motschmann, and J.-P. Lebreton, 2006. Magnetic field investigation of the Venus plasma environment: expected new results. *Planet. Space Sci.*, in press.

4 Moon's Origin And Evolution: Alternatives and Implications

Harrison H. Schmitt

Abstract. The origin of the Moon 4.56 Gyr ago, its subsequent evolution, and the implications of both relative to the Earth remain subject to lively debate. Because the internal geochemistry and geophysics of the Moon does not appear consistent with an origin by the giant impact of a Mars-sized asteroid on the Earth, this hypothesis is challenged by one that proposes the capture of an independently formed planetesimal. The Moon's internal structure also indicates that it and all the terrestrial planets initially had relatively cool, chondritic proto-cores prior to formation of metallic cores. Evidence exists that these proto-cores delayed formation of metallic cores for periods that correlate with the final mass of a planet.

The impact history of the inner solar system has been broadly outlined by the modern investigation of the Moon. First, soon after the formation of a coherent lunar crust and against an intense background of smaller cratering events, the Moon was subjected to extremely large impacts that formed basins up to 3200km in diameter. On Earth, the melt sheets from these continental-scale impacts may have been responsible for the formation of the first continental crust at ~4.4 Gyr. Second, ~50 impact basins >300 km in diameter formed between 4.5 and 3.8 Gyr probably by pulses of impactors produced during the migration and interaction of the giant planets within a structured solar disk of planetesimal rings. The last of these pulses at about 3.85 Gyr, producing the ~14 mascon basins, resurfaced most of the Moon, and suggested an apparent "cataclysm" at that time. This period of 700 Myr may have been one of "punctuated cataclysm" as one or more giant planets encountered separate planetesimal rings and gaps during outward orbital migration. Finally, the implication of this violent impact history in the inner solar system prior to 3.8 Gyr relative to the surfaces of the hydrous terrestrial planets, that is, Earth, Mars and probably early Venus, is that clays were the dominant mineral species. These clays, as well as volcanic sulfides, may have provided the templates for the formation of complex organic precursors that made up the first living cells.

4.1 Introduction: The Consensus

Debate over the origin and evolution of the Moon, and implications relative to the Earth, has remained lively and productive since the last human exploration of that small planet in 1972 (Schmitt 1973, Fig. 4.1). More recently, issues related to the evolution of Mars, Venus and Mercury have received increasing attention from the planetary geology community and the public as many robotic missions have augmented the earlier lunar investigations. A number of generally accepted hypotheses relative to the history of the Moon and the terrestrial planets (e.g., Sputis 1996, Canup and Righter 2000, Taylor

Fig. 4.1. The author investigating a large boulder of impact derived rocks in the Valley of Taurus-Littrow during the Apollo 17 Mission, December 12, 1972 (NASA Photo).

2001, Warren 2003, Palme 2004), however, deserve intense questioning as data from many of the lunar samples suggest alternatives to some of these hypotheses. Further, how much do the known, probable and possible events in lunar history constrain what may have occurred on and in Mars? Or, indeed, do such events constrain what may have happened on and in the Earth?

4.1.1 Origin

Strong agreement exists that at 4.567 ± 0.001 Gyr[1] (T_0) (Carlson and Lugmair 2000, Alexander et al. 2001, Taylor 2001, Jacobsen 2003, Amelin et al. 2004) a portion of an interstellar molecular cloud collapsed to form the solar nebula (Taylor 2001, Hester et al. 2004). A sizable consensus also exists that a few tens of millions of years after T_0, the Moon belatedly came into existence as a result of a "giant impact" between the very young Earth and a chondritic asteroid that was 11 to 14 percent the mass of the Earth and gave a combined angular momentum of 110–120 percent of the current Earth-Moon system (Hartmann and Davis 1975, Cameron and Ward 1976, Hartmann 1986, Cameron 2002, Canup 2004, Palme 2004). This angular momentum was subject to later tidal dissipation to reach today's still unusually high value. Increasingly detailed hydrodynamic simulations (Canup 2004) of such a hypothetical collision indicates that, if it occurred, the impact angle and velocity of the impactor were around 45° and less than 4 km/sec, respectively. Under these conditions, about half of the ejected material would

[1] Gyr: billions of years before present.

reach the velocities and trajectories outside the Roche limit[2] that would be necessary for a stable orbit once aggregated as the Moon. The remaining material would re-impact the Earth.

The giant impact consensus holds that at the time of such a collision, both the Earth and the impactor would have been at least partially differentiated from their chondritic parent material by the separation of core-forming, iron-rich liquid. Some chemical differentiation in non-core material may also have taken place by a least partial crystallization of silicate-rich magma oceans created during the earlier accretion of the two bodies. This hypothetical lunar origin by giant impact primarily offers an explanation for the unusually high angular momentum of the Earth-Moon system (see Taylor 2001), and allows many of the lunar geochemical characteristics to be attributed to those of the mantle of the original, at least partially differentiated, impactor. For example, the fifty percent higher iron content of the Moon relative to the Earth's mantle requires that at least 80–90 percent of the Moon be derived from the impactor rather than the core-depleted mantle of the Earth (Taylor 2001, Canup 2004).

The major, hypothetical consequences of such a Moon-forming, giant impact are presumed to be as follows: Most core-forming material in both bodies remained with or was quickly re-acquired by the Earth. The Moon formed through the rapid orbital re-aggregation of about one-half of the material ejected by the giant impact. Most of that material would have reached temperatures between 3000° and 5000°K, according to Canup's recent computer simulations (Canup 2004). 10–30 percent of material that formed the Moon vaporized during ejection and passed through a molten stage during condensation and re-aggregation into a coherent body, all in just a few days. Soon after or during re-aggregation, the lunar core formed (Agee 1991, Neal et al. 2000b, Neal and Ely 2002) and a magma ocean existed above this core (Wood et al. 1970, Smith et al. 1970, Warren 1985).

4.1.2 Evolution

The analysis of the isotopic system for extinct ^{182}Hf (decays with a 9-Myr half-life to a stable ^{182}W daughter) indicates that the average age of the lunar core forming material is about 30 Myr (Kleine et al. 2002) and the lunar magma ocean largely crystallized within 40 Myr (Shearer and Newsom 2000) of T_0. Core forming material, a dense, iron and sulfur-rich, immiscible liquid, would have separated from the silicate magma ocean as it formed (Agee 1991). As the magma ocean cooled, sequential crystallization of various silicate and oxide minerals occurred under the constraints of phase equilibria that pertain to complex, multi-element silicate systems rich in magnesium, calcium and iron but devoid of water. The magma ocean differentiated chemically into mantle and crust due to gravitational settling or floating of minerals based on the densities of those minerals relative to remaining liquid. This fractional crystallization of the magma ocean led to a lunar mantle dominated by olivine-pyroxene cumulates (accumulated crystals) and a 60–70 km thick Ca-feldspar-rich crust (Agee 1991). The large basin-dominated portions of the near-side crust, however, now appear to be only about to about 30 km

[2] Roche limit: limit inside of which tidal stress would exceed the internal strength of an orbiting body.

thick while the far-side crust reaches a thickness of 120 km in a broad region between two of the largest basins (Neumann et al. 1996, Lognonné et al. 2003). At the end of the fractional crystallization process, a residual magma rich in potassium, rare earth elements, phosphorous, and thorium known as urKREEP (Warren and Watson 1978) accumulated beneath the crust, largely in the region below what is now the very large, near-side, Procellarum basin (Haskin et al. 1999, Feldman et al. 1999, Wieczorek et al. 1999, Korotev 1999, Joliff et al. 1999). Late ilmenite-rich cumulates sank asymmetrically into the mantle of less dense olivine and pyroxene, carrying some urKREEP material with them (Spera 1992, Hess and Parmentier 1995, Parmentier and Hess 1999, Parmentier et al. 2002).

Heat from radioisotopic decay in the mantle of the Moon, critically assisted by the development of insulating impact debris on the lunar crust (Schmitt 2003), triggered intrusive and extrusive magmatic activity (Shearer and Papike 1999). This partial melting of the mantle began soon after crystallization of the majority of the magma ocean with the main sequence of partial melting and surface eruptions of mare basalt beginning at about 3.8 Gyr. The early portion of this magmatic activity produced the magnesium-rich suite (Mg-suite) of plutonic rocks (Heiken et al. 1991, Neal and Taylor 1992, Shearer and Floss 2000), extrusive KREEP-rich basalts (Taylor and Warren 1989), and the cryptomaria (Bell and Hawke 1984, Head et al. 1993, Antonenko et al. 1995, Williams et al. 1995). The cryptomaria are now, by definition, covered by impact generated, late large basin ejecta and exposed only where later impact craters have ejected dark material from beneath this ejecta.

Most modern workers have concluded that a concentrated bombardment of the lunar crust took place for about 100 Myr around 3.85 Gyr ago (Tera et al. 1974, Ryder 1990, Hartmann et al. 2000, Ryder 2001), producing most if not all of the ~50 basins greater than 300km in diameter visible today (Wilhelms 1987, Sputis 1993) as well as most other observed cratering effects in the lunar crust (Fig. 4.2). The effect of this late lunar "cataclysm" was to reset the potassium-argon ages of nearly all crustal impact breccias yet studied, including those from the Apollo sample and lunar meteorite collections (Ryder et al. 2000, Cohen et al. 2000, Cohen et al. 2004). Some of the geochemical characteristics of the Earth's mantle and the Moon's crust are attributed to the accretion of material associated with this cataclysm (Morgan et al. 2001) although other evidence related to rare gases and PGE (platinum group elements) abundances does not appear compatible with significant late accretion (Porcelli and Pepin 2000, Neal et al. 2000a).

According to the consensus hypothesis about lunar history, the ~2500 km diameter South Pole-Aitken basin is probably the only basin significantly greater than 1000 km in diameter to form during late bombardment (Spudis et al. 1994). The much larger, ~3200 km diameter Procellarum basin is considered an accidental consequence of the superposition of several smaller basins (Spudis 1996, Neumann et al. 1996). A global magnetic field, and presumably a circulating dynamo in a fluid metallic core, 350–450 km in radius (Goins et al. 1981, Williams et al. 2001, Hood et al. 1999, Hood and Zuber 2000, Khan et al. 2004), was present at least between 3.9 and 3.6 Gyr (Lin et al. 1998, Lin et al. 1999, and Mitchell et al. 2000). Between 3.9 and about 1.3 Gyr, a chemically highly diverse suite of mare basalts and basaltic pyroclastic materials erupted at the lunar surface, largely on the near side of the Moon (Apollo 11, 12, 15 and 17 Mission Science Reports 1969–1973, Schmitt 1973, McGetchin et al. 1981, Wilhelms 1987, Hiesinger

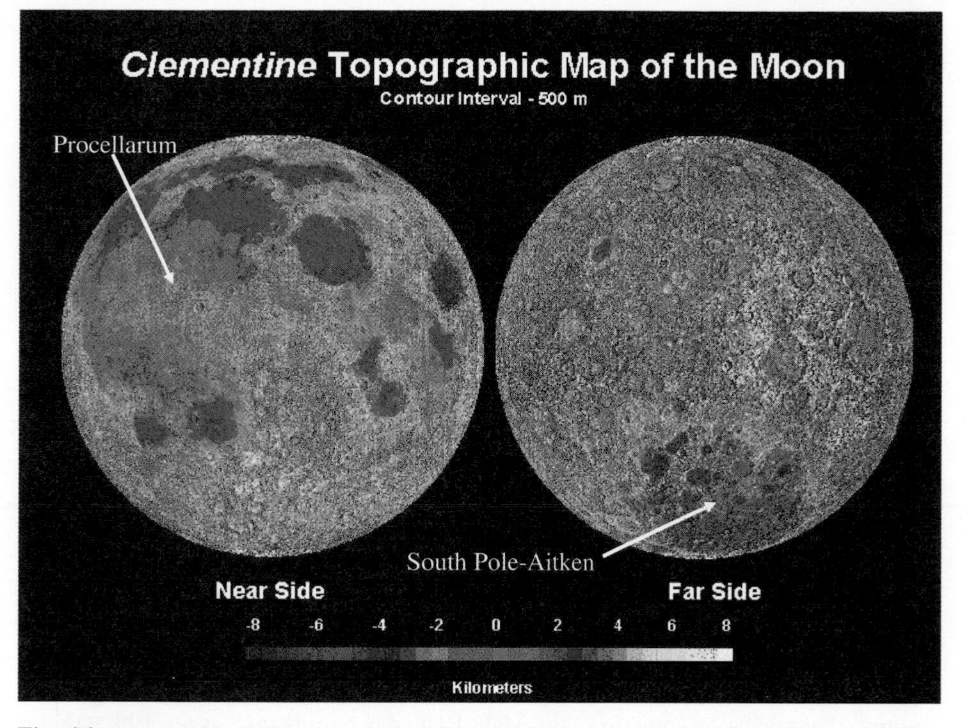

Fig. 4.2. Topographic differences produced largely by impacts on the Moon after ~4.56 Gyr. Derived from Clementine mission data (Image processing by Brian Fessler and Paul Spudis, LPI).

et al. 2000, and Schmitt 2003). Subsequent to about 3.8 Gyr, features on the Moon have been modified by meteor and cometary impact and space radiation. This maturation of the surface included the development of several meters of impact-generated regolith on most surfaces (Shoemaker et al. 1968, Schmitt 2003 and 2006).

Although many aspects of the above broad hypothesis for lunar origin and evolution are clearly correct, numerous difficulties exist in reconciling some of its major tenets with everything known about the geology of the Moon (Schmitt 1991 and 2003, Jones and Palme 2000, Stewart 2000).

4.2 Moon Origin by Giant Impact Unlikely

The giant impact hypothesis for the origin of the Moon summarized in Sect. 4.1.1, above, has become entrenched in the scientific literature, the popular scientific press, and as dogma in most academic institutions. Students and faculty alike are remarkably unquestioning about its basic premises. It is an attractive hypothesis, but too often implied as being more established than just a "hypothesis" by being called a "theory" (see also Wilson 2005). "Giant impact" should be considered a hypothesis rather than a theory as it is based on computer modeling and simulations rather than also being objectively tested against all the geochemical data available from Apollo samples. Giant impact, however, is an implicit if not explicit assumption that underlies most interpretations of new geochemical and geophysical data reported in the lunar science literature. The impact on the young Earth by a planetesimal less than fifteen percent its mass, of course, would be one way to explain the unusually high angular momentum of the Earth-Moon system. Further, the large-scale physical and orbital characteristics of the Earth and Moon can be reproduced convincingly in computer simulations (e.g. Canup 2004).

The timing of this proposed event now is constrained to have occurred within the first ~30 Myr of solar system history, that is, ~30 Myr after T_0. This constraint has been established by the isotopic systematics[3] of extinct ^{182}Hf (9-Myr half-life) and its daughter stable isotope ^{182}W (Yin et al. 2002, Kleine et al. 2002). (^{182}Hf differentiated between mineral phases during the fractional crystallization of the lunar magma ocean.) The estimated iron content of the Moon (Taylor and Esat 1996) as well as hafnium/tungsten systematics (Jones and Palme 2000) further constrain about ninety percent of the Moon's parent to be the impactor rather than the differentiated Earth's mantle, assuming, of course, that the impactor effectively had the composition of the Moon. This latter constraint essentially turns the giant impact hypothesis into an "impact assisted capture" hypothesis as the modern simulations clearly indicate. An additional critical constraint on this computer-based hypothesis is that the impactor needed to have evolved in the same oxygen isotopic reservoir as did the Earth (Jones and Palme 2000). Although chromium isotopic considerations suggest the same co-reservoir accretion (Palme 2001), broader chromium relationships, as well as oxygen isotope equivalence, suggest that the Moon's material may be more closely related to that of the HED meteorites[4] that appear to come from the Main Belt of asteroids (McSween 1999, Taylor 2001). Clearly, the original source of lunar materials is a subject for significant further study that takes into account all major variables, including thermal and dynamic accretionary history as a function of accreted planetesimal mass.

The major problem with the giant impact hypothesis lies with geochemical and geophysical information we have about the interior of the Moon below about 500 km. Specifically, both the original and more recent analyses of Apollo seismic data indicate a velocity discontinuity below about 550km (Goins et al. 1981, Kahn et al. 2000). Kahn et al., as well as Neal (2001), interpret these velocity data as suggesting that the lower lunar mantle is significantly more aluminous than overlying upper mantle. If this interpretation is correct, then melting and fractional crystallization in Moon was limited to the highly differentiated upper mantle. An initially cool lower mantle also is suggested by the delay in lunar core formation discussed in Sect. 4.3, below.

Additionally, the lower lunar mantle, based on analyses of the non-glass component of Apollo 17 orange pyroclastic[5] glass (Fig. 4.3) and Apollo 15 green pyroclastic glass, has chondritic isotopic and elemental signatures for tungsten (Lee et al. 1997), lead (Nunes et al. 1994), and siderophile and chalcophile elements (Neal 2001, Walker et

[3] Systematics: geochemical relationships in a family of isotopes.

[4] The HED meteorites have a highly distinctive spectro-photometric signature and appear to be derived from the large, Main Belt asteroid, 4 Vesta, based on spectro-photometric surveys of the large Main Belt asteroids (McSween 1999).

[5] Pyroclastic: volcanic debris derived from highly energetic eruptions of magma at a planetary surface.

al. 2004). Samarium-neodymium (Synder et al. 2000) and rhenium-osmium (Walker et al. 2004) isotopic systems also are consistent with relatively undifferentiated, that is, chondritic sources for this component. Volatiles associated with coatings on beads of both orange and green glasses are enriched over associated basalts by factors greater than 100 in Cl, F, Br, Zn, Ge, Cd, Tl, and Ag and by factors greater than 10 in Pb, Ga, Sb, Bi, In, Au, Ni, Se, Te, and Cu (Wasson et al. 1976, Krähenbühl 1980, Meyer 1989). These geochemical characteristics of the non-glass components are not consistent with volatile retention during the 1000° to 5000° degree temperatures to which proto-lunar materials would have been subjected during a giant impact event. Some workers (Walker et al. 2004) interpret chondritic material associated with pyroclastic glasses as being derived from post-magma ocean accretion (~150 Myr after T_0). This possibility of a late accretion source for the non-glass component appears highly unlikely based on the probable deep sources for these pyroclastic materials (Delano 1980, Heiken et al. 1991) and the probable loss of such volatiles from the Moon because of extreme heating during any late accretionary impacts. An internal source also is supported by the close spatial association of the eruptions with deeply penetrating graben faults at the margins of large basins such as Serenitatis (see Fig. 4.3).

In addition to the above data from the pyroclastic glasses that are inconsistent with a giant impact origin, the ratios of rare earth elements and other refractory elements (Taylor 2001) also are inconsistent with the extremely high temperatures that modeling

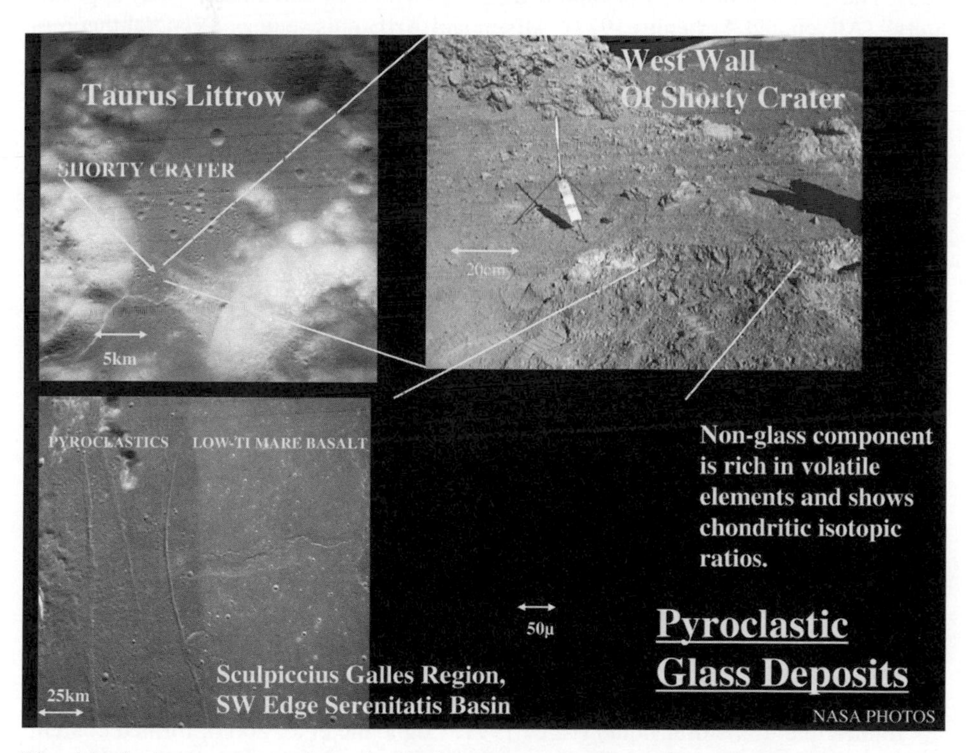

Fig. 4.3. Geologic context of the Apollo 17 orange pyroclastic glass and like-deposits of the Sculpiccius-Galles Region of the Moon (NASA photos).

indicates would accompany such an origin. The rare earth elements show none of the fractionation that would be expected if temperatures in the proto-lunar materials reached significantly over 1100° and are present in their expected solar nebular ratios except as modified by the fractional crystallization of the magma ocean (Taylor and Esat 1996). A lack of temperature fractionation in potassium isotopes (Humayun and Clayton 1995) also eliminates exposure to the high temperatures inherent in a giant impact.

If the giant impact hypothesis cannot explain the spectrum of geological evidence about the nature of the lower mantle outlined above, alternatives to that hypothesis should be considered. The most plausible alternative to the giant impact hypothesis appears to be the non-catastrophic capture of the Moon as an independently evolved planetesimal (Alfvén and Arrhenius 1969 and 1972; Singer 1970, Schmitt 1991; Schmitt 2003). Capture of one body by another appears to have been common around the giant planets (Taylor 2001, Johnson and Lunine 2005). Those satellites with retrograde, inclined orbits, Triton around Neptune and Phoebe around Saturn, for example, appear to have been captured so a capture mechanism for the Moon probably exists that is not as improbable as many have stated (Kaula 1971, Spudis 1996, Cameron 2000, Taylor 2001).

General agreement exists that for any one encounter of two bodies in orbits around the Sun, the probability is low that capture will occur. On the other hand, after each such encounter the orbit of the smaller body changes significantly and moves in an ellipse that will create new encounters twice for every complete orbit of the Sun by the larger body. The overall probability of capture increases with time and, in fact. "may approach unity" (Alfvén and Arrhenius 1972). Alfvén and Arrhenius state as a general theorem, "if two bodies move in crossing orbits and they are not in resonance, the eventual result will be either a collision or a capture." There will be exceptions to this "theorem" but there are also many factors to be considered (Alfvén and Arrhenius 1972, Singer 1986). For example, how would tidal interaction and accretion of possible pre-existing Earth satellites stabilize otherwise unstable post-capture orbits?

Although the internal geochemical and geophysical characteristics of the Moon suggest that capture is required, and physics appears to permit it, no modern modeling or simulation studies of such a dynamic interaction of the Earth and Moon have been published. In the capture alternative to giant impact, the Moon, like all the terrestrial planets, would probably have begun its existence in heliocentric orbit in the inner solar system, possibly at one of the Earth-Sun stable libration points, L4 or L5 (Belbruno and Gott 2004), with the initial, relatively slow accretion of a cool, chondritic core. Independent accretion at a libration point, of course, would take place precisely in the same reservoir of oxygen isotopes that fed the Earth, as required by analyses of lunar and terrestrial rocks (Jones and Palme 2000). As the accreting proto-Moon reached a radius of about 1200 km, roughly the current radius of the lower lunar mantle, a magma ocean formed from accreting material through a rapid rise in the conversion of kinetic and potential energy to heat. The ultimate mass of the Moon in the Earth's solar "feeding zone," as well as the Moon's ability to accrete the denser core material in the zone, would be limited by its competition with the more massive and gravitationally more attractive Earth.

Immiscible iron-sulfur liquid (Agee 1991, Lognonné et al. 2003), formed contemporaneously with silicate liquid in the growing magma ocean, would sink rapidly to the bottom of the ocean and then migrate slowly downward through the cooler lower mantle as it warmed due to radiogenic heat production. In this migration process of broadly disseminated iron-rich liquid, any water encountered would be decomposed to hydrogen and FeO (Schmitt 2003). At about 3.9 Gyr, the remaining iron-sulfur liquid coalesced to form a core capable of sustaining a fluid dynamo and a global magnetic field (see Sect. 4.3).

Devolatilization of the Moon's magma ocean (upper mantle), over and above that of the more massive terrestrial planets (Taylor 2001), would take place at the magma ocean's surface as the effectively infinite number of impacts heated portions of magma to many thousands of degrees. The cool lower mantle, however, would retain much of its chondritic character as it would remain largely unmelted as well as non-convecting in the Moon's relatively low gravitational environment. This remnant, chondritic material would later provide the non-glass, volatile component that combined with partial melts in the upper mantle to produce pyroclastic surface eruptions. Both the release of volatiles from the lower mantle and the partial melting of the upper mantle appears to have been triggered when graben forming, extensional fracturing relieved some lithostatic pressure deep within the Moon.

Geologically, a pure "capture" hypothesis appears to make a lot of sense when viewed in the total context of our knowledge about the Moon. Modern computer modeling and simulation studies of this hypothesis should be undertaken to see if the angular momentum constraint for the Earth-Moon system can be satisfied without an actual "impact assist" to capture.

4.3 Terrestrial Planets Had Early Chondritic Cores

Seismic, isotopic and elemental evidence, as discussed in Sect. 4.2 above, indicates that the lower mantle of the Moon, material below ~500km, never melted significantly and is relatively undifferentiated from its chondritic beginnings. The modern hypothesis for planetary accretion (Wetherill 1990, Weidenshilling and Cuzzi 1993, Kortenkamp et al. 2000, Taylor 2001, Alexander et al. 2001) favors slow initial accretion of many planetesimals followed by runaway accretion by the largest of these at the expense of the smaller. This, then, would suggest that the Moon and other terrestrial planets, up to some point in size, had solid chondritic cores similar in composition and size to that of the lower mantle of the Moon, that is, they had proto-cores on the order of 1200 km in radius (Schmitt 2003). In the case of the Moon, at least, a 1200 km radius appears to be the size beyond which the combination of impact energy and frequency of accreting, co-orbiting objects resulted in both the melting of newly accreted material and its retention as a magma ocean. Since 3.8 Gyr, however, impact velocities, and therefore impact energies, of asteroidal and cometary objects have been so high (10–70 km/s) that less than 0.3 percent of their mass appears to be retained by the Moon (Ryder 1999).

The existence of a relatively cool proto-core beneath the early magma oceans of the terrestrial planets would delay, to varying degrees, the migration of immiscible iron-sulfur liquid to form metallic cores. In the case of the Moon, remnant magnetic fields at the antipodes of the four youngest large impact basins (Anderson and Wilhelms 1979; Lin et al. 1988; Lin et al. 1998) suggest that core formation and an active dynamo-driven magnetic field were delayed by its proto-core until after about 3.92 Gyr, the suggested age of Nectaris, the oldest of the mascon basins (Wilhelms 1987, Chapters 9–10). Mascon basins with remnant magnetic anomalies at their antipodes (Crisium, Serenitatis, Imbrium and Orientale, in order of relative age) are the four youngest large basins. Orientale, the youngest of the four is estimated by Wilhelms to have been formed at 3.8 Gyr with the measured ages of Serenitatis and Imbrium, 3.86–3.87 and 3.84–3.86 Gyr, respectively, falling in between the ages of Nectaris and Orientale. Because of larger masses of dense core forming material and greater gravitational potential to drive core formation through displacement of a chondritic proto-core, and greater internal heating as the core forming material moved downward and released its potential energy, one would expect that progressively decreasing delay in core formation would follow a sequence based on increasing mass, that is, Moon, Mercury, Mars, Venus, and Earth. Mars, indeed, appears to have had a working dynamo in its core earlier than the Moon but after the formation of the early Martian crust. This is apparent because the remnant magnetic "striping" observed in Mars' Southern Uplands (Connerney et al. 1999) has been destroyed by the formation of very large basins (greater than 2000 km in diameter) such as Hellas. These very large basins probably formed prior to about 4.2 Gyr (Schmitt 2003), but clearly before 3.9 Gyr, the age that lunar mascon basins first recorded a magnetic field for the Moon.

This is contrary, however, to the conclusions reached by Kleine (2002) and Yin (2002) in their analyses of ^{182}Hf/^{182}W systematics for samples potentially representative of the mantles of the Earth and Mars. These workers believe that core formation in the terrestrial planets was complete within 30 Myr of T_0. Their analysis, however, assumes that the **separation** of core forming material from material that formed planetary mantles was equivalent in time to the formation of a core. As indicated by the apparent delay in lunar core formation, this assumption probably is not valid. The Kleine and Yin analyses indicate that **separation** of core forming material in Mars occurred at about 13 Myr, in the Earth at about 25 Myr, and in the Moon at between 25 and 30 Myr after T_0. An additional factor to be considered in the interpretation of the hafnium and tungsten isotopic data includes the possibility of an earlier accretion of Mars relative to planets closer to the sun, as was clearly the case with the parent bodies of the Main Belt Asteroids that were melted and differentiating prior to 2–10 Myr after T_0 (Carlson and Lugmair 2000, Baker et al. 2005). Incomplete mixing in planetary mantles after upward displacement of chondritic proto-cores is also a possible confounding factor to the Kleine and Yin analyses.

4.4 Melt from Huge Impacts Produced First Continents

Impacts of objects from space clearly have played a major role in the evolution of the Earth's crust throughout the ~1 Gyr long Hadean Era (Taylor 2001, Glikson 2001, Schmitt 2003). The 2500 km diameter basin on the far-side of the Moon (Wilhelms 1987), known as South Pole-Aitken, records an impact of an extraordinarily energetic object near the end of the period of smaller scale saturation cratering that followed the solidification of the lunar crust. As recorded in the southern highlands and much of the far-side of the Moon, that saturation cratering reached crater diameters of 60–70 km and created a mega-regolith of repetitively broken and mixed crustal rock and solar

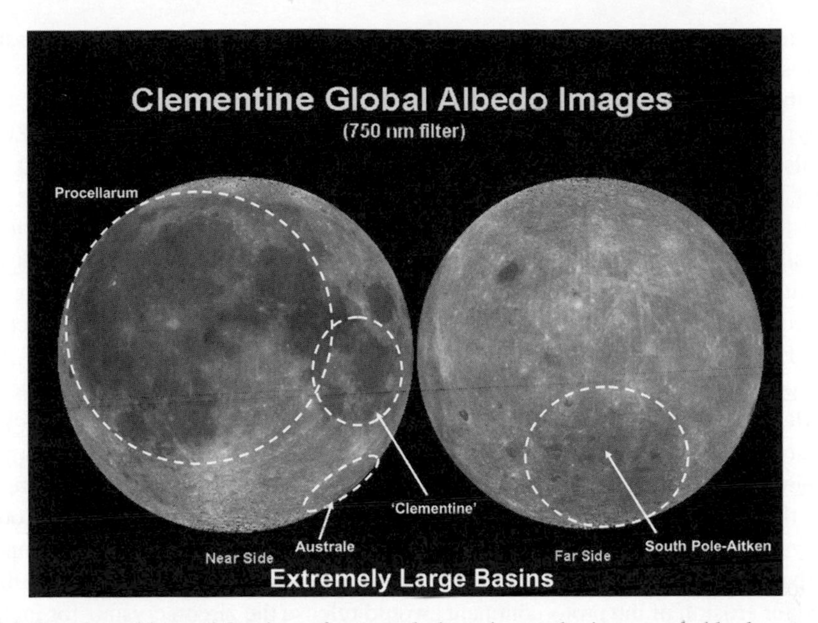

Fig. 4.4. Rough positions of the rims of extremely large impact basins recorded by lunar crustal features (Modified NASA photo).

wind volatiles at least 20–25 km deep[6] (see Wilhelms 1987, Schmitt 2003). Saturation cratering is responsible for providing the thermal insulation that resulted in the global partial re-melting of the upper mantle to produce basaltic volcanism (Schmitt 1991 and 2003) between about 4.2 and 1.3 Gyr ago.

South Pole-Aitken is probably just the most obvious manifestation of possibly four or five other such extremely large impacts (Fig. 4.2), including the 3200km diameter front-side basin, Procellarum (Whitaker 1981, Wilhelms 1987, Schmitt 2003). On the basis of the degree to which background saturation cratering affected their basin rim morphologies, Schmitt estimates that the more highly degraded Procellarum basin formed at about 4.3 Gyr and South Pole-Aitken at about 4.2 Gyr. The ~4.3 Gyr age for Procellarum also may be supported by zircon crystallization ages near 4.3 Gyr for some of the Apollo 14 KREEP-related samples from the Procellarum region (Meyer et al. 1989 and 1996) and possibly related to the Procellarum event. The formation of these old, extremely large impact basins requires an early source of large planetesimals or highly energetic smaller bodies. The interaction of the early-formed Jupiter with the parent planetesimals of the Main Belt asteroids is one possible source of impactors. Another source might be the Moon's post-capture accretion of original satellites of the Earth that have been

[6] Mega-regolith: The broken and regionally mixed, rocky crust material that would extend to depth of the saturation craters or about one-third the diameter of the crater saturation size of 60–70 km. This depth is also the depth where seismic velocities reach those of solid rock (Goins et al 1981). The "mega-regolith" contrasts with the lunar surface "regolith" that is made up largely of very fine-grained debris.

estimated to be between 2 and 10 in number based on extrapolation from the satellite systems of other planets (Alfvén and Arrhenius 1972).

If the formation ages for South Pole-Aitken and Procellarum are about right at 4.2 and 4.3 Gyr, respectively, an explanation is suggested for the 4.4 Gyr, detrital zircon ($ZrSiO_4$) crystals recently discovered in very old sedimentary rocks on Earth (Wilde et al. 2001, Turner et al. 2004). Zircon crystallizes from igneous magmas in the very late stages of crystallization when zirconium concentrations become sufficiently high due to most other silicate minerals having crystallized – minerals with crystal structures with which zirconium is incompatible. Late stage crystallization also tends to produce other, much more common silicate minerals, such as quartz, feldspar and mica that are characteristic of the Earth's continents. Early impacts on the continental scale of South Pole-Aitken and Procellarum, occurring in water-rich environments such as the surfaces of the Earth and Mars (Abe et al. 2000), would create thick sheets of impact generated rock melt thousands of kilometers across and many kilometers thick. As these magma sheets crystallized, zirconium concentrations could be expected ultimately to have reached levels that crystallized zircon. Indeed, recent analyses (Watson and Harrison 2005) of the crystallization temperature of these old zircon crystals (~700? C) are compatible with their formation late in the differentiation of water-rich silicate magma. Weathering and water erosion of the proto-continents would release the zircon crystals for inclusion as sand grains in ancient sediments. As zircons are extremely hard and durable, they can survive several cycles of erosion. The very old terrestrial zircons that have been dated and had their oxygen isotopic ratios determined apparently formed in the presence of water (Wilde et al. 2001, Mojzsis et al. 2001), consistent with this hydrous impact melt sheet hypothesis for their origin.

4.5 Lunar Cataclysm at 3.9 Gyr Faked by Late Basin-Forming Impacts

The suggestion (Tera et al. 1974) that a "cataclysm" of impacts at about 3.9 Gyr was responsible for the vast majority of craters visible on the lunar surface has gained increasing adherents in recent years (Ryder 1990, Ryder et al. 2000, Cohen et al. 2000, Cohen et al. 2004). In this hypothesis's most extreme manifestation, essentially all pre-mare basalt impact craters are attributed to such a cataclysm. More modest proposals concentrate on the timing of the formation of fifty or so craters greater than a few hundred kilometers in diameter (Fig. 4.5). The primary rationale for the cataclysm hypothesis comes from the rarity of impact breccias in the Apollo sample collection and in lunar meteorites with impact dependent, potassium-argon ages older than ~3.9 Gyr.

The primary argument against this hypothesis is, of course, the high degree of inherent sampling bias in both the Apollo suite and the lunar meteorites (Schmitt 1991, 2001 and 2003, Chapman et al. 2002). The need to consider sampling bias comes from the strong evidence that the Moon has been effectively resurfaced by debris thrown from the fourteen youngest of the fifty or so largest impacts. Also, Schmitt (1989) has discussed the clear temporal and geological distinctions between the young and old large basins. He points out, specifically, that the formation of the oldest basins preceded a period of remarkable strengthening of the lunar crust so that mass concentrations in younger

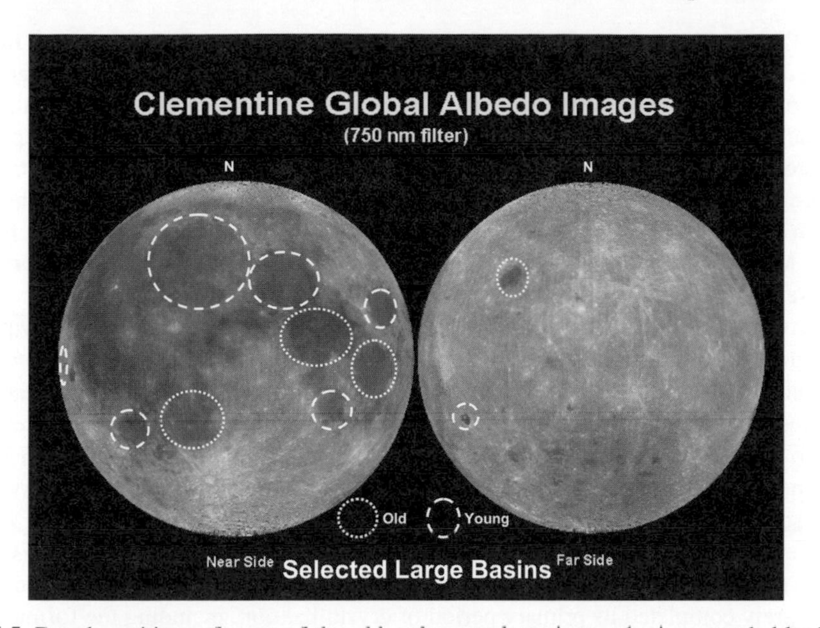

Fig. 4.5. Rough positions of some of the old and young large impact basins recorded by lunar crustal features (Modified NASA photo).

basins could be supported indefinitely. The fresher appearing so-called "mascon" basins (Muller and Sjogren 1968, Wilhelms 1987, Neumann et al. 1996) now represent the 14 youngest large impacts and are basins that have undergone little isostatic adjustment since they formed. Those young basins for which reasonable formation dates have been assigned (Wilhelms 1987), that is, Nectaris, Crisium, Serenitatis, Imbrium and Orientale, range in age between 3.9 and 3.8 Gyr, coinciding with the proposed period of "cataclysm."

Photo and telescopic geologic mapping in the 1960s and 1970s (see summary by Wilhelms 1987) established that ejecta blankets and the effects of secondary ejecta from the fourteen youngest impacts are distributed widely around the Moon. This fact has been more recently emphasized by the lunar-wide identification of "cryptomaria" (Bell and Hawke 1984, Antonenko 1999) through mapping the global distribution of small impact craters surrounded by dark ejecta and that have penetrated overlying, lighter colored deposits of impact debris. These apparent basaltic lava eruptions clearly preceded the formation of the fourteen young large basins, or they would not have been covered by ejecta from them. The cryptomaria eruptions may have been temporally associated with and immediately related to the formation of the ~35 old large basins through local pressure release partial melting in the upper mantle after each impact. As the upper mantle was approaching partial melting temperatures prior to the main period of mare basalt eruptions, beginning at about 3.8 Gyr, limited pressure release melting would have been a likely occurrence beneath each new large basin.

Whether there was a 100 Myr cataclysm at about 3.85 Gyr or a 400 Myr period of large basin formation between about 4.2 and 3.8 Gyr as proposed by Schmitt (2003), it is clear that a discrete new source of impactors appeared in the solar system (Schmitt,

1999, Dones, 2002). Of particular interest as potential new sources would be the Jupiter-induced break-up of the proto-planetesimals of the Main Asteroid Belt, the interaction of the outwardly migrating giant planets with Kuiper Belt objects (Levinson et al. 2001, Morbidelli 2004, Kerr 2004), and the possible disturbance of the Öort Cloud by a passing stellar object. The identification of the impactor source or sources is not only an intriguing challenge but also one with many implications to unraveling the evolution of the solar system and the terrestrial planets. Continuing analysis of all data on the impact history of the Moon is critical to providing constraints on the possible source of a discrete set or of sets of travelers into the inner solar system.

Recent modeling of the interaction and migration of the giant planets as they passed through "low-order mean motion resonances" relative to each other and interacted gravitationally with random planetesimals in the disk of the outer solar system suggests that a process such as this would be short lived (Tsiganis et al. 2005), that is, complete on the order of 40 Myr after T_0. Although the results of this modeling suggests that the ejection of planetesimals from the disk would be a source for a "cataclysm" of impactors in the inner solar system, the modeled interactions occur far too early to explain the various major episodes of impacts on the Moon, the last of which occurred around 3.85 Gyr. In fact, as discussed in Sect. 4.2, above, in that 40 Myr period, the lunar magma ocean would have barely completed its primary period of crystallization, including the formation of a stable crust that could record later impacts. Further, most observers estimate that the giant planets formed within 10–20 Myr after T_0 (Liu 2004, Greaves 2005, Telesco et al. 2005). As modeled, therefore this process of giant planet interaction and migration does not explain either a short cataclysm in the inner solar system at about 3.85 Gyr or a prolonged one between 4.5 and 3.8 Gyr. This process may provide, however, a dynamic mechanism for the re-introduction of volatile components, such as water, carbon dioxide and nitrogen, into the inner solar system. Re-introduction would be subsequent to the early clearing of volatiles out to $\sim$5 AU (Taylor 2001) but prior to a significant portion of the accretion of the terrestrial planets. Volatile re-introduction would be particularly likely during and immediately following the formation of Jupiter.

The new modeling related to the giant planets also suggests that a more complex scenario of interactions might provide multiple episodes of large crater formation. Models do not yet take into account the probability of significant gaps in the initial distribution of disk planetesimals after the early volatiles clearing event and the initial accretion of the giant planets beyond the planetary "snow line" at $\sim$5AU. At the very least, accretion of each giant planet would have created structural variations in the planetesimal disk (Liu 2004, Okamoto et al. 2004, Greaves 2005, Telesco et al. 2005), probably even relatively depopulated gaps. Such gaps have been reported in CoKu Tau 4, imaged by the Spitzer Space Telescope (NASA 2004). In addition, some structural variations may have been inherited from the original spiral structure of the solar nebulae. Given these additional possible complexities, it seems likely that giant planet migration would be variable in rate and timing as entry into depleted zones slowed that migration down and entry into planetesimal-rich rings speeded it up and increased the supply of objects forced into the inner solar system.

With a scenario of what might be called "punctuated cataclysm," the following sequence of impact episodes, based on the current evidence from the Moon (Schmitt 2003, Fig. 4.6), should be considered for the inner solar system:

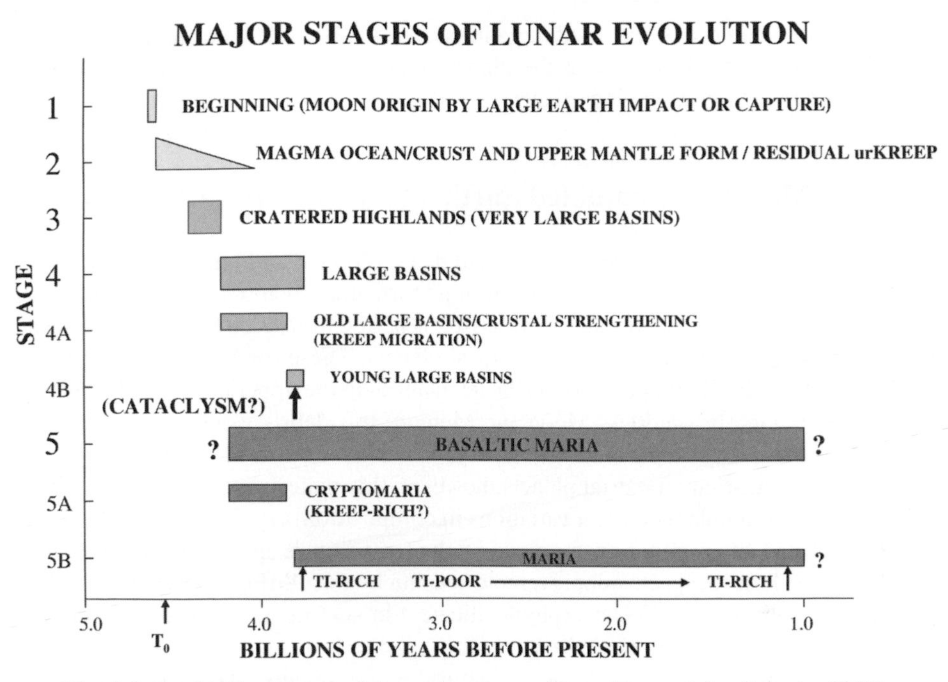

Fig. 4.6. Graphical summary of the major stages of lunar history (after Schmitt, 2003).

1. $T_0 - T_0 + 20$ Myr: Re-introduction of volatile components into the inner solar system as the giant planets formed and interacted with each other and with the icy planetesimals of the outer disk of the solar system.
2. 4.5–4.2 Gyr: A few extremely large impact events in a background of smaller cratering, both produced by the effects of Jupiter's resonance disruption of the parent bodies of the Main Belt Asteroids (McSween 1999, Taylor 2001). Saturn may have played a role in this as well (Tsiganis et al. 2005). The gradual sweeping up of leftover rocky debris in the inner solar system may have been superposed on these giant planet effects. There remains the possibility that the extremely large impacts resulted from the Moon's, post-capture sweeping up of Earth's original satellites (Alfvén and Arrhenius 1972).
3. 4.2–3.9 Gyr: One or more episodes of large basin formation as the outward migration of giant planets other than Jupiter, and possibly only Neptune, encountered concentrated rings of planetesimals in the outer solar system.
4. 3.9–3.8 Gyr: A final episode of large basin formation as the continued outward migration of giant planets encountered the final ring of concentrated planetesimals in the outer solar system.

Future modeling of the possible migration and interactions of the giant planets and the planetesimal disk might work backward from the apparent impact record on the Moon to see if it results in a reasonable, non-uniform structure for the planetesimal disk. For this new modeling effort, the episodes of large impacts might be centered within the four periods outlined above. Further, episodes (3) and (4) vary by a factor of 2.5 ($\sim$14 vs. ~35) in the number of recorded large impacts (>300 km diameter basins) and may partially quantify the difference in the planetesimal densities in the zones of the disk encountered by a migrating giant planet.

4.6 Clay Minerals Dominated Earth's Crust Before Life Appeared

The exploration of the Moon has documented that prolonged and intense impact cratering took place in the inner solar system from its formation to about 3.8 Gyr. Following the solidification of magma oceans, records of this period have been left in the crusts and upper mantles of the Moon and the terrestrial planets. These impacts would have produced abundant glassy and pulverized silicate material in the upper portions of planetary crusts. On the largely waterless Moon and Mercury this debris would just accumulate as increasingly deep and increasingly glassy, pulverized and solar wind impregnated regolith. On water-rich terrestrial planets, however, this material would alter rapidly to clay minerals, minerals with great variations in composition, crystal structural patterns, and mineral stability niches as determined by the temperature and composition of local environments. Indeed, spectroscopic remote sensing by the OMEGA spacecraft in Mars orbit has recently identified clays (phyllosilicates) in surface materials exposed in old Martian highlands (Poulet et al. 2005).

In their broad adaptability to various natural environments, clay species resemble biological species. Variable crystal structural patterns on the surfaces of the clay mineral grains, possibly along with those on the surfaces of sulfide minerals (Huber and Wächtershäuser 1998), may have assisted in the aggregation of complex organic molecules, possible precursors to the first replicating forms of such molecules (Ferris 1996). Indeed, replication may have first been symbiotic with increased diversity in the forms, growth, and/or expansion of clay mineral structures. Tubular forms of clay minerals also may have been initially incorporated in the earliest single cell organisms to assist in the movement of fluid, only to be replaced later by organic compounds.

Evidence, still controversial, of isotopic fractionation by organic processes associated with ~3.8 Gyr sedimentary rocks in Greenland (Mojzsis et al. 1996, Moorbath 2005) may not be a timing coincidence. No older rock exposures of clear sedimentary beginnings in stable water dominated environments are as yet known on Earth (Westfall 2005). As discussed in Sect. 4.5, above, the end of the large basin forming events in the inner solar system also appears to be at about 3.8 Gyr. Prior to that time, preservation of either sedimentary basins or evidence of life forms would have been unlikely. Although simple organic replication, and possibly even single cell organisms, may have existed on Earth prior to ~3.8 Gyr, the catastrophic effects of large impacts may have prevented extensive or permanent biological activity until after that time.

The cratering history of the Moon, therefore, has alerted us to the potential pervasiveness of clays on the early, water-rich crusts of Earth, Mars, and probably Venus. Mars, in this context and in the absence of the geological recycling effects of plate tectonics, may be the arrested crucible of early organic processes now only vaguely apparent on Earth and probably totally lost on Venus due to later, nearly total geological resurfacing (Taylor 2001).

4.7 Conclusion

Lunar origin and evolution remains incompletely understood and any strategy to return to the Moon should provide for a continuation of the scientific exploration of that small planet. A return to the Moon for the purpose of exploration, as proposed for the United States by President George W. Bush (2004) would inherently include lunar science, and particularly exploration science, as a major objective as it was for the majority of Apollo missions. Alternatively, the author (Schmitt 2006) proposes that private investors take the lead in financing a return to the Moon and the development of helium-3 fusion power in order to take advantage of the helium-3 resources present in the lunar regolith. A private energy resource initiative would lead directly to routine access to the Moon and early settlement and therefore to establishing a relatively low cost capability to support long-term lunar science. A combination of the two approaches may eventually be the most likely to succeed politically and economically.

The available broad outlines of lunar history and alternatives to prevailing hypotheses, however, give insights and generate questions related to the origin and evolution of the other terrestrial planets that should not be ignored. If early proto-cores delayed core formation in these planets, what evidence of this might be found during our continuing studies of the present mantles of Earth and Mars (Smith and Zuber, 1996, Albarède et al. 2000; Meibom and Frei, 2002, Bizzarro et al. 2002; Ballentine 2002, Boyet and Carlson 2005, Caro et al. 2005, McCammon 2005, Class and Goldstein 2005)? If large impact basin formation took place early in the history of terrestrial planets, would these basins be reflected in the distribution and timing of long duration seas that appear to have existed on Mars (Head et al. 1999, Squyres et al. 2004)? Whether or not giant impact played a role in the origin of the Moon, could such events explain other planetary anomalies such as the large core of Mercury relative to its mantle (Taylor 2001) and the possible giant impact relationship between Pluto and its moon, Charon (Canup 2005)?

The exciting and dynamic field of comparative planetology is searching for answers to the above questions within existing data sets and anticipating the acquisition of more as lunar and planetary exploration continues to lay the groundwork for eventual human settlement of the Moon and Mars.

References

Abe, Y., Obtani, E., Okuchi, T., Fighter, K., and Drake, M., 2000. Water in the early Earth. In *Origin of the Earth and Moon*, eds. R.M., Canup and K. Righter, University of Arizona Press, Tucson, and Lunar and Planetary Institute, Houston, pp. 413–433.

Agee, C.B., 1991. High-pressure melting of carbonaceous chondrites. In *Workshop on the Physics and Chemistry of Magma Oceans from 1 Bar to 4 Mbar*, eds. C. B. Agee and J. Longhi, Technical Report Number 92–03, Lunar and Planetary Institute, Houston, pp. 11–12.

Albarède, F., Blichert-Toft, J, Vervoort, J. D., Gleason, J. D., and Rosing, M., 2000. Hf-Nd isotope evidence for a transient dynamic regime in the early terrestrial mantle. *Nature* **404**, 488–490.

Alfvén, H. and Arrhenius, G., 1969. Two alternatives for the history of the Moon. *Science* **165**, 11–17.

Alfvén, H. and Arrhenius, G., 1972. Origin and evolution of the Earth Moon system. *The Moon* **5**, 216.

Alexander, C.M. O'D, Boss, A.P., and Carlson, R.W., 2001, The early evolution of the inner solar system: A meteoritic perspective. *Science* **293**, 64–68.

Amelin, Y, Krot, A.N., and Twelker, E., 2004. Pb isotopic age of the CB chondrite Gujba, and the duration of the chondrule formation interval. *Geochimica et Cosmochimica Acta* **68**, Abstract E958.

Anderson, K.A., and Wilhelms, D.E., 1979. Correlation of lunar farside magnetized regions with ringed impact basins. *Earth and Planetary Science Letters* **46**, 107–112.

Antonenko, I, 1999. Global estimates of cryptomare deposits: Implications for lunar volcanism. Lunar and Planetary Science Conference 30, Abstract #1703.

Antonenko, I., Head, J.W., Mustard, J.F., and Hawke, B.R., 1995. Criteria for the detection of lunar cryptomaria. *Earth, Moon and Planets* **69**, 141–172,

Baker, J., Bizzarro, M., Wittig, N., Connelly, J., and Haack, H., 2005. Early planetesimal melting from an age of 4.5662 Gyr for differentiated meteorites. *Nature* **436**, 1127–1131.

Ballentine, C.J., 2002. Tiny tracers tell tall tales. *Science* **296**, 1247–1248.

Belbruno, E., and Gott, J.R., III, 2004. Where did the Moon come from? *Astronomical Journal* **129**, 1724.

Bell, J., and Hawke, B., 1984. Lunar dark-haloed impact craters: Origin and implications for early mare volcanism. *Journal of Geophysical Research* **89**, 6899–6910.

Bizzarro, M, Simonetti, A, Stevenson, R.K., David, J., 2002. Hf isotope evidence for a hidden mantle reservoir. *Geology* **30**, 771–774.

Boyet, M. and Carlson, R.W., 2005. 142Nd evidence for early (>4.53 Ga) global differentiation of the silicate Earth. *Science* **309**, 576–581.

Bush, G. W., 2004. President Bush Delivers Remarks On U.S. Space Policy, *NASA Facts*, January 14.

Boynton, W.V., Feldman, W.C., Squyres, S.W., Prettyman, T.H., Brückner, J., Evans, L.G., Reedy, R.C., Starr, R., Arnold, J.R., Drake, D.M., Englert, P.A.J., Metzger, A.E., Mitrofanov, I., Trombka, J.I., d'Uston, C., Wänke, H., Gasnault, O., Hamara, D.K., Janes, D.M., Marcialis, R.L., Maurice, S., Mikheeva, I., Taylor, G.J., Tokar, R., and Shinohara, C., 2002. Distribution of hydrogen in the near-surface of Mars: Evidence for sub-surface ice deposits. *Science* **297**, 81–85.

Cameron, A.G.W., 2002. Birth of a solar system, *Nature* **418**, 924–925.

Cameron, A.G.W., 2000, Higher-resolution simulations of the giant impact. In *Origin of the Earth and Moon*, eds. R.M., Canup and K. Righter, University of Arizona Press, Tucson, and Lunar and Planetary Institute, Houston, pp. 133–144

Cameron, A.G.W., and W.R. Ward, 1976. The origin of the Moon. Lunar Science Conference 7, Lunar Science Institute, Houston, pp. 120–122.

Canup, R.M., 2004. Simulations of a late lunar-forming impact. *Icarus* **168**, 433.

Canup, R.M., and Agnor, C.B., 2000. Accretion of the Terrestrial Planets and the Earth Moon System. In *Origin of the Earth and Moon*, eds. R.M., Canup and K. Righter, University of Arizona Press, Tucson, and Lunar and Planetary Institute, Houston, pp. 113–129.

Canup R.M., and Righter, K. (eds.), 2000. *Origin of the Earth and Moon*, University of Arizona Press, Tucson, and Lunar and Planetary Institute, Houston, 555 pp.

Canup, R.M., 2005. A giant impact origin of Pluto-Charon. *Science* **307**, 546–550.

Carlson, R.W., and G.W. Lugmair, 2000. Timescales of planetesimal formation and differentiation based on extinct and extant radioisotopes. In *Origin of the Earth and Moon*, eds. R.M., Canup and K. Righter, University of Arizona Press, Tucson, and Lunar and Planetary Institute, Houston, pp. 25–44.

Caro, G., Bourdon, B., Wood, B.J., and Corgne, A., 2005. Trace-element fractionation in Hadean mantle generated by melt segregation from a magma ocean. *Nature* **436**, 246–249.

Carr, M. H., 1996, *Water on Mars*, Oxford University Press, Oxford, 229 p.

Chapman, C.R., Cohen, B.A., and Grinspoon, D.H., 2002, What are the real constraints on commencement of the late heavy bombardment? Lunar and Planetary Science Conference 33, Abstract #1627.

Class, C., and Goldstein, S.L., 2005. Evolution of helium isotopes in the Earth's mantle. *Nature* **436**, 1107–1111.

Cohen, B.A., Swindle, T.D. and Kring, D.A., 2000. Support for the lunar cataclysm hypothesis from lunar meteorite impact melt ages. *Science* **290**, 1754–1756.

Cohen, B.A., James, O.B., Taylor, L.A., Nazarov, M.A., and Baruskova, L.D., 2004. Lunar highland meteorite Dhofar 026 and Apollo sample 15418: Two strongly shocked, partially melted, granulitic breccias. *Meteoritics and Planetary Science*, **39**, 1419–1447.

Connerney, J.E.P., Acuña, M.H, Wasilewski, P.J., Ness, N.F., Rème, H., Mazelle, C., Vignes, D., Lin, R.P., Mitchell, D.L., and Cloutier, P.A., 1999. Magnetic lineations in the ancient crust of Mars. *Science* **284**, 794–798.

Delano, J.W., 1980. Chemistry and liquidus relations of Apollo 15 red glass: Implications for the deep lunar interior. Proceedings of Lunar and Planetary Science Conference 11, pp 251–288.

Dones, L., 2002. Dynamics of possible late heavy bombardment impactor population. Lunar and Planetary Science Conference 33, Abstract #1662.

Feldman, W.C., Lawrence, D.J., Maurice, S., et al., 1999. Classification of lunar terranes using neutron and thorium gamma-ray data. Lunar and Planetary Science Conference 30, Abstract #2056.

Ferris, J.P., Hill, A.R. Jr., Liu, R. and Orgel, L.,E., 1996. Synthesis of Long Prebiotic Oligomers on Mineral Surfaces. *Nature* **381**, 59–61.

Glikson, A.Y., 2001. The astronomical connection of terrestrial evolution: Crustal effects of post 3.8 Ga mega-impact clusters and evidence for major 3.2 ± 0.1 Ga bombardment of the Earth-Moon system. *Journal of Geodynamics* **32**, 205–229.

Goins, N.R., Dainty, A.M., and Toksöz, M.N., 1981. Lunar seismology: The internal structure of the Moon. *Journal of Geophysical Research* **86**, 5061–5074.

Greaves, J.S., 2005. Disks around stars and the growth of planetary systems. *Science* **307**, 68–75.

Hartmann, W.K., 1986. Moon origin: The impact trigger hypothesis. In *Origin of the Moon*, eds. W.K. Hartmann, R.J. Phillips, and G.J. Taylor, Lunar and Planetary Institute, Houston, p. 579–608.

Hartmann, W.K., and D.R. Davis, 1975. Satellite-sized planetesimals and lunar origin. *Icarus* **24**, 504–515.

Hartmann, W.K., Ryder, G., Dones, L, and Grinspoon, D, 2000. The time-dependent intense bombardment of the primordial Earth/Moon System. In *Origin of the Earth and Moon*, eds. R.M., Canup and K. Righter, University of Arizona Press, Tucson, and Lunar and Planetary Institute, Houston, pp. 493–512.

Haskin, L.A., Gillis, J.J., Korotev, R.L., and Jolliff, B.L., 2000. The materials of the lunar Procellarum KREEP Terrane: A synthesis of data from geomorphological mapping, remote sensing, and sample analyses. *Journal of Geophysical Research* **105**, E8, 20403–20416.

Head, J.W, 1999. Possible ancient oceans on Mars: Evidence from Mars Orbiter laser altimeter data. *Science* **286**, 2134–2137.

Head, J.W., Murchie, S., Mustard, J.F., et al., 1993. Lunar impact basins: New data for the western limb and farside (Orientale and South Pole-Aitken basins) from the first Galileo flyby. *Journal of Geophysical Research* **98**, E9, 17,149–17,181.

Heiken, G.H., Vaniman, D.T., and French, B.M., 1991. *Lunar Sourcebook*, Cambridge University Press, Cambridge, pp. 225–228.

Hess, P.C., and Parmentier, E.M., 1995. A model for the thermal and chemical evolution of the Moon's interior: Implications for the onset of mare volcanism. *Earth and Planetary Science Letters* **134**, 501–514.

Hester, J.J., Desch, S.J., Healy, K.R., and Leshin, L.A., 2004. The cradle of the Solar System. *Science* **304**, 1116–1117.

Hiesinger, H., Head, J.W., Jaumann, R., and Neukum, G., 1999. Lunar Mare Volcanism. Lunar and Planetary Science Conference 30, Abstract #1199.

Hood, L.L., Mitchell, D.L., Lin, R.P., Acuña, M.H. and Binder, A.B., 1999. Initial measurements of the lunar induced magnetic dipole moment using Lunar Prospector magnetometer data. *Geophysical Research Letters* **26**, 2327–2330.

Hood, L.L., and Zuber, M., 2000. Recent refinements in geophysical constraints on lunar origin and evolution. In *Origin of the Earth and Moon*, eds. R.M., Canup and K. Righter, University of Arizona Press, Tucson, and Lunar and Planetary Institute, Houston, pp. 397–412.

Huber, C., and Wächtershäuser, G., 1998. Peptides by activation of amino acids with CO on (Ni,Fe)S surfaces: Implications for the origin of life. *Science* **281**, 670–672.

Humayun, M., and Clayton, R.N., 1995. Precise determination of the isotopic composition of potassium: Application to terrestrial rocks and lunar soils. *Geochimica et Cosmochimica Acta* **59**, 2131.

Jacobsen, S.B., 2003. How old is planet Earth? *Science* **300**, 1513–1514.

Johnson, T.V., and Lunine, J.I., 2005. Saturn's moon Phoebe as a captured body from the outer Solar System. *Nature* **435**, 69–71.

Jolliff, B.L., Gillis, J.J., Haskin, L.A. Korotev, R.L. and Wieczorek, M.A., 2000. Major lunar crustal terranes: Surface expressions and crust-mantle origins. *Journal of Geophysical Research* **105**, E2, 4197–4216.

Jones, J. and Palme, H., 2000. Geochemical constraints on the origin of the Earth and the Moon. In *Origin of the Earth and Moon*, eds. R.M., Canup and K. Righter, University of Arizona Press, Tucson, and Lunar and Planetary Institute, Houston, pp. 197–216.

Kahn, A., Mosegaard, K, and Rasmussen, K.L., 2000. Lunar models obtained from a Monte Carlo inversion of the Apollo seismic P and S. waves. Lunar and Planetary Science Conference 31, Abstract #1341.

Kaula, W.M., 1971. The dynamical aspects of lunar origin. *Reviews of Geophysics and Space Physics* **9**, 217–238.

Kahn, A., Mosegaard, K., Williams, J.G., and Lognonné, P., 2004. Does the Moon possess a molten core? Probing the deep lunar interior using results from LLR and Lunar Prospector. *Journal of Geophysical Research* **109**, E9, E09007, 10.1029/2004JE002294.

Kerr, R.A., 2004. Did Jupiter and Saturn team up to pummel the inner solar system? News Focus, *Science* **306**, 1676.

Kleine, T., Münker, C., Jezger, K., and Palme, H., 2002. Rapid accretion and early core formation on asteroids and the terrestrial planets from Hf-W chronometry. *Nature* **418**, 952–955.

Korotev, R.L., 1999. The "Great lunar hot spot" and the composition and origin of "LKFM" impact-melt breccias. Lunar and Planetary Science Conference 30, Abstract #1305.

Kortenkamp. S.J., Kokubo, E., Weidenschilling, S.J., 2000. Formation of planetary embryos. In *Origin of the Earth and Moon*, eds. R.M., Canup and K. Righter, University of Arizona Press, Tucson, and Lunar and Planetary Institute, Houston, pp. 85–100.

Krähenbühl, U., 1980. Distribution of volatile and non volatile elements in grain-size fractions of Apollo 17 drive tube 74001/2. Lunar and Planetary Science Conference 11, pp. 1551–1564.

Lee, D., Halliday, A.N., Snyder, G.A., and Taylor, L.A., 1997. Age and origin of the Moon. *Science* **278**, 1098–1103.

Levinson, H.F., Dones, L., Chapman, C.R., Stern, S.A., Duncan, M.J., and Zahnle, K, 2001. Could the lunar "late heavy bombardment" have been triggered by the formation of Uranus and Neptune? *Icarus* **151**, 286–306.

Lin, R. P., Mitchell, D. L., Harrison, L., et al., 1999. Miniature magnetospheres on the Moon and their relation to albedo swirls. Lunar and Planetary Science Conference 30, Abstract #1930.

Lin, R.P., Anderson, K.A., and Hood, L., 1988. Lunar surface magnetic field concentrations antipodal to young large impact basins. *Icarus* **74**, 529–541.

Lin, R.P., Mitchell, D., Curtis, D., Anderson, K, Carlson, C. McFadden, J., Acuña, M., Hood, L., and Binder, A., 1998. Lunar surface magnetic fields and their interaction with the solar wind: Results from Lunar Prospector. *Science* **281**, 1480–1484.

Liu, M.C., 2004. Substructure in the circumstellar disk around the young star AU Microscopii. *Science* **305**, 1442–1444.

Lognonné, P., Gagnepain-Beyneix, J., and Chenet, H., 2003. A new seismic model of the Moon: implications for structure, thermal evolution and formation of the Moon. *Earth and Planetary Science Letters* **211**, 27–44.

McCammon, C., 2005. The paradox of mantle redox. *Science* **308**, 807–808.

McGetchin, T.R., Pepin, R.O., and Phillips, R. J., 1981. *Basaltic Volcanism on the Terrestrial Planets*, Pergamon Press, New York, pp. 236–267

McSween, H.Y., Jr., 1999. *Meteorites and Their Parent Planets*, Cambridge University Press, p. 166.

Meibom, A., and Frei, R., 2002. Evidence for an ancient osmium isotopic reservoir in Earth. *Science* **296**, 516–518.

Meyer, C., 1989. A brief literature review of observations pertaining to condensed volatile coatings on lunar volcanic glasses. In *Workshop on Lunar Volcanic Glasses: Scientific and Resource Potential*, eds. J.W. Delano and G. H. Heiken, Technical Report #90–02, Lunar and Planetary Institute, pp. 50–51.

Meyer, C., Williams, I.S., and Compston, W., 1989. Zircon-containing rock fragments within Apollo 14 breccia indicate serial magmatism from 4350 to 4000 million years. In *Workshop on Moon in Transition: Apollo 14, KREEP, and Evolved Lunar Rocks*, eds. G.J. Taylor and P.H. Warren, Technical Report # 89–03, 1989.

Meyer, C., Williams, I.S., and Compston, W., 1996. Uranium-lead ages for lunar zircons: Evidence for a prolonged period of granophyre formation from 4.32 to 3.88 Ga. *Meteoritics and Planetary Science* **31**, 370–387.

Mitchell, D.L., Lin, R.P., Harrison, L., et al., 2000. Solar wind interaction with lunar crustal magnetic fields: relation to albedo swirls. Lunar and Planetary Science Conference 31, Abstract #2088.

Mojzsis, S.J., Arrhenius, G., McKeegan, K.D., Harrison, T.M., Nutman, A.P., and Friend, C.R.L., 1996. Evidence for life on Earth before 3800 million years age. *Nature* **384**, 55–59.

Mojzsis, S.J., Harrison, T.M., and Pidgeon, R.T., 2001. Oxygen-isotope evidence from ancient zircons for liquid water at the Earth's surface 4,300 Myr ago. *Nature* **409**, 178–181.

Moorbath, S., 2005. Dating earliest life. *Nature* **434**, 155.

Morbidelli, A., 2004. How Neptune pushed the boundaries of our solar system. *Science* **306**, 1302–1304.

Morgan, J.W., Walker, R.J., Brandon, A.D., Horan, M.F., 2001. Siderophile elements in Earth's upper mantle and lunar breccias data synthesis suggests manifestations of the same late influx. *Meteoritics and Planetary Science* **36**, 1257–1275.

Muller, P.M., and Sjogren, W.L., 1968. Mascons: Lunar mass concentrations. *Science* **161**, 680–684.

NASA, 2004. NASA telescope revels clues to newborn planet, *Space News* **165**, June 5, 2004, p. 355.

Neal, C.R., 2001. Interior of the Moon: The presence of garnet in the primitive deep lunar mantle. *Journal of Geophysical Research* **106**, E11, 27865–27885.

Neal, C.R., and Taylor, L.A., 1992. Petrogenesis of mare basalts: A record of lunar volcanism. *Geochimica et Cosmochimica Acta* **56**, 2177–2211.

Neal, C.R., Ely, J.C., and Jain, J.C., 2000a. PGE abundances in Martian meteorites: No need for a "late veneer". Lunar and Planetary Science Conference 31, Abstract #1975.

Neal, C.R., Ryan, J., Jain, J.C., and Chazey, W., 2000b. The nature of the lunar mantle: Generally chondritic of the Mare basalt sources, but with Garnet in the source of the volcanic glasses. Lunar and Planetary Science Conference 31, Abstract #1944.

Neal, C.R., and Ely, J.C., 2002. Sulfide immiscibility in the lunar magma ocean: evidence for a primitive lunar lower mantle and the origin of high-μ mare basalts. Lunar and Planetary Science Conference 31, Abstract #1821.

Neumann, G.A., Zuber, M.T., Smith, D.E. and Lemoine, F.G., 1996. The lunar crust: Global structure and signature of major basins. *Journal of Geophysical Research* **101**, E7, 16,841–16,863.

Nunes, P.D., Tatsumoto, M., and Unruh, D.M., 1974. U-Th–Pb systematics of some Apollo 17 lunar samples and implications for a lunar basin excavation chronology. Lunar Science Conference 5, 1487–1514.

Okamoto, Y.K, Kataza, H., Honda, M., Yamnashita, T., et al., 2004. A early extrasolar planetary system revealed by planetesimal belts in Beta-Pictoris. *Nature* **431**, 660–663.

Palme, H., 2001. *Philosophical Transactions of the Royal Society of London*, Series A, **359**, 2061.

Palme, H., 2004. The giant impact formation of the Moon. *Science* **304**, 977–979.

Parmentier, E.M., and Hess, P.C., 1999. On the chemical differentiation and subsequent evolution of the Moon. Lunar and Planetary Science Conference 30, Abstract #1289.

Parmentier, E.M., Zhong, S., and Zuber, M.T., 2002. Gravitational differentiation due to initial chemical stratification: origin of lunar asymmetry by the creep of dense KREEP? *Earth and Planetary Science Letters* **201**, 473–480.

Porcelli, D., and Pepin, R.O., 2000. Rare gas constraints on early Earth history. In *Origin of the Earth and Moon*, eds. R.M., Canup and K. Righter, University of Arizona Press, Tucson, and Lunar and Planetary Institute, Houston, pp. 435–458.

Poulet, F., Bibring, J.-P., Mustard, J.F., Gendrin, A., Mangold, N., Langevin, Y., Arvidson, R.E., Gondet, B., Gomez, C, and the Omega Team, 2005. Phyllosilicates on Mars and implications for early Martian climate. *Nature* **438**, 623–627.

Ryder, G., 1990. Lunar samples, lunar accretion, and the early bombardment history of the Moon. EOS AGU, **71**, 313 and 322–323.

Ryder, G., 1999. Meteoritic abundances in the ancient lunar crust. Lunar and Planetary Science Conference 30, Abstract #1362.

Ryder, G., 2001. Mass flux during the ancient lunar bombardment: The cataclysm. Lunar and Planetary Science Conference 32, Abstract #1326.

Ryder, G., Koeberl, C., and Mojzsis, S.J., 2000. Heavy bombardment of the Earth at ~3.85 Ga: The search for petrographic and geochemical evidence. In *Origin of the Earth and Moon*, eds. R.M., Canup and K. Righter, University of Arizona Press, Tucson, and Lunar and Planetary Institute, Houston, pp. 475–492.

Schmitt, H.H., 1964. Petrology and Structure of the Eiksundal Eclogite Complex, Hareidland, Sunmöre, Norway, Ph.D. Dissertation, Harvard University.

Schmitt, H.H., 1973. Apollo 17 report on the Valley of Taurus-Littrow. *Science* **182**, 681–690.

Schmitt, H.H., 1989. Lunar crustal strength and the large basin-KREEP connection. In *Workshop on Moon in Transition: Apollo 14, KREEP, and Evolved Lunar Rocks*, eds. G.J. Taylor and P.H. Warren, Technical Report #89–03, Lunar and Planetary Institute, Houston, p. 111–112.

Schmitt, H.H., 1991. Evolution of the Moon: The Apollo Model. *American Mineralogist* **76**, 775–776.

Schmitt, H.H., 1999. Early lunar impact events: terrestrial and solar system implication. Geological Society of America Annual Meeting, Abstract #50440.

Schmitt, H.H., 1999. Origin and evolution of the Moon: Apollo 2000 Model. Workshop on New Views of the Moon II, Lunar and Planetary Institute, Contribution No. 980, p. 57.

Schmitt, H.H., 2001. Lunar cataclysm? Depends on what "cataclysm' means. Lunar and Planetary Science Conference, Abstract #1133.

Schmitt, H.H., 2003. Apollo 17 and the Moon. In *Encyclopedia of Space*, ed. H. Mark, Wiley, New York, Chapter 1 (on line).

Schmitt, H.H., 2006. *Return to the Moon*, Copernicus-Praxis, New York, 335 pp.

Shearer, C.K., and Papike, J.J., 1999. Magmatic evolution of the Moon. *American Mineralogist* **84**, 1469–1494.

Shearer, C.K., and Fløss, C., 2000. Evolution of the Moon's mantle and crust as reflected in trace-element microbeam studies of lunar magmatism. In *Origin of the Earth and Moon*, eds. R.M., Canup and K. Righter, University of Arizona Press, Tucson, and Lunar and Planetary Institute, Houston, pp. 339–360.

Shearer, C.K., and Newsom, H.E., 2000. W-Hf isotope abundances and the early origin and evolution of the Earth-Moon System. *Geochimica et Cosmochimica Acta* **64**, 3599–3613.

Shoemaker, E.M., Morris, E.C,. Batson, R.M., et al., 1968. Television observations from Surveyor, In *Surveyor Project Final Report, Part II. Science Results*, NASA Technical Report 32–1265, pp. 21–108.

Singer, S.F., 1970. Origin of the Moon by capture and its consequences. *EOS Transactions of the American Geophysical Union* **51**, 637–641.

Singer, S.F., 1986. Origin of the Moon by capture. In *Origin of the Moon*, ed. W. Hartmann, Lunar and Planetary Institute, Houston, pp. 471–485.

Smith, D.E., and Zuber, M.T., 1996. The shape of Mars and the topographic signature of the hemispheric dichotomy. *Science* **271**, 184–188.

Smith, J.V., Anderson, A.T., Newton, R.C., Olsen, E.J., Wyllie, P.J., Crewe, A.V., Isaacson, M.S., and Johnson, D., 1970. Petrologic history of the moon inferred from petrography, mineralogy, and petrogenesis of Apollo 11 rocks. Proceedings Lunar Science Conference 1, pp. 897–925.

Snyder, G.A. Borg, L.E. , Nyquist, L.E., and Taylor, L.A., 2000. Chronology and isotopic constraints on lunar evolution. In *Origin of the Earth and Moon*, eds. R.M., Canup and K. Righter, University of Arizona Press, Tucson, and Lunar and Planetary Institute, Houston, p. 381.

Spera, F.J., 1992. Lunar magma transport phenomena. *Geochimica et Cosmochimica Acta* **56**, 2253–2266.

Spudis, P.D., 1993, *Geology of Multi-Ring Impact Basins*, Cambridge University Press, New York, 263 pp.

Spudis, P.D., Reisse, R.A., and Gillis, J.J., 1994. Ancient multiring basins on the Moon revealed by Clementine laser altimetry. *Science* **266**, 1848–8151.

Spudis, P.D., 1996. *The Once and Future Moon*, Smithsonian, 308 pp.

Squyres, S.W., Grotzinger, J.P., Arvidson, R.E., et al., 2004. In situ evidence for an ancient aqueous environment at Meridiani Planum, Mars. *Science* **306**, 1709–1714.

Stewart, G.R., 2000. Outstanding questions for the giant impact hypothesis. In *Origin of the Earth and Moon*, eds. R.M., Canup and K. Righter, University of Arizona Press, Tucson, and Lunar and Planetary Institute, Houston, pp. 217–226.

Taylor, S.R., 2001. *Solar System Evolution*, Cambridge University Press, Cambridge, 460 pp.

Taylor, S.R., and Esat, T.M., 1996. Geochemical constraints on the origin of the Moon. In *Earth Processes: Reading the Isotopic Code*, eds. A. Basu and S. Hart, Geophysical Monograph 95, American Geophysical Union, pp. 33–46.

Taylor, G.J., and Warren, P.H. (eds.), 1989. Workshop on Moon in Transition: Apollo 14, KREEP, and Evolved Lunar Rocks, Technical Report Number 89–03.

Telesco, C.M., Fisher, R.S., Wyatt, M.C., Dermott, S.F., et al., 2005. Mid-infrared images of Beta Pictoris and the possible role of planetesimal collisions in the central disk. *Nature* **433**,

133–136.

Tera, F., Papanstassiou, D.A., and Wasserburg, G.J., 1974. Isotopic evidence for a terminal lunar cataclysm. *Earth and Planetary Science Letters* **22**, 1–21.

Tsiganis, K., Gomes, R., Morbidelli, A., and Levison, H.F., 2005. Origin of the orbital architecture of the giant planets of the Solar System. *Nature* **435**, 459–469.

Turner, G., Harrison, T.M., Holland, G., Mojzsis, S.J., and Gilmour, J., 2004. Extinct 244Pu in ancient zircons. *Science* **306**, 89–91.

Walker, R.J., Horan, M.F., Shearer, C.K., and Papike, J.J., 2004. Depletion of highly siderophile elements in the lunar mantle: evidence for prolonged late accretion. *Earth and Planetary Science Letters* **224**, 399–413.

Warren. P.H., 1985. The magma ocean concept and lunar evolution. *Annual Reviews in Earth and Planetary Science* **13**, 201–240.

Warren, P.H., 2003. The Moon. In *Treatise on Geochemistry, 1*, ed. A. Davis, Elsevier, Amsterdam, pp. 559–599.

Warren, P.H., and J.T. Watson, 1978. Compositional-petrographic investigation of pristine nonmare rocks. Lunar and Planetary Science Conference 9, pp. 185–217.

Wasson, J.T, Boynton, W.V., Kallemeyhn, G.W., Sundberg, L.L., and Wai, C. M., 1976. Volatile compounds released during lunar lava fountaining. Proceedings Lunar Science Conference 7, pp. 1583–1595.

Watson, E. B., and Harrison, T. M., 2005. Zircon thermometer reveals minimum melting conditions on earliest Earth. *Science* **308**, 841–844.

Weidenshilling. S. J., and J. N. Cuzzi, 1993. Formation of planetesimals in the solar nebula. In *Protostars and Planets III* , eds. E.H. Levy and J.I. Lunine, University of Arizona Press, Tucson, Arizona, pp 1031–1060.

Westfall, F., 2005. Life on the early Earth: A sedimentary view. *Science* **308**, 366–367.

Wetherill, G.W., 1990. Formation of the Earth. *Annual Reviews of Earth and Planetary Science* **18**, 205–256.

Wharton, D.A., 2002. *Life at the Limits: Organisms in Extreme Environments*, Cambridge University Press, Cambridge, 320 p.

Whitaker, E.A., 1981. The lunar Procellarum basin, Lunar and Planetary Science Conference 12, pp. 105–111.

Wieczorek, M.A., and Phillips, R.J., 2000. The "Procellarum KREEP Terrane": Implications for mare volcanism and lunar evolution. *Journal of Geophysical Research* **105**, E8, 20,417–20,430.

Wilde, S.A., Valley, J.W., Peck, W.H., and Graham, C.M., 2001. Evidence from detrital zircons for the existence of continental crust and oceans on the Earth 4.4 Gyr ago. *Nature* **409**, 175–178.

Wilhelms, D. E., 1987. The Geologic History of the Moon: U.S. Geological Survey Professional Paper 1348, U.S. Government Printing Office, Washington, 302 p.

Williams, D.A., Greeley, R., Neukum, G., et al., 1995. Multispectral studies of western limb and farside maria from Galileo Earth-Moon encounter 1. *Journal of Geophysical Research* **100**, E11, 23,291–23,299.

Williams, J.G., Boggs, D.H., Yoder, C.F., and Ratcliff, J.T, 2001. Lunar rotational dissipation in solid body and molten core. *Journal of Geophysical Research* **106**, E12, 27933–27968.

Wilson, T. J., 2005. Theory vs. hypothesis, Wall Street Journal, A12.

Wood, J.A., Dickey, J.S, Marvin, U.B, and Powell, B.N., 1970. Lunar anorthosites and a geophysical model of the Moon, *Proceedings Apollo 11 Lunar Science Conference*, pp. 965–988.

Yin, Q., Jacobsen, S.B., Yamashita, K., Blichert-Toft, J., Télouk, P. and Albarède, F., 2002. A short timescale for terrestrial planet formation from Hf-W chronometry of meteorites. *Nature* **418**, 949–952.

5 Evidence for Climate Change on Mars

Stephen R. Lewis and Peter L. Read

Abstract. One of the most striking differences between the present day climates of Earth and Mars is the ubiquitous and abundant presence of liquid water on Earth and the extremely dry atmosphere and surface of Mars. Features on the surface of Mars, discovered by early spacecraft missions in the 1970s and apparently caused by flowing water on the surface in the past, have lead to much speculation concerning the early Martian climate and the possibility that the planet was once relatively warm and wet. Such speculation is fuelled by the search for life on Mars, either in the present or as a fossil record. Until recent missions, however, there has been little direct evidence for the existence of large water deposits, other than in the form of ice, largely around the northern polar cap. During the past 2 years, however, NASA's Mars Odyssey and ESA's Mars Express spacecraft have discovered evidence for considerable amounts of ice lying at relatively shallow depths in the Martian regolith. The NASA Mars Rovers have also found considerable *in situ* evidence for ancient water in the nearby rocks and landscape.

It still seems unclear, despite various attempts to model the ancient Martian climate, whether Mars had a sustained warm, wet climate, with liquid water flowing on the surface, or whether it has remained mostly in a frozen state, interrupted by occasional melting events for short periods of time. Climate change on more recent timescales (10^4–10^6 years BP) has perhaps been less dramatic, but more amenable to systematic modelling. There is strong evidence of changes in Mars' climate on these timescales in the polar-layered deposits, associated with the obliquity cycle, and Mars GCMs have started to make progress in modelling climate change associated with varying astronomical parameters. We will briefly review such studies as well as the limited observational evidence for more dramatic climate change since the early epochs of the planet's history.

5.1 Introduction

The climate of Mars has long fascinated observers. Huygens, Cassini, Maraldi and Herschel, amongst others, observed the advance and retreat of the Martian polar ice caps, with Maraldi making the first detailed studies between 1672 and 1719. Transient brightenings were also seen, perhaps evidence of dust and water clouds. Giovanni Schiaparelli first claimed to see linear features on the surface during the favourable perihelic opposition of 1877. These "canali" were later interpreted by Percival Lowell and others as artificial canals (Lowell 1895, 1906, 1908). This contributed to the popular belief in an alien civilization, perhaps attempting to transport water from the polar regions toward the equator in an increasingly dry and hot desert climate. Changes in albedo, now known to be due to movement of dust, were interpreted as seasonal changes in vegetation.

Although improved telescopic observations during the mid-20th century made this belief less tenable, it was emphatically dispelled by the NASA Mariner 4 spacecraft fly-by in 1965. This showed a rocky, cratered and apparently lifeless landscape, at the same time demonstrating that the seasonal polar caps were mostly composed of frozen carbon dioxide, at temperatures of 150 K or below, rather than water. With the advent of successful orbiting spacecraft from Mariner 9, which arrived at Mars in 1971, Viking 1 and 2 in the late 1970s, and more recently Mars Global Surveyor (MGS), Mars Odyssey (MO), ESA's Mars Express (MEx) and NASA's Mars Reconnaissance Orbiter (MRO), launched on 12th August 2005, a much more detailed picture of the present day climate of Mars has been accumulated. This includes the discovery of water ice in the permanent polar caps and most recently in the upper layers of the surface over much of the planet.

Intriguingly, these spacecraft missions have produced evidence that Mars has not always been the dry, dusty and cold planet we observe today, but has undergone dramatic climate change on a variety of timescales since its creation. This evidence includes geological formations and rock types observed from space, the discovery of icy polar layered terrains, and the more detailed geology conducted *in situ* by landed spacecraft including Viking Landers 1 and 2, Mars Pathfinder and most recently the Spirit and Opportunity Mars Exploration Rovers (MER).

Interpreting and understanding Martian climate change is not only a considerable challenge in itself, but stimulates intriguing comparisons and contrasts with climate changes in the Earth's history and with present day concerns about possible future climate change. The ability to model and to understand the present and past climates of both planets is an important prerequisite for the ability to predict future changes with greater confidence. Investigating the possibility of past climates on Mars which might have been more hospitable to life than that of the present day is also essential to guiding exobiology studies, whether searching for evidence of present life or for a fossil record of primitive organisms.

In the following sections we first briefly summarize some observations which have led to our present understanding of the current state of the Martian climate. The following section then discusses the most recent evidence for climate change on Mars over a variety of timescales. The extent to which climate change on Mars can be understood and modelled is reviewed and we conclude with a discussion of future prospects for some important forthcoming observations which may lead to a better understanding of these problems.

5.2 Martian Climate in the Present Epoch

The Martian climate is governed by dynamical processes that are in many ways similar to those of the Earth, despite the obvious differences in mean surface pressure and atmospheric composition (e.g., Read & Lewis 2004). Both planets receive an excess of heating near the equator, mainly through visible radiation, and radiate to space in the thermal infrared, emitted from the surface and atmosphere. Atmospheric motions act to transport heat away from the equator toward the poles. At low latitudes on both planets, heat transport is through a mean over-turning circulation, though the Martian meridional circulation is at many times dominated by a single, huge, equator-crossing Hadley Cell. Warm air rises near the hottest latitude, diverges at high levels and sinks at higher latitudes, with a low-level return flow. The vertical extent of this motion is limited to the lowest 10–15 km by the strong tropopause on Earth, but on Mars may extend to much greater heights because of the lack of a clear transition to a stratosphere. In mid-latitudes, the primary form of heat transport becomes an almost horizontal type of convection, baroclinic instability (e.g., Andrews 2001), in the form of waves rather than a longitudinally uniform circulation. This instability gives rise to mid-latitude weather systems of cyclones and anticyclones. The similarities between Earth and Mars in rotation rate, stratification and vertical scale height, mean that the typical horizontal scale for mid-latitude weather systems on both planets, the radius of deformation, is close to 1000 km in both cases.

Perhaps the most notable difference between the planets is in the major role that the oceans, and the atmospheric water cycle, play on Earth. In contrast, present day Mars is largely dry, and the little water vapour that has been observed is insufficient to play any thermodynamical role in the climate system. It might also be thought that the lack of oceans on Mars, and the relatively short radiative timescales of one or two days associated with the atmosphere, would mean that there is little or no interannual variability on Mars. In fact, this is not the case, but variability on Mars is closely linked to the dust cycle, with regional storms taking place at different times and places each year, and global storms occurring in some years and not others, e.g. as recently observed by MGS (Smith et al. 2002, Smith 2004).

For a long time it has been suspected that large water deposits must exist somewhere on Mars, based on its proximity to the Earth, presumably similar origins, and simple models of the planet's composition. Until recent missions, however, there has been little direct evidence for the existence of large amounts of water, other than in the form of polar ice.

The presence of water near the Martian north pole was observed by the Viking Orbiters in the late 1970s, largely through the detection by the Mars Atmospheric Water Detector (MAWD) instrument of enhanced atmospheric water vapour released in the region during northern springtime. After sunlight returns to northern polar regions, and the seasonal CO_2 ice has sublimed, water ice can sublime directly into the atmosphere reaching levels of around 100 μm precipitable in the total atmospheric column compared to annual average levels of around 10 μm (both of which are extremely dry compared to the amounts of water in the Earth's atmosphere, where precipitable cm are more common). It now seems likely that there is a layer of water ice under the more permanent CO_2 cover near the south pole as well, though the measurements here are more recent and were initially less direct. These include those from the Mars Orbiter Laser Altimeter (MOLA) aboard MGS, which measured the height of the south polar cap and concluded that the most likely composition was 'dirty' water ice (ice mixed with Martian dust), and the Gamma Ray Spectrometer (GRS) aboard MO, which found large amounts of hydrogen within 1m of the Martian surface, most concentrated in polar regions. Water ice is the most likely explanation for the presence of the hydrogen.

In the last couple of years, the MO Thermal Emission Imaging System (THEMIS) has seen evidence for exposed water ice around the southern polar cap (Titus, Kieffer & Christensen 2003; Byrne & Ingersoll 2003), when combined with data from the MGS Thermal Emission Spectrometer (TES). TES has also measured an enhancement in the atmospheric water vapour in southern hemisphere spring, although this is variable from year-to-year and less dramatic than the release of water vapour from the northern polar cap (Smith 2002, 2004).

These discoveries have been confirmed and enhanced by new observations from the ESA MEx orbiter in 2004–5. The Observatoire pour la Minéralogie, l'Eau, les Glaces et l'Activité (OMEGA) instrument (Bibring et al. 2004) has used near-infrared reflection spectroscopy to show the presence of water ice at the edge of the perennial southern polar cap. As in the case of the previous measurements, this discovery only relates to the exposed ice at or near the surface, and the total inventory of water ice in the cap remains difficult to assess.

5.3 Evidence for Climate Change on Mars

One of the most striking differences between the present day climates of Earth and Mars is the ubiquitous and abundant presence of liquid water on Earth in contrast to the extremely dry atmosphere and surface of Mars. Such differences would appear to suggest analogies between the present climate on Mars and that of the dry, cold valley regions of Antarctica or the Canadian Arctic.

5.3.1 The ancient martian landscape

A number of features on the surface of Mars, discovered by the Mariner 9 and Viking spacecraft missions in the 1970s and apparently caused by flowing water on the surface in the past, however, have led to much speculation concerning the early Martian climate and the possibility that the planet was once relatively warm and wet (e.g. see Kargel 2004 for a detailed discussion). These features particularly include the variety of sinuous and dendritic channels, found especially in the geologically old southern uplands (see Figs. 5.1 and 5.2). Such features must clearly have been produced by the action of a flowing liquid, which most authorities now accept is most likely to have been liquid water, though the evidence that this water arrived at the surface in the form of rain (or any other kind of precipitation) is at best fragmentary.

Numerous other features, especially near the boundaries of the northern plains, testify to the effects of large flows of water, sometimes produced in catastrophic flooding events. An example of such terrain can be seen in Fig. 5.3, showing flood outflow regions near Chryse Planitia. The presence of various indicators of the age of such landscapes, however, clearly imply that, in most cases, these channels and flood features are extremely old in geological terms, sometimes up to 4 Gyrs (1 Gyr = 10^9 years). Despite these factors, speculation concerning the past presence of liquid water on Mars continues to be fuelled by the search for life on Mars, with the suggestion that more hospitable conditions in the past might have led to the evolution of simple organisms on Mars.

It is convenient to divide Martian history into a few epochs, defined by the times at which various geological features were laid down. The earliest epoch is the Noachian, which is taken to last for roughly the first Gyr of the planet's existence, 4.6–3.5 Gyr before the present (BP). This is followed by the Hesperian epoch, associated with extensive lava plains, from about 3.5–1.8 Gyr BP. Finally, the Amazonian epoch extends from 1.8 Gyr

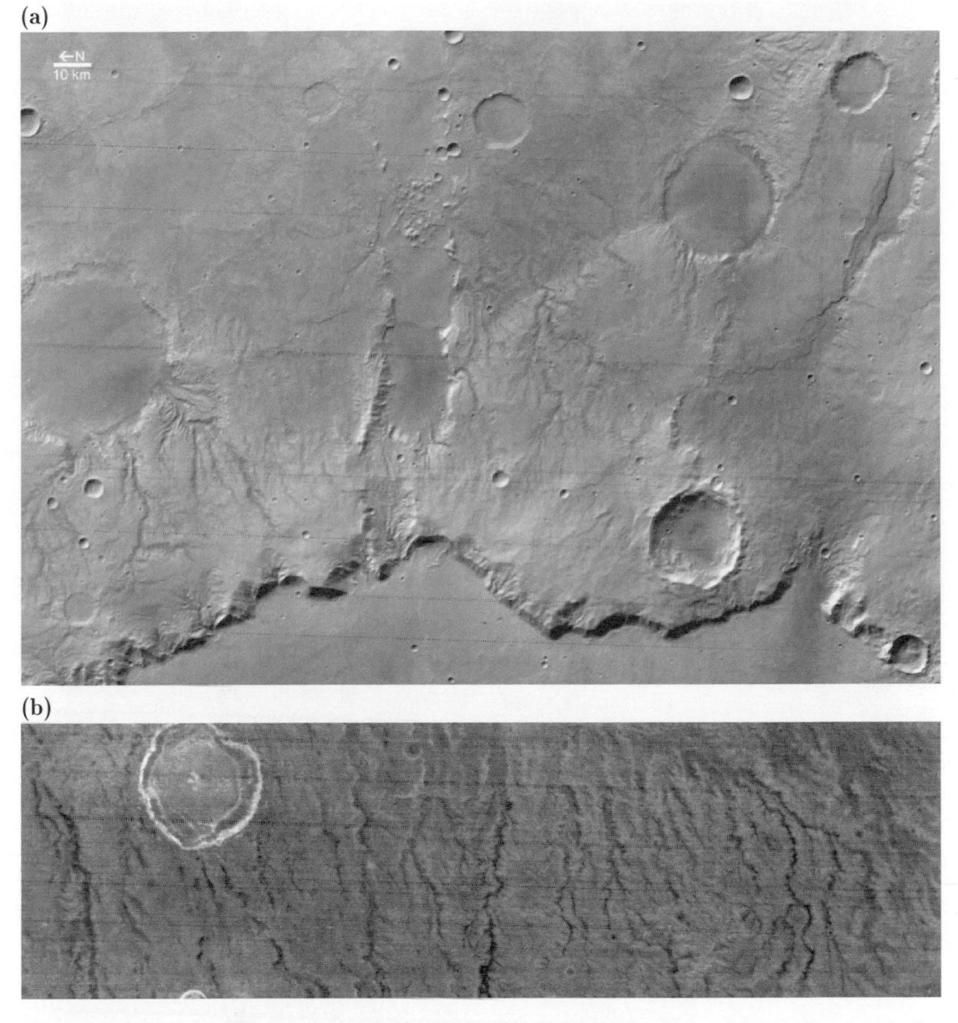

Fig. 5.1. Two images of valley networks in the southern highlands of Mars **(a)** near the eastern rim of Huygens crater, around 14° S, 61° E, from the HRSC camera on MEx (credit: ESA/DLR/Freie Universität Berlin (G. Neukum)) and **(b)** along Parana Vallis around 24.6° S, 10° W from a night-time THEMIS image (credit: NASA/JPL). The resolution of each image is around 70–100 m per pixel, so that **(a)** covers an area roughly 240 × 170 km and **(b)** approximately 95 × 32 km. North is to the left in each case.

BP to the present, and has seen the formation of varied, younger terrains, including the Olympus Mons volcano.

The prevailing view of Mars after the Viking spacecraft missions was of a planet which may once have had a warm, wet climate in the distant past, most likely up to the late Noachian or early Hesperian epoch, but which has since been in a very cold, dry state for most of its history. This picture is now being challenged by the most recent spacecraft observations (e.g. Head et al. 2003), which suggest a more dynamic climate, varying

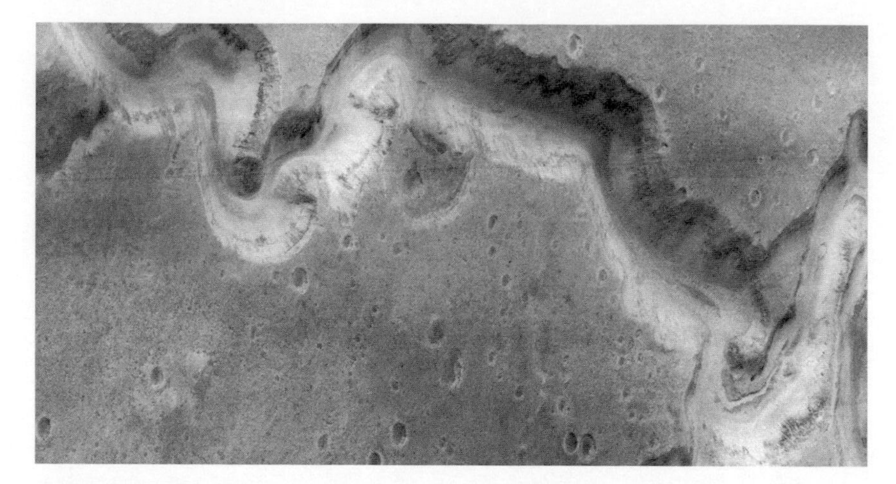

Fig. 5.2. Nanedi Vallis (at 28° S, 42° W) from the Mars Observer Camera (MOC) on Mars Global Surveyor (MGS). The picture covers an area 18.5×9.8 km. Nanedi is the word for "planet" in Sesotho, the national language of Lesotho, Africa. Credit: NASA/JPL/Malin Space Science Systems.

Fig. 5.3. Topography in the region of Chryse Planitia. Credit: Image adapted from the high resolution Mars Orbiter Laser Altimeter (MOLA) topographic map.

on much shorter timescales of thousands to millions of years up to the present day. In particular, rather than having been lost to space or absorbed deep within the planet, it appears possible that there remain reservoirs of water which are quite close to the surface and which could be released or redeposited under different climatic conditions.

Evidence for such a hypothesis comes from various geomorphological features, such as the presence of lobate ejecta blankets suggestive of mud-slides following major impact events (see Fig. 5.4). The effects of sub-surface permafrost can also be inferred from recent high resolution images from the Mars Observer camera (MOC) aboard MGS, sometimes in the form of small-scale polygonal ridges and fissures in the surface produced when water-rich soil freezes and expands (see Fig. 5.5) and a variety of other features (Head et al. 2003).

It is notable that the mean surface pressure on Mars is presently close to 610 Pa, a pressure which corresponds roughly to the triple point of water at typical Martian surface temperatures. At lower pressures and temperatures, water sublimes directly from ice to vapour and condenses directly from vapour to ice, with no liquid state being possible. Hence it is only at restricted locations and times of year that it is even possible to have liquid water on the surface of Mars, although if water is in the form of a strong brine solution this will expand the range of conditions at which a liquid is possible (Haberle

Fig. 5.4. Lobate ejecta around Yuty Crater (diameter 18 km) from a Viking image, seeming to indicate fluidized movement. Credit: NASA/JPL.

Fig. 5.5. Small scale polygonal structures in the floor of a crater in the northern plains of Mars, 67.5° N, 312.5° W (the scale bar indicates 200 m). Credit: NASA/JPL/Malin Space Science Systems.

et al. 2001). The implication is that the formation of fluid erosion features must imply a climate which was much warmer and with a significantly higher atmospheric pressure than is observed today.

5.3.2 The Mars Exploration Rovers

While some of the main issues concerning the past and present presence of groundwater on Mars can be addressed with high resolution imagery from orbit, certain questions can only be tackled effectively using the techniques of modern *in situ* geology. In practice, this means sending robotic spacecraft, equipped with sophisticated instrumentation, to study Martian rocks and minerals at close proximity, especially in key locations identified from orbit.

In 2004, therefore, NASA landed two highly successful rovers on Mars, with a major objective of searching for evidence for the former presence of water in the Martian geology. The Spirit Rover landed in Gusev Crater on 4 January 2004, a location specifically targetted as a possible site of a former lake based on remotely-sensed observations (Squyres et al. 2004a). In fact, Spirit did not find the evidence of lacustrine sedimentation that had been anticipated, but a mainly volcanic basalt lithology with evidence of subsequent impact events and aeolian transport. Some rocks, however, were coated in a way which suggests aqueous alteration and perhaps might have been deposited as a result of a watercharged debris flow, although there seems to be little direct evidence for this. Ventifacts (rocks whose shape has been modified by the "sandblasting" effect of the wind) and deposits seen by Spirit were all consistent with afternoon winds from the northwest, as predicted for this location by models of the present day Martian atmosphere (Rafkin & Michaels 2003). It is possible that rocks were partially buried, and then exhumed by wind deflation of the surface, and that the coatings on the rocks were formed during periods in which they were buried. As water migrates upwards through the regolith, it would finally evaporate, leaving the salts as a coating. Another possibility is that the rocks were covered by frost deposits in periods of high planetary obliquity, when the climate was substantially different from that of the present day. In conclusion, while Spirit found evidence for some of the effects of water in the geology of that location, it has not to date found unambiguous evidence for a large body of standing water in Gusev Crater or for substantial climate variability in the past few billion years.

The Opportunity Rover landed on Meridiani Planum on 24 January 2004 (Squyres et al. 2004b), and, in contrast to Spirit, found extensive geological evidence for liquid water having flowed on the Martian surface in the more recent past. Under a thin layer of sand-like regolith, Opportunity found flat sedimentary rocks, with small-scale cross laminations which indicate that deposition occurred in flowing water. Figure 5.6 shows a part of one Martian rock, known as "Upper Dells," which displays fine laminae. The laminae are somewhat disordered, lying at angles to each other, and many are truncated. Squyres et al. (2004b) interpret these rocks as a mixture of sediments formed during episodic events of shallow surface water, evaporation, exposure and dessication. Haematite-rich spherules (informally known as "blueberries"), eight of which can also be seen in Fig. 5.6, were also found embedded in these deposits, and many more were scattered across the surface in this region. These objects were a surprise to many Martian geologists, but appear to represent nodules of iron oxide which have precipitated out from solution in liquid groundwater as it evaporated and dried out.

Fig. 5.6. The sinuous curves seen in the face of this rock, known as "Upper Dells," provide strong evidence that liquid water was flowing on the Martian surface in the past, either toward or away from the viewer in this orientation. Image from NASA's Mars Exploration Rover Opportunity. Credit: NASA/JPL/Cornell/USGS.

Based on MOC images, the region explored by Opportunity is not atypical of extensive regions of Mars, and rather similar types of rock occur right across Meridiani Planum, an area of several 10^4 km^2. The age of the rock is uncertain, but could be several Gyr, overlying middle to late Noachian cratered terrains. The sand may have been the result of ejecta from craters, erosion of overlying features, or could have been transported to the region by winds. Despite the considerable uncertainties in age, the results of Opportunity do make clear that liquid water was present in significant quantities, perhaps many times, at the surface in Meridiani, and thus the Martian climate in this region must at times have been substantially different from that of the present day.

These results from Spirit and Opportunity emphasise the complexity and variety of the Martian system. As on Earth, different regions of the planet have simultaneously experienced very different climates over the course of their history. Thus, Spirit appears to have landed in an area that has been remarkably dry for the past 3 Gyr, whereas Opportunity arrived in an area which has seen repeated flooding with liquid water, perhaps as recently as a few Myr BP (Squyres et al. 2004b).

5.3.3 Water ice in equatorial regions

Several new observations have highlighted relatively recent ice-related activity on Mars. The MEx High Resolution Stereo Camera (HRSC) has revealed evidence for both active volcanism, perhaps multiple times, in the past several million years and glacial activity and erosion in equatorial regions (Neukum et al. 2004, Hauber et al. 2005).

Head et al. (2005) describe morphological evidence for tropical and midlatitude snow and ice accumulation and glacial-like flow in debris aprons, sloping away from highland regions. Features reminiscent of debris-covered, ice-rich rock glaciers are observed at the base of Olympus Mons (18°N) and debris aprons east of the Hellas basin (at 39°–43°S; see Fig. 5.7) suggest the former presence of ice. Both features are geologically very young, perhaps just a few million years old. Present conditions do not favour ice accumulation at low latitudes on Mars, and indeed there is evidence also of sublimation and retreat. But these features do suggest that at the time they were formed the climate must have instead favoured equatorward water transport and ice accumulation. The debris aprons near Hellas also show evidence for multiple resurfacing, suggesting cyclical change, perhaps related to the orbital configuration, with Mars presently in a low-obliquity, interglacial phase. It is also possible that climate change could be associated with different levels of volcanic activity, which could inject large amounts of aerosols into the atmosphere and also modify surface albedos.

A particularly intriguing discovery from HRSC images is that of evidence for a "frozen sea" close to the equator at 5° N, 150° E (Murray et al. 2005). Figure 5.8 is a HRSC image of the region. This plain, in Elysium Planitia, is remarkably flat (with slopes less than 0.005° over 60 km distances) and covered with irregular blocks, with a morphology reminiscent of floating sea ice on Earth. The area is about 800 km $\times$ 900 km in total and, based on typical crater heights, is estimated at 45 m deep. Murray et al. (2005) suggest that the block-like shapes are the result of floating pack ice, broken up into rafts, and subsequently covered in volcanic ash and dust. Other possible explanations, such as lava flows and mud rafts, appear less likely owing to the relative densities of these materials and the detailed morphology including evidence of drifts and the formation

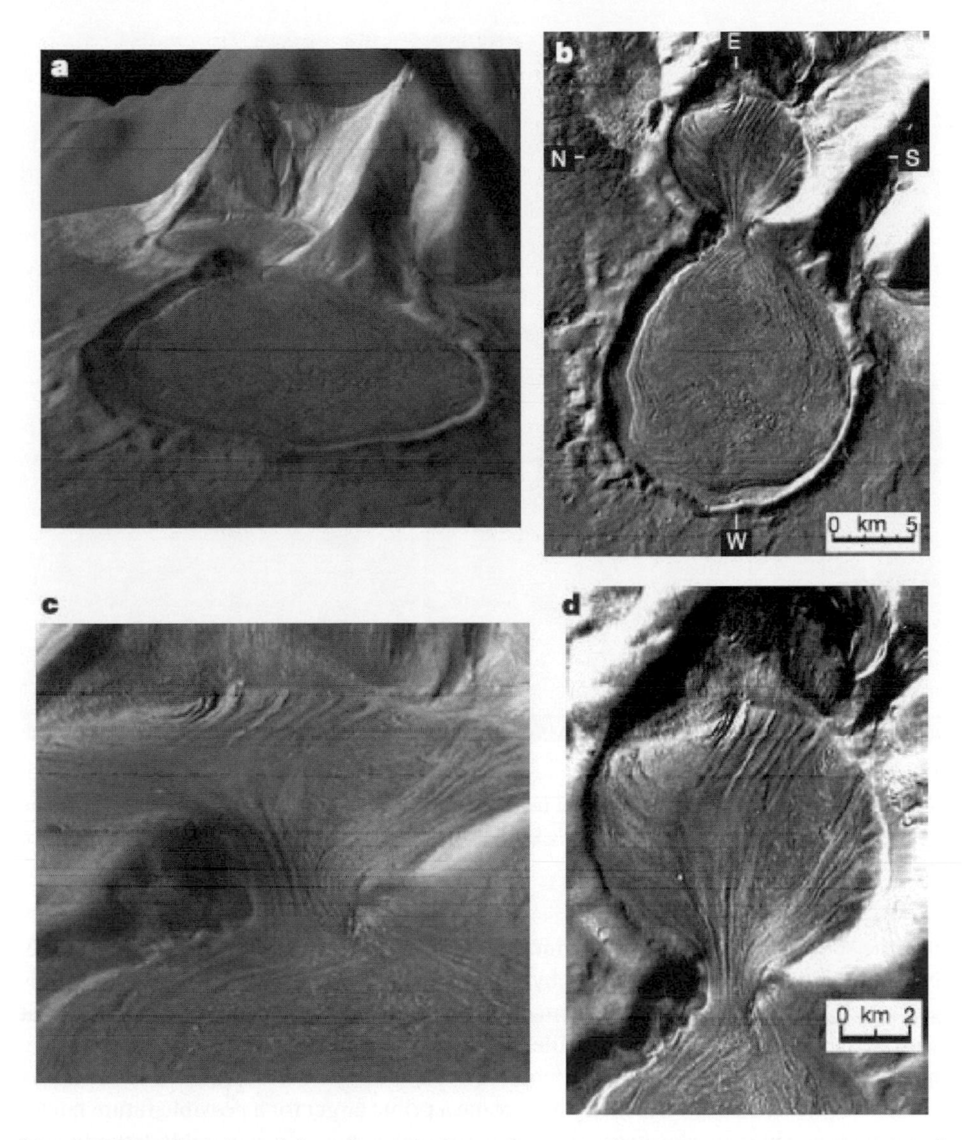

Fig. 5.7. Hourglass-shaped deposits at the base of a mountainous area near the eastern rim of the Hellas basin (257° W, 39.2° S) showing (a) a perspective view of the 3.4 km high massif and viscous flow of material from a 9 km diameter crater, (b) a vertical view, (c) an enlargement of the notch between the two craters and (d) a perspective view. Images were derived from MEx HRSC orbit 248 by Head et al. (2005) (*Nature* with permission).

of downstream leads. The surface in this area can be dated to about 5 ± 2 million years BP, although there is some indication that the brighter regions between the plates are 1 million years younger than the darker plates themselves (Murray et al. 2005). The region of water probably formed as the result of a flood from an underground reservoir, the sea of water eventually freezing, with some loss to evaporation and seepage back

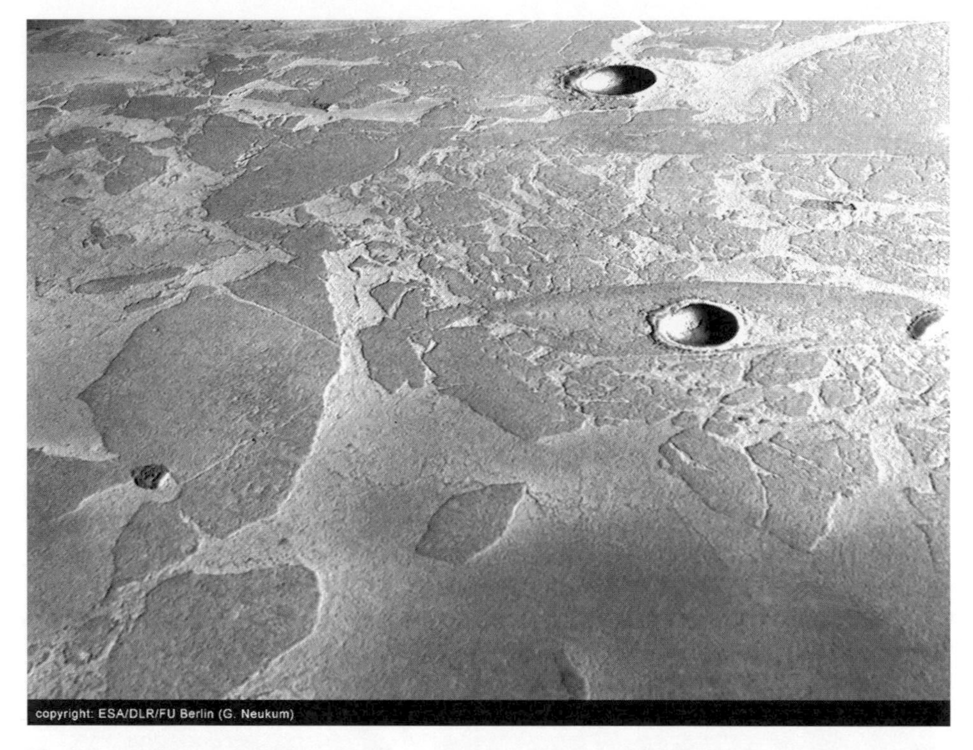

Fig. 5.8. This image, taken by the High Resolution Stereo Camera (HRSC) on the MEx spacecraft, shows what appears to be a dust-covered frozen sea near the Martian equator. The image is a few tens of km across and centred near 5° N, 150° E. Credit: ESA/DLR/Freie Universität Berlin (G. Neukum).

into the substratum. In the present climate, the ice would rapidly sublime at this latitude, but it appears to have been protected by a covering of dust. It seems possible that there is still much of the water left in the form of ice under the dust, which would be consistent with the level to which craters are filled and the lack of height variation in the surface, but confirmation may require new observations, perhaps with a radar instrument as described later. This region is likely to become a prime target for a possible future lander investigation, since its closeness to the equator makes it easily accessible from typical spacecraft orbits.

5.3.4 Polar layered deposits

The polar regions are substantial areas on Mars where there is strong evidence of changes in Mars' climate on timescales of between 10^4–10^6 years. In particular, in the regions surrounding the permanent ice caps lie the so-called polar-layered deposits, within which can be seen thin layers of dusty water ice of differing properties, each around 5–10 m thick (see Fig. 5.9). Each layer is thought to be composed of water ice, with differing amounts of dust mixed into each layer, perhaps reflecting atmospheric conditions at the

(a)

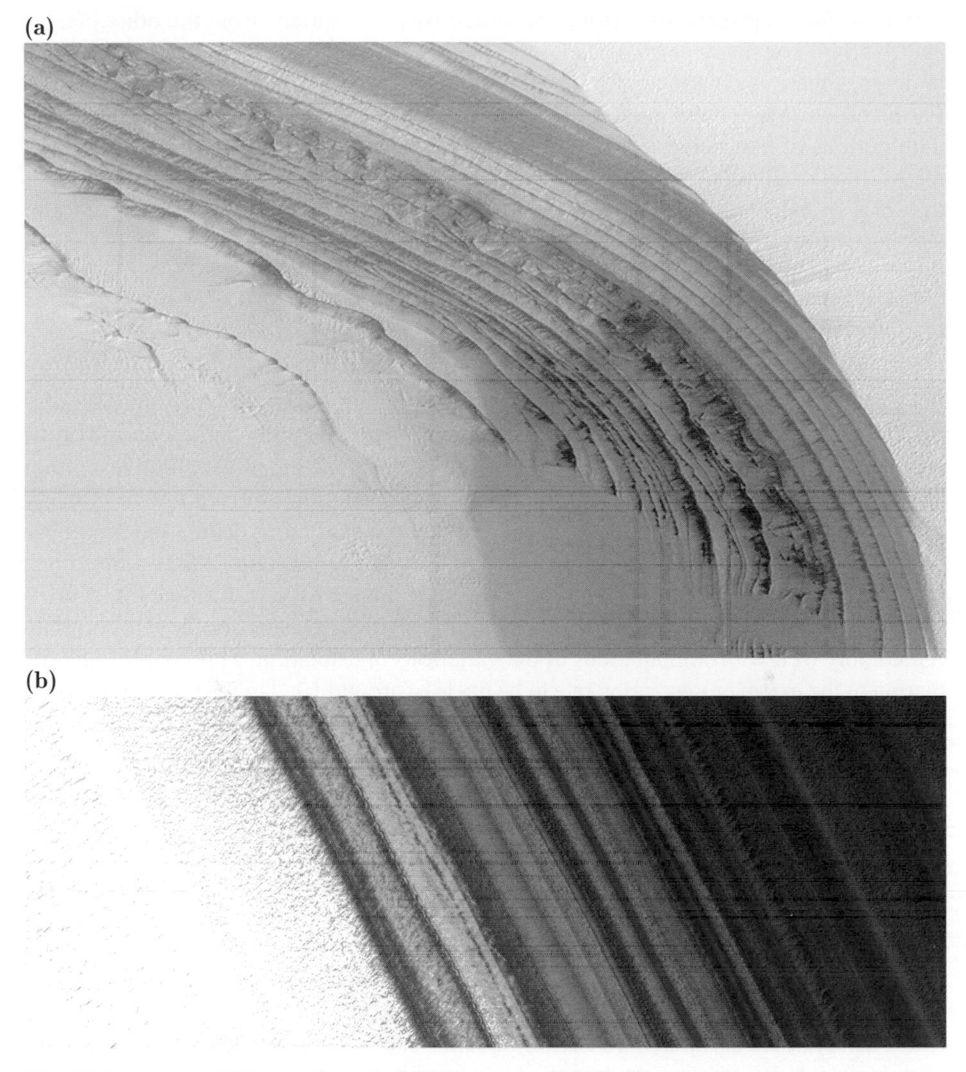

(b)

Fig. 5.9. Narrow-angle images from the MOC camera on MGS of **(a)** a small region of the southern layered terrains centred at 73° S, 224.5° W, and **(b)** a region of the northern layered terrains (part of image M00-02100) centred at 86.5° N, 279.5° W. The image in **(a)** shows a region 1.5 km × 4.6 km long, at a resolution of 1.2 × 2.2 m per pixel, while **(b)** shows a region 1.7 km × 4.3 km at a resolution of 1.84 m per pixel. North is to the left in (b). Credit: NASA/JPL/Malin Space Science Systems.

time each was laid down. Polarlayered deposits have been photographed near both poles from orbiters utilizing many different instruments, including MGS/MOC, although the layering can sometimes be obscured by frosts, fogs and clouds.

These layers are now widely thought to indicate varying rates of deposition of ice and dust, modulated on timescales over which the orbit and rotation of Mars are believed to vary in oscillations (actually chaotic) produced by perturbations from the other planets and the Sun. For many years celestial mechanics computations have shown that Mars undergoes these oscillations of its orbital eccentricity and obliquity, like those of the Earth responsible for the long-term (Milankovitch) cycles of glaciations or ice-ages, with periods of between 5×10^4 and 10^6 years. A particular feature of interest about the Martian oscillations is that, especially for obliquity, their amplitude is much larger than that of the Earth, ranging in the past 10 Myr period from around 14° to 48° and varying chaotically (Touma &Wisdom 1993; Laskar et al. 2004: see Fig. 5.10). Such large variations in obliquity lead to strong variations in insolation at polar latitudes, which are thought to exert a strong influence on the atmospheric conditions and climate, including the mean rate of ice deposition and its dust content.

Recent work has sought to establish whether the detailed structure of the observed polar layered deposits can be associated with the obliquity cycle in Mars' geologically recent past, using observations, detailed computations of obliquity variations and, most recently, Mars GCMs (as reviewed in Sect. 5.4). With the advent of very high resolution images from MGS and other recent missions, with resolution of just a few m in many cases, exquisite detail within these layers can now be studied in detail, and a number of new results are beginning to emerge. Laskar, Levrard & Mustard (2002) showed one intriguing case in the northern hemisphere (in MOC image M00-02100, see Fig. 5.9(b)) where the layer-by-layer variation in albedo across a series of layers appeared to correlate remarkably well with computations of the variations in insolation. This enabled an estimation of the mean ice deposition rate to be made of around 10^{-4} m yr^{-1} at that location. More recently, Milkovich & Head (2004) have shown that the pattern of variation across the layers in M00-02100 is also reproduced reasonably well at several other locations far removed in longitude. Such observations provide a strong clue that the broad pattern of layers is coherent across large areas of the planet, and is likely, therefore, to represent effects modulated by the large-scale climate.

Somewhat intriguingly, the variations seen in the northern hemisphere are not reflected in the south, where the polar layered deposits seem to reflect an era of deposition significantly older than those in the north. This again appears to indicate the importance of the hemispheric asymmetry in the topography of Mars in currently favouring systematic transport of water from the south towards the north, ablating and evaporating ice from the south polar cap and depositing it in the north. However, the detailed mechanism by which these layers are deposited and modified by subsequent processes are not well understood (e.g. see Pathare & Page 2005). There remains, therefore, a strong need for *in situ* studies of the polar layered deposits to study their structure and geology in more detail. Two landers were sent to the Martian polar regions in recent years, but failures during landing mean that there are still no *in situ* observations of these fascinating features.

5.3.5 Martian Climate Change in the Present

It should not be assumed that the Martian climate is now in a static state, even on relatively short timescales. Although Mars' obliquity is inferred to have varied widely over the past 5–10 Myrs, the amplitude of this variability seems to have gone through a relative minimum in the past ~200 kyrs (e.g. see Fig. 5.10(a)). Such a minimum seems

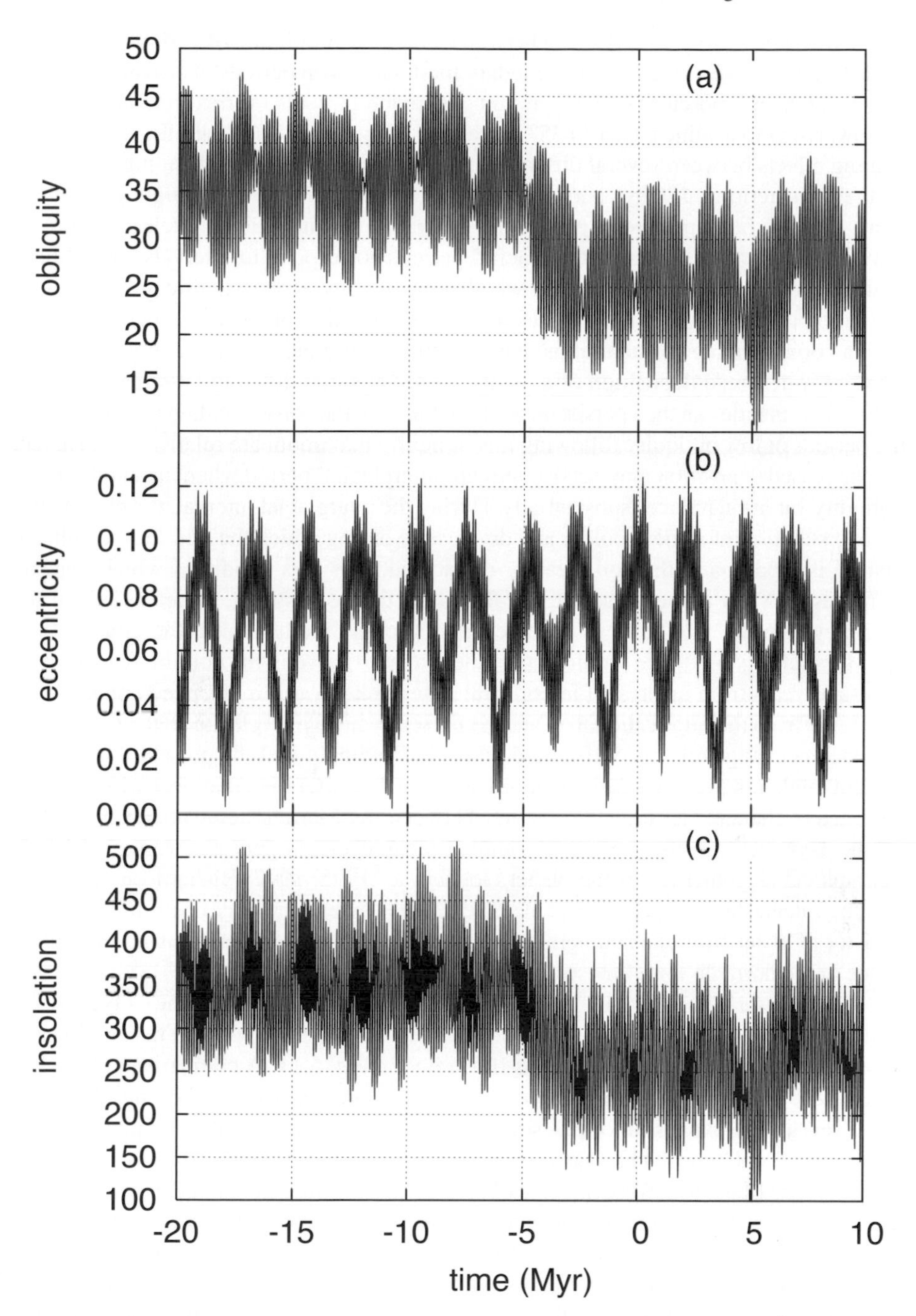

Fig. 5.10. Martian **(a)** obliquity (degrees), **(b)** eccentricity and **(c)** insolation (W m^{-2}) at the north pole surface at summer solstice (LS = 90°) from the solution La2004 for −20 Myr to +10 Myr (Laskar et al. 2004, Elsevier with permission).

to occur roughly every 2–3 Myrs, in between which the amplitude of variation rises to a peak-to-peak range of ~20° with secondary modulations on periods of around 1.3 Myrs. On even longer timescales (>10 Myrs) the mean obliquity also varies from around 25°, as now, rising to a value closer to 35° more than 5 Myrs ago. Such variations arise from beating effects between several different modes of oscillation. The changing amplitude of the short-term (~120 kyr period) obliquity modulation is now thought (e.g. Head et al. 2003) to be important in producing 'glacial' and 'interglacial' cycles on Mars at intervals of around 2.5 Myrs. The 'glacial' periods correspond to intervals when Mars' obliquity reaches values ~35°, during which polar regions receive significantly increased insolation during the summer. This leads to strong melting of the main water ice caps, with a corresponding substantial increase in atmospheric humidity, which in turn leads to more widespread deposition of a mantle of icy material down to latitudes as low as ±30°. This mantle can then persist throughout much of the corresponding summer. Since the periods of low obliquity following this obliquity maximum are relatively shortlived, this increased deposition may persist until an 'interglacial' period when the amplitude of obliquity variation reduces substantially. During the interglacial interval, the climate has time to stabilise into a new, colder and dryer, state during which the mid-latitude glacial mantle is eroded and the soil steadily dessicated. It is this condition which appears to represent Mars at the present time. Based on this assessment, therefore, given the tendency of Mars' obliquity to become more strongly modulated in the coming 0.2–0.5 Myrs (cf Fig. 5.10), we would expect the planet to move in the coming few hundred thousand years from its present interglacial state back towards a glaciated condition.

Aside from the implication that Mars is presently in an interglacial state as its orbital parameters change (cf Fig. 5.10), modelling studies imply that the present day climate system might not even be steady on timescales much shorter than that over which orbital parameters change. Net transfer of water between the Martian hemispheres is implied by the asymmetry in the present day annual mean cycle, largely as a result of the topographical asymmetries of the planet (Joshi et al. 1995, 1997; Richardson & Wilson 2002b; Takahashi 2003; Böttger et al. 2005). This asymmetry may result in a slow net transfer of water from the southern to the northern hemisphere, and thus accumulation of ice in the north on long timescales, rather than a quasi-steady state.

Recent observations, even from only one Mars year apart, show that the CO_2 ice in the Martian polar caps is also not in a steady state, but that there is consistent net erosion and ablation of this ice in some regions, as illustrated in Fig. 5.11. A comparison of images around the southern polar cap in two successive Mars years (separated by 687 Earth days) reveals the expansion of pits several hundred m across and the disappearance of small mesas and buttes. The scarps in these images have retreated by about 3 m in the course of one Mars year.Were this rate of erosion to continue, and not be balanced by net deposition elsewhere, the total atmospheric mass of Mars would increase significantly at the expense of the surface ice reservoir, virtually doubling on a thousand year timescale. Of course, over such a short record as is available from spacecraft observations to date, it is impossible to say how much of this change is a longterm trend and how much is simply due to internal interannual variability, which has been observed in the atmosphere over the course of the MGS mission (Smith 2004).

As Fig. 5.10 indicates, the Martian obliquity can be expected to rise again in the future to values as high as 35° and this implies potential change towards a warmer, higher

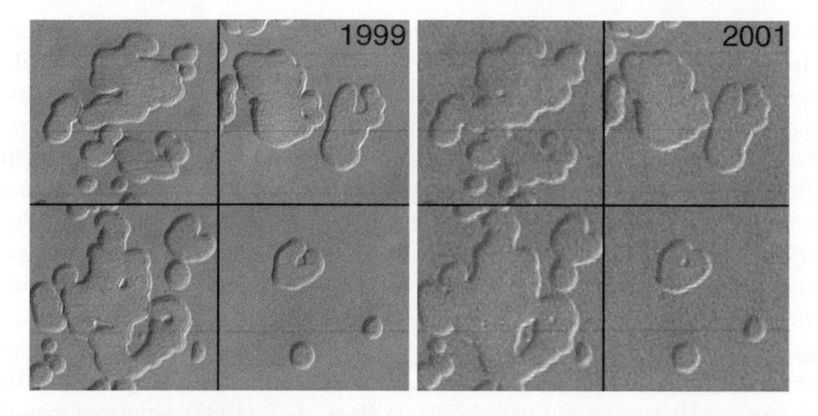

Fig. 5.11. Martian south polar pits at four locations in a layer of frozen CO_2 at the same areocentric longitude in two successive Mars years. Each frame in each image is about 250 m across. Credit: NASA/JPL/Malin Space Science Systems.

mean pressure climate on geologically-short timescales of 10,000–50,000 yr. The more extreme, higher obliquities of >45° seen before about 5 Myr BP may not return for over 10Myr, according to Laskar et al. (2004). These predictions are subject to increasing errors the further in time they are from the present, owing to the chaotic orbit of Mars. It seems likely that very high obliquities will occur again, as may very low obliquities of 10° or less, which would imply a particularly cold and dry climate and much lower surface pressure than today.

5.4 Martian Climate Change Modelling

Numerical modelling and simulation has led directly to much of our present knowledge of the Martian atmosphere and its circulation, and has significantly enhanced the interpretation of spacecraft and telescopic observations. This process has mirrored developments in understanding of terrestrial weather and climate processes, making use of a hierarchy of models, from analytical work with simplified equations to large-scale numerical simulations. Among the most powerful tools for weather and climate prediction for the Earth are so-called general circulation models (GCMs). These are attempts to model all the processes important to the large-scale atmosphere, from global scales down to the smallest horizontal scales which can be represented within computational constraints. GCM resolution is typically limited to around 100 km, depending on the problem in hand and the time scale of the simulation required, and resolution of finer features generally requires the use of a limited-area, mesoscale model.

5.4.1 Climate change and early Mars

It still seems unclear whether Mars had a very sustained early warm, wet climate, with liquid water flowing on the surface, or whether it has always remained mostly in a frozen state, interrupted by occasional melting events for shorter periods of time. The present day Martian climate could only support liquid water, most likely in the form of a strong brine solution, in a few locations at limited times of year. The geological flow features, described in the previous section, have naturally led to attempts to model a Martian climate which would permit running water on the surface for significant periods of time. Owing to the considerable uncertainties in modelling the early Martian climate, most previous attempts have involved the use of simple global models, one-dimensional column models or two-dimensional models with latitudinal variations, rather than the more complex, and computationally expensive, GCMs now used to model the present day climate. A more detailed review of previous early Mars modelling is provided by Haberle (1998).

Initial attempts to produce a warmer Mars climate focussed on a more massive greenhouse atmosphere consisting of CO_2, H_2O and sometimes SO_2 as an additional greenhouse gas (Postawko & Kuhn, 1986). Pollack et al. (1987) estimated that a 5 bar (5×10^5 Pa) CO_2 atmosphere, with H_2O as a greenhouse gas, would be required to maintain a global mean surface temperature close to 273 K. It should be noted that solar luminosity would have been about 70% of the present value towards the end of the Noachian epoch, meaning that a strong atmospheric greenhouse would be required to maintain surface temperatures much higher than observed on Mars now. The 5 bar CO_2 estimate is subject to many uncertainties, but may be reasonable to within a factor of about two. Such a high atmospheric mass is not unreasonable in the early history of the planet, but it is harder to account for how this atmosphere might have evolved to the presently observed, 610 Pa CO_2 atmosphere, three orders of magnitude less massive. This is especially true since very few, if any, carbonate rocks have been detected by recent spacecraft missions, whereas a large amount should have formed in a wet, CO_2-rich climate. It is possible that much of the atmosphere was lost to space, or that carbonate rocks have been somehow buried or processed, but the almost complete apparent absence of carbonates needs to be explained.

Kasting (1991) pointed out a flaw with the very dense atmosphere model, which is that CO_2 condensation could occur within the atmosphere and limit the atmospheric mass to about 2.5 bar CO_2, not enough to raise surface temperatures to more than about 220 K according to the Pollack et al. (1987) model. Since that time, absorption of shortwave radiation by SO_2 (Yung, Nair & Gerstell 1997) and the greenhouse effect of CO_2 clouds (Forget & Pierrehumbert 1997) have both been suggested as possible mechanisms which might produce a warmer climate with a less dense atmosphere of around 2 bar CO_2. The details remain speculative and there has been little recent progress in three-dimensional modelling or detailed microphysical modelling of the Noachian Martian atmosphere. It is at least possible that Mars never had a consistently warm climate, but instead passed through transitory warmer states linked to impact melting, volcanic eruptions and cycles in orbital parameters.

5.4.2 Climate change with orbital parameters

Although full Mars GCMs have seen little use to date in tackling the extremely challenging problem of demonstrating the stability of a warm, wet, high atmospheric density Martian climate in the distant past, there have recently been a series of studies with different GCMs investigating Martian climate change on shorter timescales (Richardson &

Wilson 2002a; Mischna et al. 2003; Haberle, Murphy & Schaeffer 2003; Armstrong & Leovy 2005; Newman, Lewis & Read 2005; Segschneider et al. 2005). Climate change on more recent timescales (10^4–10^7 years BP) has been less dramatic than the massive changes required for a proposed high density early atmosphere, but has proved more amenable to detailed modelling using GCMs validated against present day conditions.

Figure 5.10 indicates the possible range of Martian orbital parameters over recent times. The eccentricity of the Martian orbit, coupled with the time of perihelion (not shown), will determine the relative intensity of solar forcing in northern and southern winter. It appears that Mars is presently in a state of relatively high eccentricity ($\approx$0.093) which, with perihelion in the northern hemisphere winter ($L_S \approx 251°$ [1]), tends to result in stronger inter-hemispheric thermal contrasts, and hence winds, at that time of year. Clearly, lower eccentricity would tend to give a more symmetric seasonal cycle and changes in the time of perihelion might alter the time of year at which the strongest zonal winds occur, but Richardson & Wilson (2002b) conclude that altering the time of perihelion does not change the hemispheric bias in the annual-mean tropical zonal mean circulation in their GCM, which is governed principally by topography.

The most significant parameter for climate change appears to be the Martian obliquity, o. Presently $o \approx 25.2°$, but o seems to have reached at least 45° around 5–10 Myr BP and values as high as $o = 60°$ do not seem impossible in the more distant past. Martian obliquity can also fall to $o = 10°$ and below. GCM studies at different obliquities (Haberle, Murphy & Schaeffer 2003; Armstrong & Leovy 2005; Newman, Lewis & Read 2005; Segschneider et al. 2005) consistently demonstrate that low obliquity states are associated with permanent polar ice caps, low mean temperatures and relatively weak winds. In contrast, at high obliquities, the summer poles become much warmer, the inter-hemispheric temperature contrasts are enhanced and the peak winds generally become stronger. These features are illustrated in Figure 5.12, which shows the zonal-mean surface temperature as a function of latitude and time of year for four different obliquites, two lower and two higher than at present, from experiments performed with the Oxford Mars GCM (Newman, Lewis & Read 2005). It is clear that warmer temperatures and much stronger latitudinal temperature gradients occur at higher obliquities, with the summer poles becoming the warmest places on the planet and the polar CO_2 ice caps (though not necessarily the water ice, see below) subliming completely in summer, but extending to lower latitudes in winter, at high obliquities.

Most of the GCM experiments described have used a constant atmosphere plus seasonal ice CO_2 budget. Newman, Lewis & Read (2005) show that for low obliquity experiments, the global mean atmospheric surface pressure equilibrates after about twenty Martian years at lower values than at present, e.g. around 350 Pa at $o = 15°$, as more CO_2 is held in the permanent polar ice caps. High obliquity states might have increased mean surface presure as CO_2 is released from high latitude, subsurface reservoirs, although it is unclear how much CO_2 is available and how much of that released might be stored again in seasonal polar caps. Armstrong & Leovy (2005) show a few simulations at

[1] The time of year on Mars is measured in areocentric longitude, L_S, with $L_S = 0°$ being northern hemisphere spring equinox, $L_S = 90°$ being northern hemisphere summer solstice, etc.. This angle does not advance exactly linearly with time owing to the significant eccentricity of the Martian orbit.

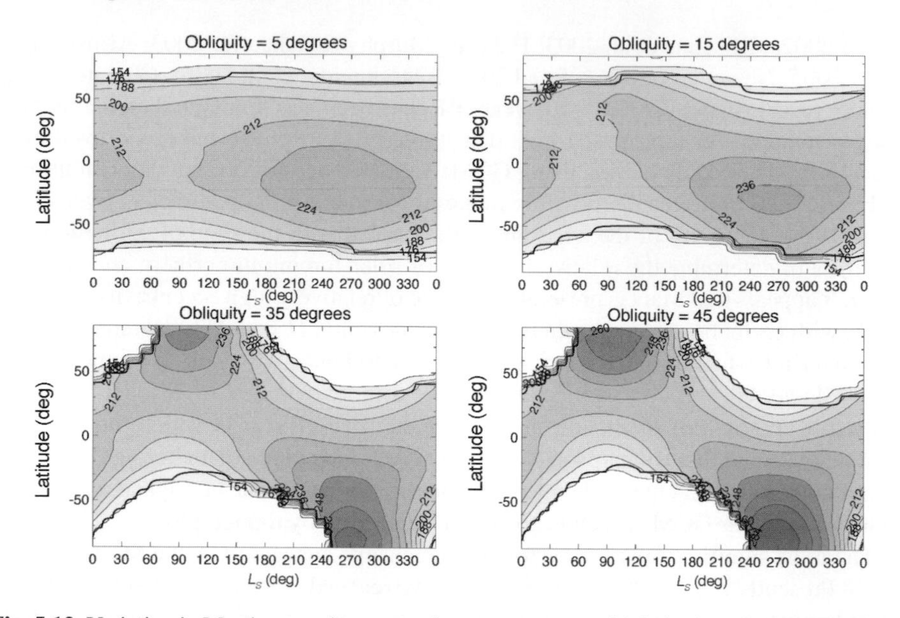

Fig. 5.12. Variation in Martian zonal-mean surface temperature with latitude and season (LS) over a range of obliquities, as simulated by the Oxford Mars GCM. The black solid line shows the mean extent of the CO_2 ice cap.

high obliquity with arbitrarily higher surface pressure, double and quadruple the present value, which does indeed increase wind erosion in the model.

The enhanced thermal contrasts at high obliquites tend to lead to stronger peak winds which, possibly in combination with higher atmospheric densities, would be expected in a fully coupled model to result in much greater dust lifting by near-surface wind stress, making the atmosphere dustier and warmer still. Several studies (Haberle, Murphy & Schaeffer 2003; Armstrong & Leovy 2005; Newman, Lewis & Read 2005) have tried to predict dust lifting at states of higher obliquity and there is now some evidence that regions of dust lifting at high obliquity $o = 45°$ correlate with regions of erosion and relatively high thermal inertia seen on the present day Martian surface.

Richardson & Wilson (2002a) and Mischna et al. (2003) demonstrate the effect of obliquity changes on a GCM with a water cycle included. Although the water cycle is modelled in a relatively simple way, with many uncertainties, the results indicate that at times of high obliquity the atmosphere may become much wetter than is currently observed and the winter seasonal water ice cap may extend to remarkably low latitudes, perhaps even crossing the equator into the summer hemisphere. Such results are particularly interesting in the light of recent observations of evidence for low latitude snow and ice deposits on Mars (Head et al. 2003; Head et al. 2005; Murray et al. 2005).

5.5 Future Prospects

All previous spacecraft missions, whether remote sensing or surface rovers, have only been able to examine the Martian surface at relatively low latitudes, or penetrate no more than about 1 m into the topsoil, in the search for water and geological evidence of climate change. Amongst other missions planned for the coming few years, NASA's Phoenix lander, due for launch in 2007, is intended to emulate aspects of the two MER rover missions (Spirit and Opportunity) but at high northern latitudes. Although at the time of writing the landing site has not been decided, it is likely that this will be in the vicinity of the north polar layered terrains discussed above, hopefully enabling a more detailed study of these fascinating markers of Mars' climate in the recent past.

More immediately, two new radar instruments will, in the near future, make observation possible of liquid water or ice in the top few km of the Martian crust, effectively making a three-dimensional map of much of the Martian surface for the first time. If successful, this should revolutionize our understanding of the present day Martian water cycle and perhaps provide many clues about past climates.

5.5.1 The MARSIS experiment

The Mars Advanced Radar for Subsurface and Ionosphere Sounding (MARSIS) is one of the instruments aboard ESA's MEx spacecraft. MARSIS's main scientific objective (e.g. Picardi et al. 2004) is to map the distribution of both liquid water and ice in the upper portion of the Martian crust, in addition to various other subsurface and ionospheric investigations. As of August 2005, this instrument has only just been turned on and results are not yet available, the delay being a result of concerns about the deployment of the radar antennae. MARSIS offers intriguing possibilities for the extension of the subsurface water ice mapping observations at horizontal scales of 5–10 km. The real advance over previous instruments is that the radar will penetrate up to about 5km below the surface, well down to possible permafrost layers and water reservoirs, compared with the relatively shallow depths of around a metre, observed so far by other instruments such as GRS onboard MO. The radar signal would reflect particularly strongly from liquid water, if this is present, but will detect any interface, such as layers of rock and ice interspersed near the surface. Vertical resolution will be on the order of 100m, so very thin layers, or features smaller than about 10km in the horizontal will not be detected. MARSIS will also be able to characterize material composition and surface roughness, perhaps detecting layers of sediment in areas which might be the sites of ancient lakes or seas.

5.5.2 The SHARAD experiment

The Shallow Subsurface Radar (SHARAD) is an instrument which is being flown on NASA's Mars Reconnaissance Orbiter (Seu et al. 2004), launched on 12th August 2005. Similar in some ways to MARSIS, SHARAD operates at higher frequency, 15–25 MHz compared to 1.3–5.5 MHz. This means that the signal will not penetrate as deeply below the surface, roughly 1 km, but the instrument will have higher spatial resolution, 300 m in the horizontal and 15 m in the vertical. The data will thus be complementary to

MARSIS in being able to detect and map much smaller features close to the surface. As GRS has already detected signs of water ice only 1m below the surface it seems likely that SHARAD will be able to enhance this picture considerably.

Both MARSIS and SHARAD offer the opportunity of a much more direct measurement of water, both in solid and liquid form than has been possible to date, and should be able to offer unambiguous confirmation of previous interpretations of morphology, such as the glacial features described earlier.

5.6 Conclusion

The picture which seemed to emerge from the Mariner 9 and Viking Orbiters was of a long dead, dessicated and cold Mars. There was evidence of an enormous change from a warm, wetter climate, which had only existed early in the planet's history, several billion years ago. That overall scenario still seems possible, although the detailed evidence is limited and ambiguous. Recent spacecraft results, described here, are building up a much more complicated picture of ongoing, though less drastic, climate change on a large variety of timescales, with a substantially different Martian climate as recently as a few million years ago and with the prospect of future change. Adding to the complexity, different landing sites appear to show different climate records, as would, of course, be expected on Earth. What remains clear, however, is that new observations are still required to better characterize and understand the present day Martian climate and to begin to be able to unravel its history.

References

Andrews, D. G., 2001. *An Introduction to Atmospheric Physics*. Cambridge University Press, Cambridge, UK.

Armstrong, J. C., Leovy, C. B., 2005. Long term wind erosion on Mars. *Icarus* **176**, 57–74. doi:10.1016/j.icarus.2005.01.005.

Bibring, J.-P., Langevin, Y., Poulet, F., Gendrin, A., Gondet, B., Berthé, M., Soufflot, A., Drossart, P., Combes, M., Bellucci, G., Moroz, V., Mangold, N., Schmitt, B., The OMEGA Team, 2004. Perennial water ice identified in the south polar cap of Mars. *Nature* **428**, 627–630.

Böttger, H. M., Lewis, S. R., Read, P. L., Forget, F., 2005. The effects of the Martian regolith on GCM water cycle simulations. *Icarus*, in press.

Byrne, S., Ingersoll, A. P., 2003. A sublimation model for martian south polar ice features. *Science* **299**, 1051–1053.

Forget, F., Pierrehumbert, R. T., 1997. Warming early Mars with carbon dioxide clouds that scatter infrared radiation. *Science* **278**, 1273–1276.

Haberle, R. M., 1998. Early Mars climate models. *J. Geophys. Res.* **103** (E12), 28467–28479.

Haberle, R. M., McKay, C. P., Schaeffer, J., Cabrol, N. A., Grin, E. A., Zent, A. P., 2001. On the possibility of liquid water on present-day Mars. *J. Geophys. Res.* **106** (E10), 23317–23326.

Haberle, R. M., Murphy, J. R., Schaeffer, J., 2003. Orbital change experiments with a Mars general circulation model. *Icarus* **161**, 66–89.

Hauber, E., van Gasselt, S., Ivanov, B., Werner, S., Head, J. W., Neukum, G., Jaumann, R., Greeley, R., Mitchell, K. L., Müller, P., The HRSC Co- Investigator Team, 2005. Discovery of a flank caldera and very young glacial activity at Hecates Tholus, Mars. *Nature* **434**, 356–361.

Head, J. W., Mustard, J. F., Kreslavsky, M. A., Millikan, R. E., Marchant, D. R., 2003. Recent ice ages on Mars. *Nature* **426**, 797–802.

Head, J. W., Neukum, G., Jaumann, R., Hiesinger, H., Hauber, E., Carr, M., Masson, P., Foing, B., Hoffmann, H., Kreslavsky, M., Werner, S., Milkovich, S., van Gasselt, S., The HRSC Co-Investigator Team, 2005. Tropical to midlatitude snow and ice accumulation, flow and glaciation on Mars. *Nature* **434**, 346–350.

Joshi, M. M., Haberle, R. M., Barnes, J. R., Murphy, J. R., Schaeffer, J., 1997. Low-level jets in the NASA Ames Mars general circulation model. *J. Geophys. Res.* **102** (E3), 6511–6523.

Joshi, M. M., Lewis, S. R., Read, P. L., Catling, D. C., 1995. Western boundary currents in the martian atmosphere: Numerical simulations and observational evidence. *J. Geophys. Res.* **100** (E3), 5485–5500.

Kargel, J. S., 2004. Mars: *A Warmer, Wetter Planet*. Springer-Verlag, Berlin. ISBN: 1-85233-565-8.

Kasting, J. F., 1991. CO_2 condensation and the climate of early mars. *Icarus* **94**, 1–13.

Laskar, J., Correia, A. C. M., Gastineau, M., Joutel, F., Levrard, B., Robutel, P., 2004. Long term evolution and chaotic diffusion of the insolation quantities of Mars. *Icarus* **170**, 343–364. doi:10.1016/j.icarus.2004.04.005.

Laskar, J., Levrard, B., Mustard, J. F., 2002. Orbital forcing of the martian polar layered deposits. *Nature* **419**, 375–377.

Lowell, P., 1895. *Mars*. Houghton Mifflin, Boston.

Lowell, P., 1906. *Mars and its Canals*. Macmillan, New York.

Lowell, P., 1908. *Mars and the Abode of Life*. Macmillan, New York.

Milkovich, M., Head, J. W., 2004. Characterization and comparison of layered deposit sequences around the north polar cap of Mars: Identification of a fundamental climatic signal. *Lunar Planet Sci.* **XXXV**, abstr. 1342.

Mischna, M., Richardson, M. I., Wilson, R. J., McCleese, D. J., 2003. On the orbital forcing of Martian water and CO_2 cycles: A general circulation model study with simplified volatile schemes. *J. Geophys. Res.* **108** (E6), 5062, doi:10.1029/2003JE002051.

Murray, J. B., Muller, J.-P., Neukum, G.,Werner, S. C., van Gasselt, S., Hauber, E., Markiewicz, W. J., Head III, J. W., Foing, B. H., Page, D., Mitchell, K. L., Portyankina, G., The HRSC Co-Investigator Team, 2005. Evidence from the Mars Express High Resolution Stereo Camera for a frozen sea close to Mars' equator. *Nature* **434**, 352–355.

Neukum, G., Jaumann, R., Hoffmann, H., Hauber, E., Basilevsky, A. T., Ivanov, B. A., Werner, S. C., van Gasselt, S., Murray, J. B., McCord, T., The HRSC Co-Investigator Team, 2004. Recent and episodic volcanic and glacial activity on Mars revealed by the High Resolution Stereo Camera. *Nature* **432**, 971–979.

Newman, C. E., Lewis, S. R., Read, P. L., 2005. The atmospheric circulation and dust activity in different orbital epochs on Mars. *Icarus* **174**, 135–160.

Pathare, A. V., Paige, D. A., 2005. The effects of Martian orbital variations upon the sublimation and relaxation of north polar troughs and scarps. *Icarus* **174**, 419–443. doi:10.1016/j.icarus.2004.10.030.

Picardi, G., Biccari, D., Seu, R., Marinangeli, L., Johnson, W. T. K., Jordan, R. L., Plaut, J., Safaenili, A., Gurnett, D. A., Ori, G. G., Orosei, R., Calabrese, D., Zampolini, E., 2004. Performance and surface scattering models for the Mars Advanced Radar for Subsurface and Ionosphere Sounding (MARSIS). *Plan. Space Sci.* **52**, 149–156.

Pollack, J. B., Kasting, J. F., Richardson, S. M., Poliakoff, K., 1987. The case for a wet, warm climate on early Mars. *Icarus* **71**, 203–224.

Postawko, S. E., Kuhn, W. R., 1986. Effect of greenhouse gases (CO_2, H_2O, 2) on Martian paleoclimates. *J. Geophys. Res.* **91** (suppl.), D431–D438. SO

Rafkin, S. C. R., Michaels, T. I., 2003. Meteorological predictions for 2003 Mars Exploration Rover high-priority landing sites. *J. Geophys. Res.* **108**, doi:10.1029/2002JE002027.

Read, P. L., Lewis, S. R., 2004. *The Martian Climate Revisited: Atmosphere and Environment of a Desert Planet.* Springer-Verlag, Berlin. ISBN: 3-540-40743-X.

Richardson, M. I., Wilson, R. J., 2002a. Investigation of the nature and stability of the martian sasonal water cycle with a general circulation model. *J. Geophys. Res.* **107** (E5), doi:10.1029/2001JE001536.

Richardson, M. I., Wilson, R. J., 2002b. A topographically forced asymmetry in the Martian circulation and climate. *Nature* **416**, 298–301.

Segschneider, J., Grieger, B., Keller, H. U., Lunkeit, F., Kirk, E., Fraedrich, K., Rodin, A., Greve, R., 2005. Response of the intermediate complexity Mars Climate Simulator to different obliquity angles. *Plan. Space Sci.* **53**, 659–670.

Seu, R., Biccari, D., Orosei, R., Lorenzoni, L. V., Phillips, R. J., Marinangeli, L., Picardi, G., Masdea, A., Zampolini, E., 2004. SHARAD: The MRO 2005 shallow radar. *Plan. Space Sci.* **52**, 157–166.

Smith, M. D., 2002. The annual cycle of water vapor on Mars as observed by the Thermal Emission Spectrometer. *J. Geophys. Res.* **107** (E11), 5115, doi:10.1029/2001JE001522.

Smith, M. D., 2004. Interannual variability in TES atmopsheric observations of Mars during 1999–2003. *Icarus* **167**, 148–165.

Smith, M. D., Conrath, B. J., Pearl, J. C., Christensen, P. R., 2002. Thermal Emission Spectrometer observations of martian planet-encircling dust storm 2001a. *Icarus* **157**, 259–263.

Squyres, S. W., Arvidson, R. E., Bell III, J. F., Brückner, J., andW. Calvin, N. A. C., Carr, M. H., Christensen, P. R., Clarke, B. C., Crumpler, L., Des Marais, D. J., d'Uston, C., Economou, T., Farmer, J., Farrand, W., Folkner, W., Golombek, M., Gorevan, S., Grant, J. A., Greeley, R., Grotzinger, J., Haskin, L., Herkenhoff, K. E., Hviid, S., Johnson, J., Klinigelhöfer, G., Knoll, A., Landis, G., Lemmon, M., Li, R., Madsen, M. B., Malin, M. C., McLennan, S. M., McSween, H. Y., Ming, D. W., Moersch, J., Morris, R. V., Parker, T., Rice Jr., J. W., Richter, L., Rieder, R., Sims, M., Smith, M., Smith, P., Soderblom, L. A., Sullivan, R., Wänke, H., Wdowiak, T., Wolff, M., Yen, A., 2004a. The Spirit Rover's Athena science investigation at Gusev Crater, Mars. *Science* **305**, 794–799.

Squyres, S. W., Arvidson, R. E., Bell III, J. F., Brückner, J., andW. Calvin, N. A. C., Carr, M. H., Christensen, P. R., Clarke, B. C., Crumpler, L., Des Marais, D. J., d'Uston, C., Economou, T., Farmer, J., Farrand, W., Folkner, W., Golombek, M., Gorevan, S., Grant, J. A., Greeley, R., Grotzinger, J., Haskin, L., Herkenhoff, K. E., Hviid, S., Johnson, J., Klinigelhöfer, G., Knoll, A. H., Landis, G., Lemmon, M., Li, R., Madsen, M. B., Malin, M. C., McLennan, S. M., McSween, H. Y., Ming, D. W., Moersch, J., Morris, R. V., Parker, T., Rice Jr., J. W., Richter, L., Rieder, R., Sims, M., Smith, M., Smith, P., Soderblom, L. A., Sullivan, R., Wänke, H., Wdowiak, T., Wolff, M., Yen, A., 2004b. The Opportunity Rover's Athena science investigation at Meridiani Planum, Mars. *Science* **306**, 1698–1703.

Takahashi, Y. O., Fujiwara, H., Fukunishi, H., Odaka, M., Hayashi, Y.-Y., Watanabe, S., 2003. Topographically induced north–south asymmetry of the meridional circulation in the martian atmosphere. *J. Geophys. Res.* **108** (E3), 5018, doi:10.1029/2001JE001638.

Titus, T. N., Kieffer, H. H., Christensen, P. R., 2003. Exposed water ice discovered near the south pole of Mars. *Science* **299**, 1048–1051.

Touma, J., Wisdom, J., 1993. The chaotic obliquity of Mars. *Science* **259**, 1294–1297.

Yung, Y. L., Nair, H., Gerstell, M. F., 1997. CO_2 greenhouse in the early Martian atmosphere: SO_2 inhibits condensation. *Icarus* **130**, 222-224.

6 The Habitability of Mars: Past and Present

Thomas M. McCollom

Abstract. Data collected from recent spacecraft missions to Mars are making it possible to scientifically evaluate whether habitable environments are present, or existed in the past, on the planet. Determining where, when, and how such environments occur will play a crucial role in guiding future exploration efforts. Mars appears to possess the critical elements for life as we know it, including the required chemical elements, energy sources, and the presence of water. Identification of potential habitats for life therefore depends on the convergence of these factors under favorable circumstances. Spacecraft observations are more clearly defining the history of water on Mars, and appear to indicate that significant amounts of liquid water have been present at the surface in the past. Most of this water has been lost over time or retreated underground, although there is some evidence for periodic near-surface water in recent eras. If Mars did indeed have a warmer, wetter past, the possibilities for habitable environments were substantial and diverse. On the other hand, the inhospitable nature of the present surface indicates that potential habitats for extant life are much more limited. Nevertheless, habitable environments might exist in the subsurface or at the surface during more favorable climatic periods.

6.1 Introduction

Human beings have long been fascinated with the possibility that Mars might be inhabited by extraterrestrial life, and looking for signs of life is currently one of the foremost goals of human exploration of the planet. Whether or not life might exist on Mars has been the subject of scientific debate for well over a century, and over the years opinions on the likelihood of life on Mars have spanned the entire spectrum from "There can't be anything there!" to "Life's a virtual certainty!" Around the turn of the last century, claims by Percival Lowell and others that they had sighted canals on the planet brought visions of a Mars teeming not only with life, but with civilized, sentient beings. But Lowell's claims quickly dried up under scientific scrutiny, and when the first spacecraft flybys of Mars in the early 1960's revealed an arid, heavily cratered surface, Mars seemed as cold and forever-dead as the Moon (Fig. 6.1). Within a few years, however, new spacecraft missions began to reveal abundant signs that water had played a prominent role in the planet's history, reviving the belief that Mars could harbor some form of life.

Confidence that evidence of life might be detected on Mars was high in the 1970's when NASA sent two Viking landers to the surface equipped with several experiments designed specifically to look for evidence of life. When the landers not only failed to return signs of life but also found the surface was highly oxidized, bombarded with intense UV radiation, and seemingly devoid of organic molecules, it once again appeared that life was only a distant possibility. Yet, scrutiny of the images returned by the accompa-

Fig. 6.1. Image of the martian surface returned by the Mariner 4 spacecraft in 1965. The surface, from the southern hemisphere of Mars, is dominated by impact craters, providing little evidence of a habitable world. Image is about 250 km across. Image credit: NASA/JPL.

nying Viking orbiters in the years following the missions continued to provide evidence for wetter and more clement epochs in the past. At the same time, scientists were finding that the diversity of life on Earth is much broader than had been previously recognized, and many terrestrial environments once thought to be too extreme to be inhabited by organisms, such as thermal springs and the deep subsurface, have now been found to be capable of supporting substantial microbial populations. Together, these factors renewed interest in the possibility that life could be present on Mars, even under the challenging conditions found there.

Debate over the potential for life on Mars was once again brought to the forefront in 1996 when scientists from NASA claimed they had found evidence for life in the martian meteorite ALH84001 [McKay et al., 1996]. While the majority of the scientific community appears to have concluded that all of the putative evidence for life in ALH84001 can be explained by non-biological processes and therefore provides doubtful evidence for life [Treiman, 2003; Golden et al., 2004], a minority still maintains that some features of the meteorite can only be explained by biological processes [e.g., Gibson et al., 2003]. Wherever one stands on this debate, it has brought renewed interest and research into the potential for life on Mars and how to go about looking for it.

After a drought of almost 20 years following the Viking missions, spacecraft exploration of Mars returned in earnest in the 1990's, and during the past decade the planet has been visited by an armada of highly capable spacecraft that has included both orbiters and landers. Additional spacecraft are on their way or planned for launch in the next few years. These missions are ushering in a new era in the astrobiological study of Mars. Whereas in the past the possibility of life on Mars had been largely a matter of speculation based on a modicum of fact, the enormous amount of data being collected by current spacecraft missions are more clearly defining the geologic and climatic history of the planet, and allowing more concise hypotheses about the potential for life to be formulated and tested. Unless we are particularly fortunate in our selection of future samples and landing sites, it may take many more missions and several decades to obtain compelling evidence for the presence of life on Mars (or to convince us of the lack thereof). For the present, however, we can begin to make concrete assessments about where, and when, biologically habitable environments occur, and to infer the types of organisms that might live there. These assessments will play a major role in formulating future exploration plans, and increase our prospects for recognizing any traces of life that might be present.

6.2 Mars and the Bare Necessities of Life

Life, at least as we currently know it, has three fundamental requirements: a complement of basic chemical elements, a source of metabolic energy, and the presence of liquid water (although it has frequently been speculated that exotic lifeforms might exist that are based on silicon rather than carbon and use ammonia or other solvents rather that water, as yet such lifeforms remain purely hypothetical and will not be discussed further here). On Earth, biomass is composed primarily of organic molecules constructed from the elements C, H, O, N, P, and S, with the addition of some salts (mainly K, Ca, and Mg chlorides) and a few trace metals such as Fe and Co that function as co-factors for enzymes. All of the elements that are required by terrestrial life are known to be present in abundance in the crust or atmosphere of Mars; therefore, it appears likely that the necessary elements for biomass synthesis would be available to any organisms living there and that elemental availability is not a factor constraining the potential distribution of life [Banin and Mancinelli, 1995; Fisk and Giovannoni, 1999]. Accordingly, the availability of liquid water and suitable energy sources have been the focus of most discussion about whether life could exist on Mars.

Up until a couple of decades ago, it was widely believed that life could tolerate only a limited range of physical conditions, and therefore required environments within narrow limits of temperature, pressure, pH, salinity, etc. In recent years, however, scientists have found microorganisms in many environments once thought to be too "extreme" to be inhabitable, and organisms are now known that have adapted to life at extremes of temperature, pH, salinity, dissolved metal concentrations, radiation, etc. (Table 6.1). Given the apparent flexibility of life to adapt to a broad range of physical conditions, it is likely that essentially any environment with liquid water and a sufficient source of biologically available energy can be inhabited.

6.2.1 Water on Mars

Because of its critical importance to life, the search for habitable environments on Mars is closely linked to the search for liquid water, so much so that NASA has adopted "Follow the Water" as the central theme in their current exploration strategy for the planet. The large amount of attention given to this subject over the last couple of decades has resulted in a sharp increase in scientific understanding of the history of water on Mars. It would

Table 6.1. Range of physical environments known to be tolerated by microorganisms on Earth [see Nealson, 1997; Rothschild and Mancinelli, 2001].

Parameter	Known Limits
Temperature	$< -15°C$ to $>121°C$
Pressure	<1 bar to >10 kbar
pH	1 to >10
Salinity	Distilled water to NaCl saturation
Radiation	At least 20 kGy gamma radiation & at least 1,000 J m^{-2} UV radiation
Dessication	Survival in stasis for thousands of years or more

be impossible to do justice in a few paragraphs to the tremendous progress made in this field, and only a few of the highlights with the greatest relevance to habitability will be covered here. For a more thorough discussion, the reader is referred to recent comprehensive reviews of the topic [Baker, 2001; Jakosky and Phillips, 2001; Jakosky and Mellon, 2004].

It has long been recognized that liquid water is currently unstable at the surface of Mars owing to the extremely cold conditions and low atmospheric pressure there. Nevertheless, some water is present at the surface in the form of ice and as vapor in the atmosphere. Water occurs in the ice caps at both of the planet's poles, along with solid CO_2 [Bibring et al., 2004]. In addition, recent observations made by the Odyssey spacecraft indicate that water in the form of hydrated minerals and/or ice within pores spaces comprises a significant fraction of the regolith over much of the planet (as much as 35% at high latitudes), including some areas near the equator [Boynton et al., 2002]. However, owing to the low temperatures, the water in these shallow deposits is either frozen or tightly bound to mineral structures, and thus is unlikely to be readily available to biological organisms.

By any measure, the current inventory of water at the martian surface is sparse; atmospheric and surface water inventories are only about one ten-thousandth of that on the Earth. It is widely believed, however, that the surface reservoirs represent only a fraction of the planet's current inventory of water, and that a substantially larger reservoir of water could be distributed within pore spaces in the subsurface (Fig. 6.2) [Clifford, 1993; Clifford and Parker, 2001]. At shallow depths (several hundred meters to a few kilometers beneath the surface), this water is probably frozen to form a global subsurface cryosphere. At greater depths, higher temperatures owing to geothermal heating would keep water in pores spaces liquid, suggesting that Mars may have a planet-wide, liquid water aquifer at some depth within its crust. Furthermore, atmospheric processes and hydrologic constraints might allow for continuous, gradual circulation of water through this aquifer.

In contrast to the scarcity of water on the present surface of Mars, many ancient surfaces are covered with valley networks apparently carved by flowing water as well as areas that have been interpreted by some to represent basins that contained large lakes or seas [e.g., Irwin et al., 2002]. These features indicate that there were periods in Mars' distant past when the surface was much wetter and warmer than it is at present [Kargel,

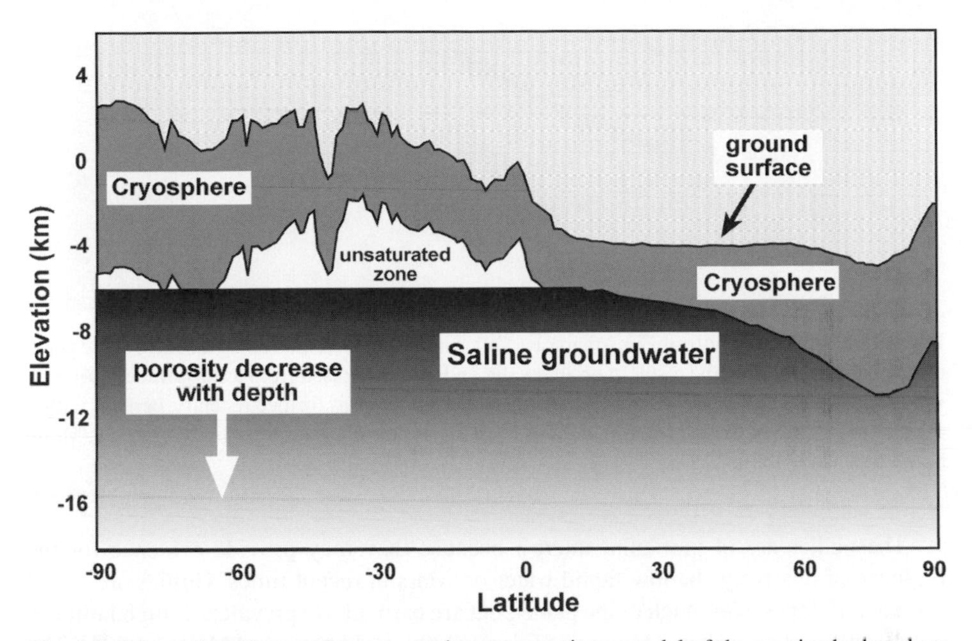

Fig. 6.2. Schematic pole-to-pole cross-section representing a model of the martian hydrosphere along 157°W longitude, showing hypothesized subsurface reservoirs of frozen (cryosphere) and liquid (saline groundwater) water [after Clifford and Parker, 2001].

2004]. Although many of the valley networks had been interpreted in the past to be fed by groundwater sapping [Carr, 1995], analysis of more recent higher resolution imagery indicates these features were instead fed by atmospheric precipitation [Hynek and Phillips, 2003; Mangold et al., 2004], implying that surface conditions were sufficiently warm and wet to allow hydrologic cycles to operate and rainfall to occur. Such periods may have been intermittent, but would have had to persist for sufficient time to carve the valley networks. Although other explanations are possible, the valley networks and putative seas may reflect an ancient Mars that once had a much larger inventory of water than it currently has, much of which has either been gradually lost to space or retreated into underground reservoirs.

In addition to the precipitation-fed valley networks, there are numerous geologic features that are believed to be formed by groundwater expelled to the surface, providing evidence for the existence of water in subsurface reservoirs (Fig. 6.3). The scale of these landforms range from large, catastrophic outflow channels (hundreds of kilometers in length) to small gullies (hundreds of meters) [Carr, 1995; Malin and Edgett, 2000]. The larger outflow channels may result from extensive melting of subsurface water ice reservoirs during thermal anomalies such as volcanic activity or large impacts [Squyres and Kasting, 1994; Cabrol et al., 2001]. While most of the outflow channels are ancient, a paucity of craters in some channels suggests they may have young ages (perhaps less than a couple of million years old), indicating that mobilization of groundwater or subsurface ice to the surface might have continued into recent history (Fig. 6.4) [Baker, 2001; Berman and Hartmann, 2002].

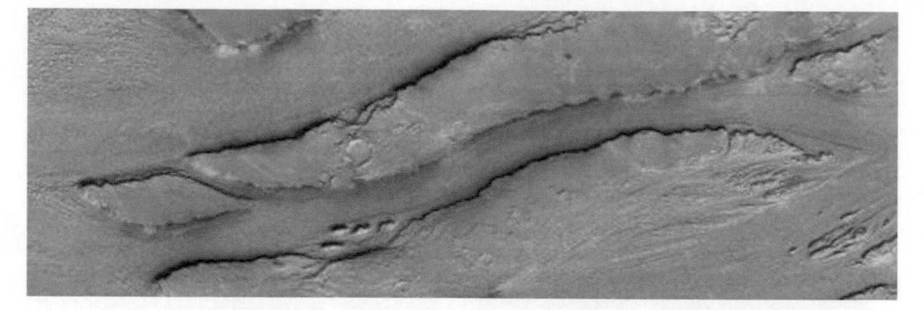

Fig. 6.3. Streamlined landforms in Athabasca Valles believed to be carved by a catastrophic flood of liquid water, possibly the result of groundwater and/or ice released from a subsurface aquifer by volcanic processes. The paucity of craters indicates the feature was formed recently, perhaps within the last million years. Scale of figure is about 3 km top to bottom. Image credit: NASA/JPL/Malin Space Science Systems.

The gullies are of particular interest because they may provide evidence for the presence of relatively shallow liquid water on Mars in recent times. Gullies are found carved into slopes over much of the planet, but are particularly prevalent at high latitudes in the southern hemisphere [Malin and Edgett, 2000; Heldmann and Mellon, 2004]. The lack of superimposed landforms and paucity of impact craters on the gullies indicates that they are mostly very recent in age. While there have been several mechanisms proposed to account for their formation, the most plausible appears to be that they are formed by water expelled from subsurface aquifers [Malin and Edgett, 2000; Mellon and Phillips, 2001; Heldmann and Mellon, 2004]. The source of the liquid water is not certain, but it appears feasible that heating from the background geothermal gradient within the planet may be sufficient to drive convection of liquid water to near the surface (i.e., within a couple hundred meters) [Travis et al., 2003]. If this interpretation is correct, it suggests that liquid water may currently be present at relatively shallow depths at some locations on Mars, and that this water periodically erupts to the surface, possibly carrying with it evidence for any biological activity occurring within the aquifer.

The largest of the outflow channels drain into the northern lowlands which cover most of the northern half of the planet. The lowlands are much younger (as indicated by lower crater counts) and smoother than the ancient, heavily cratered highlands that dominate the southern hemisphere. The presence of subdued features such as buried craters indicates that the northern plains are covered by a mantle a couple of hundred meters in thickness overlying a more ancient cratered surface [Frey et al., 2002]. One interpretation for the origin of the mantle is that it represents sediments flushed into the northern lowlands by floods that carved the outflow channels and valley networks. If so, the fluids carrying these sediments are likely to have formed large bodies of water that persisted at the surface for extended periods of time, perhaps as frozen or ice-covered liquid seas. Some researchers have even suggested that a large ocean may have covered much of the northern half of the planet at one time [e.g., Parker et al., 1989; Baker et al., 1991]. However, whether or not the northern plains were ever covered by a large sea, when that sea might have been there, and how long it may have persisted, are matters that remain unresolved (Baker, 2001).

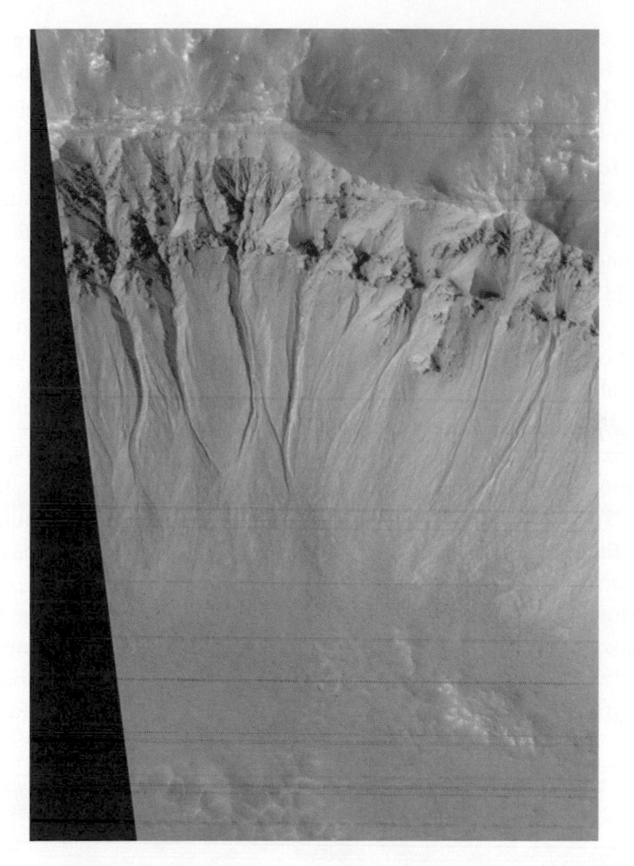

Fig. 6.4. Gullies on the edge of an impact crater on Mars that have been proposed to be carved by groundwater expelled to the surface. Image is about 1.6 km across. Image credit: NASA/JPL/Malin Space Science Systems.

The picture that emerges from our current understanding of the history of water on Mars, then, is that the planet was wetter at some periods in the past, and still retains substantial amounts of water in a frozen form at the surface and perhaps in liquid form several hundred meters or more in the subsurface. However, many of the larger questions with particularly significant implications for the habitability of the planet have not yet been answered. Foremost among these is the timing and duration of wetter periods in the planet's past. How much wetter the surface of Mars was, how long the wet eras persisted, and whether or not they were accompanied by clement conditions that allowed standing bodies of liquid water to persist for prolonged intervals are all subjects of ongoing scientific debate [Squyres and Kasting, 1994; Solomon et al., 2005]. One side of this debate contends that in its early history (perhaps for the first 500 million to a billion years), Mars was characterized by warm conditions during which hydrological cycles operated, including precipitation, and possibly even large seas were present at the surface. This scenario, of course, requires early Mars to have been much warmer than it is today, which is somewhat problematic since the intensity of sunlight is thought to have been significantly less in the early solar system than it is today. It has been suggested that early Mars may have been warmed by an atmospheric greenhouse similar to that hypothesized to warm the early Earth, but thus far a robust model for a stable atmosphere that could provide enough heat has not been proposed [Squyres and Kasting, 1994].

The opposite side of the debate suggests that the features attributed to water can be explained by several short-lived intervals during which warmer, wetter conditions were temporarily brought about by volcanic eruptions or large impacts [Baker et al., 1991; Baker, 2001; Segura et al., 2002]. These events may have ejected massive amounts of water vapor and other gases as well as heat into the atmosphere, warming the surface to the point where water was stable and leading to a temporary hydrologic cycle. Over time, this situation would be unstable, and Mars would return to colder conditions [Baker, 2001]. The number, duration, and intervals between such events that would be required to explain the surface features remains unclear.

The Mars Exploration Rovers (MERs) are currently returning the first ground observations providing direct evidence of water. The Opportunity rover was sent to Meridiani Planum, where spectral evidence had been recorded from orbit for the mineral hematite, an iron oxide that frequently forms in water. In addition to hematite, the lander has observed bedrock rich in sulfate salts and bearing layered features resembling those deposited by flowing water (Fig. 6.5). The MER team has interpreted the bedrocks to be sediments deposited by wind and water that were permeated by one or more generations of briny groundwaters that subsequently evaporated to leave behind an abundance of sulfate salts [Squyres et al., 2004; Squyres and Knoll, 2005]. Similar sulfate-rich layered

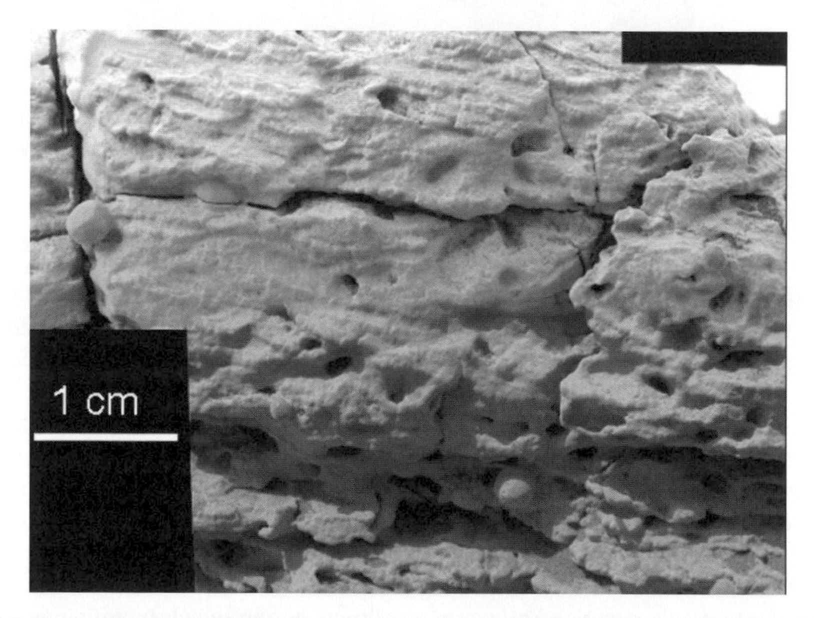

Fig. 6.5. Exposed bedrock, dubbed 'Last Chance', observed in Eagle Crater at Meridiani Planum by the Mars rover Opportunity. The rock has been interpreted to be sediments deposited by flowing surface water that were later cemented by sulfate salts precipitated by evaporation of briny groundwater. Image credit: NASA/JPL.

deposits, albeit on a smaller scale, have since been reported on the opposite side of the planet in Gusev Crater by the Spirit rover. This interpretation invokes deposition at a time when the surface environment was much wetter than it is now, and where the climate was clement enough to allow standing water to persist for extended periods of time. It should be borne in mind that we are only in the initial stages of analyses of the data from the rovers and already alternative scenarios are emerging to explain the observations by volcanic or impact processes that might not require warm surface conditions or standing bodies of water [McCollom and Hynek, 2005; Knauth et al., 2005]. Nevertheless, if the MER team's interpretation is correct, the Meridiani deposits may record a warmer, wetter era in Mars' distant past where life would have been comfortable, and that might still preserve evidence for past biological activity [Knoll et al., 2005].

6.2.2 Energy for martian life

Nearly all life on Earth relies either directly or indirectly on sunlight as a source of metabolic energy. Although Mars is further from the sun and receives less intense sunlight, the solar energy received on Earth far exceeds that required for photosynthesis and, even at the distance of Mars, the intensity of sunlight would be sufficient to drive photosynthetic metabolism. However, utilization of sunlight as an energy source requires organisms to live at, or very near, the surface of the planet. Yet, the intensely cold temperatures, absence of liquid water, and high levels of potentially damaging ultraviolet radiation make the current surface of Mars inhospitable for life. For this reason, it has been widely considered that the most likely habitats for extant life on Mars are beneath the surface [Boston et al., 1992; Shock, 1997; Fisk and Giovannoni, 1999].

Since organisms in subsurface environments cannot utilize sunlight for photosynthesis, they must instead rely on chemical sources of energy for metabolism. Furthermore, in the absence of a steady stream of organic compounds from photosynthesis or some other source, subsurface organisms would have to rely on inorganic chemical energy sources [Boston et al., 1992; Shock, 1997; Nealson, 1997; Jakosky and Shock, 1998; Fisk and Giovannoni, 1999]. Organisms that utilize inorganic chemical energy sources for metabolic energy and use this energy to synthesize their own biomass are classified by biologists as *chemolithoautotrophs*.

Chemolithoautotrophic microorganisms were first identified on the Earth well over a century ago, and are widely distributed in the biosphere. A wide variety of chemical reactions have been identified that supply metabolic energy to chemolithoautotrophs, a few examples of which are shown in Table 6.2. Not all of the reactions shown in Table 6.2 can supply energy in all environments; rather, which reactions can supply energy in a particular environment depends on whether or not the reactants are present in suitable quantities to make the reaction energetically favorable [e.g., McCollom and Shock, 1997]. All known chemolithoautotrophs exploit oxidation-reduction (redox) reactions that are out of equilibrium in the environment owing to slow reaction kinetics, and derive metabolic energy from driving the reactions towards equilibrium. In accordance with their metabolic requirements, chemolithoautotrophs frequently occupy habitats that are at an interface between relatively oxidizing and reducing environments (for example, where reducing hydrothermal fluids are exposed to oxidized surface waters or where oxidized groundwater interacts with chemically reduced rocks like basalt).

Table 6.2. Some chemical reactions that supply metabolic energy to chemolithoautotrophic organisms in terrestrial ecosystems.

Energy source	Chemical reaction
Aerobic metabolism:	
Sulfide oxidation	$HS^- + 2O_2 \rightarrow SO_4^{2-} + H^+$
Methanotrophy	$CH_4 + 2O_2 \rightarrow HCO_3^- + H^+ + H_2O$
Iron (II) oxidation	$Fe^{2+} + 1/4O_2 + H^+ \rightarrow Fe^{3+} + 1/2H_2O$
Mn (II) oxidation	$Mn^{2+} + 1/2O_2 + H_2O \rightarrow MnO_{2(s)} + 2H^+$
Hydrogen oxidation	$H_2 + 1/2O_2 \rightarrow H_2O$
Anaerobic metabolism:	
Methanogenesis	$HCO_3^- + H^+ + 4H_2 \rightarrow CH_4 + 3H_2O$
Sulfate reduction	$SO_4^{2-} + H^+ + 4H_2 \rightarrow HS^- + 4H_2O$
Anaerobic iron reduction	$2Fe^{3+} + H_2 \rightarrow 2Fe^{2+} + H^+$

Most chemolithoautotrophs live in environments like marine sediments where the chemical energy they live on arises as a by-product of the decomposition of photosynthetically produced organic matter. However, other chemolithoautotrophs inhabit environments where the chemical energy sources arise as the result of geologic processes with no direct link to photosynthesis. Examples of such environments include seafloor hydrothermal systems, geothermal springs on land, and pore spaces within igneous rocks in the deep subsurface. In these environments, the biomass produced in situ by chemolithoautotrophs forms the energetic basis of the microbial community. Because chemolithoautotroph-based microbial communities do not require any input of external organic matter from photosynthesis or other sources and derive their energy from geochemical processes, they may be particularly suitable analogs for the study of life on Mars and elsewhere.

Those metabolic reactions in Table 6.2 that utilize H_2, such as methanogenesis and hydrogen oxidation, may be especially relevant to the study of life on Mars. Hydrogen is generated in geologic systems by the reaction of water with ferrous iron-bearing minerals like olivine and pyroxene. These minerals are abundant in basalt, an iron-rich igneous rock that forms during solidification of lava flows. Basalts are widespread on the Earth, and are familiar to most people as the predominant rock formed by volcanic eruptions like those in Hawaii and Iceland. Basaltic rocks formed at mid-ocean ridges also underlay all of the ocean basins that cover two-thirds of the Earth's surface. Basalts are also the predominant rock type comprising the surface and crust of Mars. Basalts on Mars are even more iron-rich than those on Earth, and interaction of martian basalts with water would be expected to produce even greater amounts of H_2 than equivalent reactions on Earth. As a consequence, production of H_2 would be an inevitable outcome of groundwater migration in the martian subsurface. Microbial communities based on utilization of H_2 by chemolithoautotrophs are increasingly recognized in basalt-hosted hydrothermal environments on Earth that may prove to be informative analogs to life on Mars [Chapelle et al., 2002; Spear et al., 2005].

Fig. 6.6. Chemolithoautotrophic, iron-oxidizing bacteria growing in the lab on the surface of the mineral pyrite (FeS_2) as a source of Fe [Edwards et al., 2004]. These organisms, cultured from the deep seafloor, are able to live on nothing but rock, inorganic nutrients, and trace levels of oxygen. (**left**) Biofilm formed on the surface of the pyrite, with the altered mineral surface exposed where the biofilm is removed. (**right**) Stalk-like iron oxides deposited by the bacteria are visible on the exposed surface. Oxides like these could potentially be biomarkers for similar bacteria growing on Mars. Scale bars are 10 μm. Images courtesy K. Edwards, Woods Hole Oceanographic Institution.

On Earth, many chemolithoautotrophic organisms gain metabolic energy by combining reduced compounds from rocks or geologic fluids with dissolved O_2 (Fig. 6.6). In the absence of photosynthesizing organisms to supply O_2, similar reactions will play a much smaller role on Mars. Nevertheless, the atmosphere of Mars does contain a small amount of O_2 (10^{-5} bar) produced by photochemical reactions in the atmosphere, and even this small amount is sufficient to make many chemolithoautotrophic reactions energetically favorable. Consequently, any fluid in contact with the atmosphere through convection or diffusion may be able to transport O_2 to subsurface environments where it could be utilized by chemolithoautotrophs [Fisk and Giovannoni, 1999]. Atmospheric processes might also generate compounds such as CO or H_2 that could provide additional energy sources [Weiss et al., 2002].

Even on the geologically active Earth, the amount of chemical energy generated by geologic processes is order of magnitudes less than the energy available for photosynthesis, so that the total biomass produced by chemolithoautotrophic communities is only a small fraction of that produced by photosynthetic communities [see, for example, McCollom and Shock, 1997]. The amount of chemical energy available from geologic processes is likely to be even less on Mars. Consequently, the total amount of biomass that could be supported by chemolithoautotrophy on Mars is extremely low by terrestrial standards [Jakosky and Shock, 1998; Link et al., 2005]. Nevertheless, the local abundance of chemolithoautotrophic life can be very high where geologic processes focus the available chemical energy, such as occurs at hydrothermal systems or thermal springs [Varnes et al., 2003].

6.3 Habitats for Life: The Present

The discovery of extant life native to Mars would doubtlessly constitute a major milestone in the history of science, and would have repercussions well beyond science as well. But are there really places on Mars where organisms could live there now? As indicated by the foregoing discussion, there is reason to believe that Mars does indeed possess each of the fundamental requirements for life. All that is required to make a habitable environment, therefore, is to have these ingredients converge in one spot. In the current era, habitats with liquid water appear to be confined to the subsurface, and possibly only at depths of many hundreds of meters or more. However, because the crust of Mars is largely composed of iron-rich basalts, there are likely to be energy sources in the subsurface. In particular, energy within basalts and other rocks in Mars' subsurface may arise from two sources (Table 6.2): (1) transport of O_2 or other oxidants from surface sources into the deep subsurface by groundwater movements where it could be combined with ferrous iron or other reduced compounds in the rocks by organisms like sulfide- and iron-oxidizers, or (2) production of H_2 by water-rock reactions giving rise to H_2-based microbial communities (e.g., methanogens, sulfate reducers; Table 6.2).

Perhaps the most closely analogous environments on Earth would be groundwater aquifers within terrestrial basalts or within the thick basaltic layers that form the upper part of the oceanic crust [Fisk and Giovannoni, 1999]. Recent studies have indicated that both of these environments host indigenous, chemolithoautotroph-based microbial communities (Fig. 6.7) [Stevens and McKinley, 1995; Fisk et al., 1998; Chapelle et al., 2002; Fisk et al., 2003; Edwards et al., 2004, 2005]. However, owing to the difficult accessibility of these environments, little is yet known about the specific organisms that

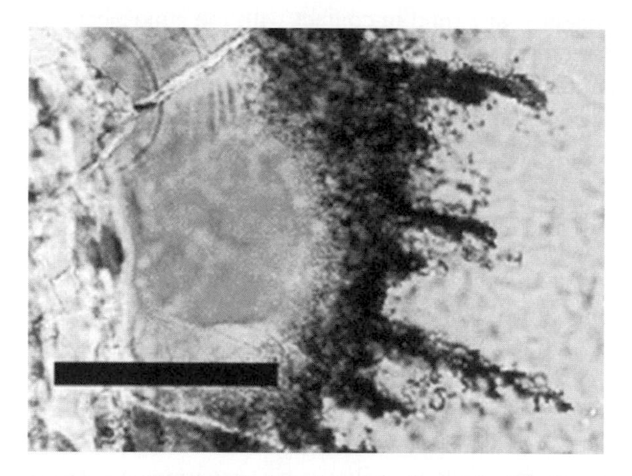

Fig. 6.7. Tubular structures within basalt from beneath the seafloor believed to be attributable to the activity of microorganisms in the subsurface [Fisk et al., 1998]. The figure shows a view through a thin-section of a piece of basalt from a hole drilled into the ocean crust. There is a fracture to the left of the figure, while the lighter gray area to the right is unaltered basalt glass. The darker gray material in the central part of the figure is altered basalt, and the tubules are thought to be channels bored into the glass by microorganisms seeking nutrients from the rock. Scale bar is 50 μm. Image courtesy M. Fisk, Oregon State University.

are present, the size of the populations, or how they interact with their environment. Nevertheless, the fact that these environments are very similar geochemically to those on Mars suggests that subsurface basaltic environments on Mars should be inhabitable.

While the deep subsurface appears to be the most likely place for life on Mars today, other environments closer to, or at, the surface may have been habitable in recent history (i.e., within the last few million years or less), and these locations may still retain a record of biological activities. Although no evidence has yet been found of current volcanic activity, Mars appears to have experienced volcanic eruptions in the near past, perhaps as recently as a million years ago or less [Neukum et al., 2004]. Intermittent volcanic activity, as well as occasional large impacts, would generate heat that would melt near-surface ice layers and lead to hydrothermal circulation, potentially bringing liquid water to the surface. Fluid-rock interactions within hydrothermal environments on Mars would create chemical gradients that provide a locally abundant source of metabolic energy, just as they do on Earth [Varnes et al., 2003]. On Earth, chemolithoautotrophic populations in hydrothermal environments are dominated by organisms such as sulfide- and hydrogen-oxidizers that utilize photosynthetically produced O_2 [e.g., McCollom and Shock, 1997; Spear et al., 2005]. Many of the same metabolic reactions used by these organisms would yield energy, albeit somewhat less, with the lower amount of O_2 in the martian atmosphere. Furthermore, many reactions such as methanogenesis and sulfate reduction can provide metabolic energy in hydrothermal environments even in the complete absence of O_2 [McCollom, 1999; Chappelle et al., 2002].

In addition, while the surface of Mars is currently too cold for liquid water to be stable, it appears that changes in the planet's obliquity may periodically allow for temperatures at the surface to exceed the melting point of briny water at high latitudes [e.g., Jakosky et al., 2004]. During periods of high obliquity when the poles are tilted towards the Sun for prolonged periods, pockets of liquid water may develop along mineral interfaces. Organisms living in liquid water within pore spaces in the rocks (termed *endoliths*) would be protected from UV radiation, but would still be able to utilize sunlight as an energy source for photosynthesis [Cockell and Raven, 2004].

Of course, inhabitation of intermittent surface environments would require that the organisms persist through long periods of dormancy in a frozen state or be continuously re-seeded from deeper habitats whenever surface conditions become favorable. However, neither of these possibilities appears to be implausible. Microorganisms have now been identified that appear to have survived for millions of years trapped within minerals [Rothschild and Mancinelli, 2001], and the rapid colonization of new hydrothermal environments in widespread locations on the Earth suggests that microorganisms can be readily transported through subsurface conduits.

6.4 Habitats for Life: The Past

The prospect that Mars had a prolonged period of warm and wet conditions early in its history provides many tantalizing possibilities for life. If conditions on early Mars were really clement enough to allow liquid water to persist at the surface for long periods, for large standing seas to form, and for a hydrologic cycle complete with precipitation to develop, it is easy to envision a variety of habitable environments. Under these conditions, large areas of the surface would be open to inhabitation by photosynthetic organisms. Increased geologic activity and impact frequency on early Mars would have provided additional energy sources and, perhaps, delivered organic nutrients that could be utilized by biological organisms.

It may be that the only barrier to life inhabiting early Mars under prolonged warm and wet conditions would be whether or not these conditions persisted long enough for life to originate and evolve. Many astrobiologists believe that life on Earth appeared very soon after conditions were sufficiently clement to allow it (possibly within 100 million years; e.g., Nisbet and Sleep, 2001) and rapidly diversified. While our knowledge of the conditions required for life to originate is still rudimentary, there doesn't appear to be any obvious reason why life could not have evolved on Mars in a comparable time period. Indeed, it can be plausibly argued that if wet, clement conditions persisted on early Mars for any lengthy period of time, a separate origin of life there would have been inevitable [Russell and Hall, 1999]. Although still a matter of debate, dating of water-related features suggests many of the major features formed over a period spanning 500 million years.

But contemplation of the possibility of life on a warm, wet, and hospitable early Mars must be tempered with consideration of the uncertainties over the early climate and history of water. If the features attributed to water can be explained by a few brief eras of warm and wet conditions separated by prolonged periods of extreme cold and dryness resembling conditions on Mars today, the prospects for habitable environments would be substantially diminished. Photosynthetic organisms in particular would have fewer surface habitats to colonize, and would have had to adapt rapidly to changing climatic conditions and to survival through long intervals of inactivity in cold, desiccating conditions.

Even without prolonged clement conditions, however, environments for life would likely have been more widespread on early Mars than they are today. Volcanic processes were much more vigorous on early Mars [Solomon et al., 2005] and, combined with more frequent impacts, may have maintained a nearly continuous presence of hydrothermal environments on the planet. Biological niches within these environments would be similar in character to those that might continue intermittently to the present day, but would have been more prolific. In addition, hydrothermal environments at the surface may have allowed oases for early photosynthetic organisms to survive between wetter periods.

6.5 Is There Anybody Out There?

It seems increasingly apparent that habitable environments very likely exist on Mars today, and may have been considerably more diverse and abundant in the past. Of course, it remains to be seen if any organisms are actually inhabiting these environments. Consideration of possible inhabitable environments, however, suggests two very daunting challenges in the search for extant life. First, the most probable environments for life today may be at hundreds of meters deep in the subsurface, requiring elaborate drilling equipment to gain access. Obtaining reliable samples for biological study from deep drilling has been problematic even on the Earth. Second, the limited availability of energy sources means that life on Mars will be substantially less abundant than what we are familiar with on Earth, compounding the difficulties in searching for life. Given these impediments, we may find more immediate success by looking for surviving evidence of past life in near surface environments on Mars that may have been inhabited in recent times. Whatever the eventual outcome of our search for life on Mars, we will gain a better understanding of the history of life in our solar system and the prospects for life beyond.

References

Baker V.R. (2001) Water and the Martian Landscape, *Nature*, **412**, 228.

Baker V.R., Strom R.G., Gulick V.C., et al. (1991) Ancient Oceans, Ice Sheets and the Hydrologic Cycle on Mars, *Nature*, **352**, 589.

Banin A. & Mancinelli R.L. (1995) Life on Mars? I. The chemical environment. *Adv. Space Sci.*, **15**, 163.

Berman D.C. & Hartmann W.K. (2002) Recent Fluvial, Volcanic, and Tectonic Activity on the Cerberus Plains of Mars, *Icarus*, **159**, 1.

Bibring J-P., Langevin Y., Poulet F., et al. (2004) Perennial water ice identified in the south polar cap, *Nature*, **428**, 627.

Boston P.J., Ivanov M.V. & McKay C.P. (1992) On the Possibility of Chemosynthetic Ecosystems in Subsurface Habitats on Mars, *Icarus*, **95**, 300.

Boynton W.V., Feldman W.C., Squyers S.W., et al. (2002) Distribution of Hydrogen in the Near Surface of Mars: Evidence for Subsurface Ice Deposits, *Science*, **297**, 81.

Cabrol N.A., Wynn-Williams D.A., Crawford D.A. & Grin E.A. (2001) Recent Aqueous Environments in Martian Impact Craters: An Astrobiological Perspective, *Icarus*, **154**, 98.

Carr M.H. (1995) The Martian Drainage System and the Origin of Valley Networks and Fretted Channels, *J. Geophys. Res.*, **100**, 7479.

Chapelle F.H., O'Neill K., Bradley P. M. et al. (2002) A hydrogen-based subsurface community dominated by methanogens. *Nature*, **415**, 312.

Clifford S.M. & Parker T.J. (2001) The Evolution of the Martian Hydrosphere: Implications for the Fate of a Primordial Ocean and the Current State of the Northern Plains, *Icarus*, **154**, 40.

Clifford S.M. (1993) A Model for the Hydrologic and Climatic Behavior of Water on Mars, *J. Geophys. Res.* (E), **98**, 10,973.

Cockell C.S. & Raven J.A. (2004) Zones of Photosynthetic Potential on Mars and the Early Earth, *Icarus*, **169**, 300.

Edwards K.J., Bach W., McCollom T.M. & Rogers D. (2004) Neutrophilic Iron-oxidizing Bacteria in the Ocean: Their Habitats, Diversity, and Roles in Mineral Deposition, Rock Alteration, and Biomass Production in the Deep Sea. *Geomicrobiol. J.*, **21**, 393.

Edwards K.J., Bach W., & McCollom T.M. (2005) Geomicrobiology in oceanography: Mineral-microbe interactions in the deep-sea. *TRENDS Microbiol.*, **13**, 449.

Fisk M.R. & Giovannoni S.J. (1999) Sources of Nutrients and Energy for a Deep Biosphere on Mars, *J. Geophys. Res.* (E), **104**, 11,805.

Fisk M.R., Giovannoni S.J. & Thorseth I.H. (1998) Alteration of Oceanic Volcanic Glass: Textural Evidence of Microbial Activity, *Science*, **281**, 978.

Fisk M.R., Storrie-Lombardi M.C., Douglas S., et al. (2003) Evidence of Biological Activity in Hawaiian Subsurface Basalts, *Geochem. Geophys. Geosys.*, **4**, 12.

Frey H.V., Roark J.H., Shockey K.M., Frey E. L. & Sakimoto S.E.H. (2002) Ancient Lowlands on Mars, *Geophys. Res. Lett.*, **29**, doi: 10.1029/2001GL013832.

Gibson Jr. E.K., McKay D.S., Thomas-Keprta K.L., et al. (2001) Life on Mars: Evaluation of the evidence within martian meteorites ALH84001, Nakhla, and Shergotty, *Precambrian Res.*, **106**, 15.

Golden D.C., Ming D.W., Morris R.V., et al. (2004) Evidence for Exclusively Inorganic Formation of Magnetite in Martian Meteorite ALH84001, *Am. Mineral.*, **89**, 681.

Heldmann J.L. & Mellon M.T. (2004) Observations of Martian Gullies and Constraints on Potential Formation Mechanisms, *Icarus*, **168**, 285.

Hynek B.M. & Phillips R.J. (2003) New Data Reveal Mature, Integrated Drainage Systems on Mars Indicative of Past Precipitation, *Geology*, **31**, 757.

Irwin R.R. III, Maxwell T.A., Howard A.D., et al. (2002) A Large Paleolake Basin at the Head of Ma'adim Vallis, Mars, *Science*, **296**, 2209.

Jakosky B.M. & Phillips R.J. (2001) Mars' Volatile and Climate History, *Nature*, **412**, 237.

Jakosky B.M., Nealson K.H., Bakermans C. et al. (2003) Subfreezing Activity of Microorganisms and the Potential Habitability of Mars' Polar Regions, *Astrobiology*, **3**, 343.

Jakosky B.M. & Mellon M.T. (2004) Water on Mars, *Physics Today*, **57**, 71.

Jakosky B.M. & Shock E.L. (1998) The Biological Potential of Mars, the Early Earth, and Europa, *J. Geophys. Res.* (E), **103**, 19,359.

Kargel J.S. *Mars: A Warmer Wetter Planet*, (Springer, London 2004).

Knauth L.P., Burt D.M. & Wohletz K.H. (2005) Impact Origin of Sediments at the Opportunity Landing Site on Mars, *Nature*, **438**, 1123.

Knoll A.H., Carr M., Clark B., et al. (2005) An Astrobiological Perspective on Meridiani Planum, *Earth Planet. Sci. Lett.*, **240**, 179.

Link L.S., Jakosky B.M. & Thyne G.D. (2005) Biological Potential of Low-temperature Aqueous Environments on Mars, *Inter. J. Astrobiol.*, **4**, 155.

Malin M.C. & Edgett K.S. (2000) Evidence for Recent Groundwater Seepage and Surface Runoff on Mars, *Science*, **288**, 2330.

Malin M.C. & Edgett K.S. (2003) Evidence for Persistent Flow and Aqueous Sedimentation on Early Mars, *Science*, **302**, 1931.

Mangold N., Quantin C., Ansan V., et al. (2004) Evidence for Precipitation on Mars from Dendritic Valleys in the Valles Marineris Area, *Science*, **305**, 78.

McCollom T.M. (1999) Methanogenesis as a Potential Source of Chemical Energy for Primary Biomass Production by Autotrophic Organisms in Hydrothermal Systems on Europa, *J. Geophys. Res.* (E), **104**, 30,729.

McCollom T.M. & Shock E.L. (1997) Geochemical constraints on chemolithoautotrophic metabolism by microorganisms in seafloor hydrothermal systems, Geochim Cosmochim. *Acta*, **61**, 4375.

McCollom T.M. & Hynek B.M. (2005) A Volcanic Environment for Bedrock Diagenesis ar Meridiani Planum on Mars, *Nature*, **438**, 1129.

McKay D.S., Gibson Jr. E.K., Thomas-Keprta K.L., et al. (1996) Search for Past Life on Mars: possible Relic Biogenic Activity in Martian Meteorite ALH84001, *Science*, **273**, 924.

Mellon M.T. & Phillips R.J. (2004) Recent Gullies on Mars and the Source of Liquid Water, *J. Geophys. Res.* (E), **106**, 23,165.

Nealson K.H. (1997) The Limits for Life on Earth and Searching for Life on Mars, *J. Geophys. Res.* (E), **102**, 23,675.

Neukum G., Jaumann R., Hoffmann H. et al. (2004) Recent and Episodic Volcanic and Glacial Activity on Mars Revealed by the High Resolution Stereo Camera, *Nature*, **432**, 971.

Nisbet E.G. & Sleep N.H. (2001) The Habitat and Nature of Early Life, *Nature*, **409**, 1083.

Parker T.J., Saunders R.S. & Schneeberger D.M. (1989) Transitional Morphology in West Deuteronilus Mensae, Mars – Implications for Modification of the Lowland Boundary, *Icarus*, **82**, 111.

Rothschild L.J. & Mancinelli R.L. (2001) Life in Extreme Environments, *Nature*, **409**, 1092.

Russell M.J. & Hall A.J., 1999, On the Inevitable Emergence of Life on Mars, in: The Search for Life on Mars, Proceedings of the 1st UK conference, J.A. Hiscox, Ed. British Interplanetary Society, pp. 26–36.

Segura T.L., Toon O.B., Colaprete A. & Zahnle K. (2002) Environmental Effects of Large Impacts on Mars, *Science*, **298**, 1977.

Shock E.L. (1997) High-temperature Life Without Photosynthesis as a Model for Mars, *J. Geophys. Res.* (E), **102**, 23,687.

Solomon S.C., Aharonson O., Aurnou J.M., et al. (2005) New Perspectives on Ancient Mars, *Science*, **307**, 1214.

Spear J.R., Walker J. J., McCollom T.M. & Pace N. R. (2005) Hydrogen and Bioenergetics in the Yellowstone Geothermal Ecosystem, *Proc. Nat. Acad. Sci.*, **102**, 2555.

Squyres S.W. & Kasting J.F. (1994) Early Mars: How Warm and How Wet?, *Science*, **265**, 744.

Squyres S.W. & Knoll A.H. (2005) Sedimentary Rocks at Meridiani Planum: Origin, Diagenesis, and Implications for Life on Mars, *Earth Planet. Sci. Lett.*, **240**, 1.

Squyres S.W., Arvidson R.E., Bell J.F. III, et al. (2004) The Opportunity rover's Athena science investigation at Meridiani Planum, Mars, *Science* **306**, 1698.

Stevens T.O. & McKinley J.P. (1995) Lithoautotrophic Microbial Ecosystems in Deep Basalt Aquifers, *Science* **270**, 450.

Travis B.J., Rosenberg N.D. & Cuzzi J.N. (2003) On the Role of Widespread Subsurface Convection in Bringing Liquid Water Close to Mars' Surface, *J. Geophys. Res.* (E), **102**, doi: 10.1029/2002JE001877.

Treiman A.H. (2003) Submicron Magnetite Grains and Carbon Compounds in Martian Meteorite AH84001: Inorganic Abiotic Formation by Shock and Thermal Metamorphism. *Astrobiology*, **3**, 369.

Varnes E.S., Jakosky B.M. & McCollom T.M. (2003) Biological Potential of Martian Hydrothermal Systems, *Astrobiology*, **3**, 407.

Weiss B.P., Yung Y.L. & Nealson K.L. (2000) Atmospheric Energy for Subsurface Life on Mars?, *Proc. Nat. Acad. Sci.*, **97**, 1395.

7 Jupiter-sized Planets in the Solar System and Elsewhere

Patrick G. J. Irwin

Abstract. The discovery of the first extrasolar planet, 51 Pegasi, in 1995 has opened up a new and exciting area of planetary astronomy. Using a number of different techniques 170 extrasolar planets have now been discovered, and this number is rapidly increasing. Due partly to observational biases, just over half of these newly discovered planets are 'Hot Jupiters' – Jupiter sized planets orbiting within 1 AU of their parent stars. However, the number of giant planets discovered in more distant orbits, as seen in our own Solar System, is increasing all the time as detection techniques improve and longer time-series are analysed. In this paper, the characteristics of the extrasolar planets discovered to date will be reviewed, together with the implications these characteristics have upon theories of the formation of planetary systems in general and our own Solar System in particular. Current and future methods for detecting extrasolar planets are also reviewed, together with prospects for detecting terrestrial planets and perhaps even searching for the signs of life itself.

7.1 Introduction

The giant planets dominate the planetary mass and total angular momentum of our Solar System. They are believed to have played an important role in the evolution of the Earth by gravitationally shielding the inner Solar System planets from cometary and asteroidal impacts, which might otherwise have inhibited the evolution of life on our planet. Unlike the terrestrial planets of the inner Solar System, which are predominantly composed of rocks, the giant planets are instead composed of ices and gas at pressures and temperatures which do not allow the formation of a surface crust. Although it is widely thought that these planets have rocky cores, these planets are entirely gaseous in their upper, observable layers.

The giant planets can actually be split into two categories (e.g. Irwin, 2003). Jupiter and Saturn are examples of gas giants whose bulk are dominated by gaseous hydrogen and helium. Uranus and Neptune, on the other hand, are mostly composed of water, ammonia and methane ice, although their upper, observable atmospheres are composed mainly of hydrogen and helium.

Until recently the giant planets were believed by most scientists to have been formed by the core accretion model (Sect. 7.2.4) whereby ice and rock in the protosolar nebula coalesced rapidly enough to form an 'embryo' of sufficient mass to gravitationally attract the hydrogen and helium of the solar nebula before it dissipated, either due to photo-evaporation from nearby, massive stars (Boss, 2002), or when the T-Tauri phase of the Sun swept away any remaining gas (e.g. Irwin, 2003). However the discovery of a 'Hot Jupiter' (Jupiter-sized planet orbiting within 1 AU) around another star (Mayor and Queloz, 1995) and the discovery since of over 100 other such planets has called this model into question and has revived the alternative gravitational instability model, whereby Jupiter-sized planets rapidly form directly from the solar nebula. In this review we will summarise the observations of extrasolar giant planets (EGPs) that have been made to date and discuss the questions they raise concerning the formation of giant planets in general, particularly the Solar System giant planets. We will then go on to explore what we hope to learn from future planned observations of extrasolar planets.

7.1.1 Gross characteristics of the giant planets

The two Solar System gas giants, Jupiter and Saturn, form a class of planet with large radius (~65,000 km) and bulk composition similar to that of the proto-solar nebula. Jupiter has a mass of 318 M_{Earth}, an equatorial radius (referenced to the 1 bar pressure level) of 71,492 km and a mean density of 1330 kg/m^3. Although Saturn is much less massive than Jupiter (95 M_{Earth}), it has a similar equatorial radius (at 1 bar) of 60,268 km leading to a density of only 700 kg/m^3. The similar radii of these planets is due to the compressibility of hydrogen gas, which is found to be little affected by temperature. Hence, a large increase in mass produces almost no change in radius and instead the characteristic radius depends almost entirely on the mean composition (Hubbard, 1997). For an approximately solar composition, the characteristic radius would be 70,000 km, close to the observed radii of both Jupiter and Saturn.

The two Solar System ice giants, Uranus and Neptune, are predominantly composed of water, ammonia and methane ices and are thus noticeably smaller than their gas giant cousins, with a characteristic radius of ~25,500 km. Uranus has a mass of 14.5 M_{Earth}, an equatorial radius (at 1 bar) of 25,559 km and a density of 1270 kg/m^3. Neptune is slightly more massive (17.1 M_{Earth}), but has a similar equatorial radius (24,764 km) and is thus somewhat denser (1760 kg/m^3).

It is found that there is a steady increase in the abundance of heavy elements in the giant planets, varying monotonically with the planets' distance from the Sun. Thus the C/H ratio, relative to the solar value of 3.311×10^{-4} (Grevesse and Sauval, 1998), increases from 3.2 × solar for Jupiter, to 7×, 29× and 40× solar for Saturn, Uranus and Neptune, respectively. Similarly the D/H ratio in the atmospheres also increases with solar distance, which is thought to be due to the increasing component of icy material, a supposition which fits well with the observed radii and densities of these planets. Hence the abundance of other heavy elements such as oxygen and nitrogen also increase with distance from the Sun. This variation in composition provides strong constraints on planetary formation theories.

7.2 Extrasolar Planets

A question that has intrigued planetary scientists for many years is how typical our Solar System is. It has generally been presumed that there are planets orbiting other stars, but until only very recently none had ever been observed and so it was impossible to determine if our Solar System was normal, unusual, or perhaps even unique. Hence, scientists began to investigate ways of detecting extrasolar planets. Consider the problem of trying to detect Jupiter about our Sun at a distance of, say, 5 parsecs. At this distance, the greatest angular separation of the Sun and Jupiter would be just 1" (i.e. one arc-second) and given that the Sun/Jupiter flux ratio is 10^9 in the visible, diffraction and scattering in a telescope would make it impossible to pick Jupiter out from the Sun's glare (Lewis, 2004). One possible solution to this problem is to search for planets around dimmer stars such as white and brown dwarfs. Once such white dwarf search is currently being undertaken (e.g. Burleigh et al., 2003) and a planet has recently actually been imaged by the Very Large Telescope (VLT) orbiting around a brown dwarf, situated 200 light years away, at a distance of $\sim$ 60 AU (Chauvin et al., 2005). Another strategy is to attempt to detect the planet at wavelengths near the peak of the planet's Planck function. Observing at 50 µm rather than 0.6 µm reduces the flux ratio to 10^4 for the Sun-Jupiter system, but at these longer wavelengths the diffraction-limited angular resolution of a telescope would be insufficient unless it was unfeasibly large. Thus direct detection of extrasolar planets appeared, at first, to be impossible.

7.2.1 Radial velocity detections

Although direct optical detection of extrasolar planets appeared very difficult, it was realised in the mid-eighties that it might be possible to indirectly detect them through their influence on the motion of the central star. For a planet of mass M_p in a circular orbit of radius a about a star of mass M_*, the star and planet will both orbit about their centre-of-mass, situated at a distance $2aM_p/\left(M_p + M_*\right)$ from the star. Equating the gravitational force with the centripetal force acting on the star, and assuming that $M_* \ggg M_p$, the maximum velocity of the star v in the line of sight of an observer may be shown to satisfy $v^2 = G\left(M_p \sin i\right)^2/2M_*a$, where i is the inclination of the planet's orbit with respect to the observer, i.e. the angle between the normal to orbital plane of the planet and the line from the star to the observer on the Earth. The radial velocity method can determine both $M_p \sin i$ and also, from the shape of the variation of v with time, the eccentricity of the planet's orbit. It is worth noting that unless the inclination can be determined from other methods such as astrometry (Sect. 7.3.3), this method only provides a lower limit on the planet's mass. If the inclination is 90°, then $M_p \sin i$ is equal to the planet's mass. However, if the inclination is small, then M_p will clearly be much larger than $M_p \sin i$. The technique is most likely to be effective for larger mass planets orbiting close to the lower mass stars (i.e. G and K type) since this gives the largest line-of-sight stellar velocity and it is crucial to be able to distinguish the radial velocity of the star due to the orbit of a planet from the naturally occurring turbulent velocities present in a stellar photosphere. Given this caveat, the technique seemed to be the most likely to detect planets and thus a number of campaigns were initiated in the early nineties. Subsequently, Mayor and Queloz (1995) made the first discovery using this technique of a planet orbiting the star 51 Pegasi. What they found however was astonishing: a planet with $M_p \sin i = 0.46M_J$, (where M_J is the mass of Jupiter) orbiting at a distance of only 0.05 AU in a period of just 4.2 days! The surface temperature of the planet, so close to its star, was calculated to be enormous ($\sim$ 1400 K) and the planet was dubbed a 'Hot Jupiter'.

Since the discovery of 51 Peg b (the exoplanetary naming convention is to list the star name followed by 'b', 'c' ... in order of the planet's discovery), there have been detections (almost all by the radial velocity technique) of well over 100 EGPs, with new planets discovered almost every week. Indeed it is now estimated that more than 6% of sun-like stars have a detectable 'wobble' due to the orbit of at least one Jupiter-mass planet. At the time of writing, the total number of EGPs listed the Extrasolar Planets Encyclopedia (http://www.obspm.fr/planets) was 170 in 146 planetary systems (including 18 multiple planet systems). Most of the recent radial velocity planet searches have been able to detect velocity variations as small as 10 m/s (Marcy et al., 2003) and so a Sun-Jupiter system (for which the Sun's radial velocity is 13.2 m/s) should have been just about detectable and, indeed, such planets are now being found. For example (Carter et al., 2003) report the discovery of HD 70642 b, a planet with mass 2 M_J, low eccentricity and an orbital radius $a = 3.3$ AU. Recent improvements have meant that current observations can now achieve even greater accuracies of 3 m/s and thus the number of planets detectable by this technique is expected to increase. In addition, the current data sets only last for ~ 10 years. As measurements continue, and the sensitivity improves, the discovery of more Jupiter-like planets orbiting far from their star with longer periods is expected.

So many planets have now been found that it is possible to consider the statistics of the mass and orbital parameter distributions, as has been done by Cameron (2002), and Marcy et al. (2003). Radial velocity measurements can only provide information on the distribution of $M_p \sin i$. However, it can be shown (Jorissen et al., 2001) that for a random distribution of planetary systems, the distribution of $M_p \sin i$ is very close to the distribution of M_p and thus statistical conclusions on the overall mass distribution can be inferred from the distribution of $M_p \sin i$ for known EGPs, shown in Fig. 7.1. Considering the selection effects of radial velocity measurements, a predominance of

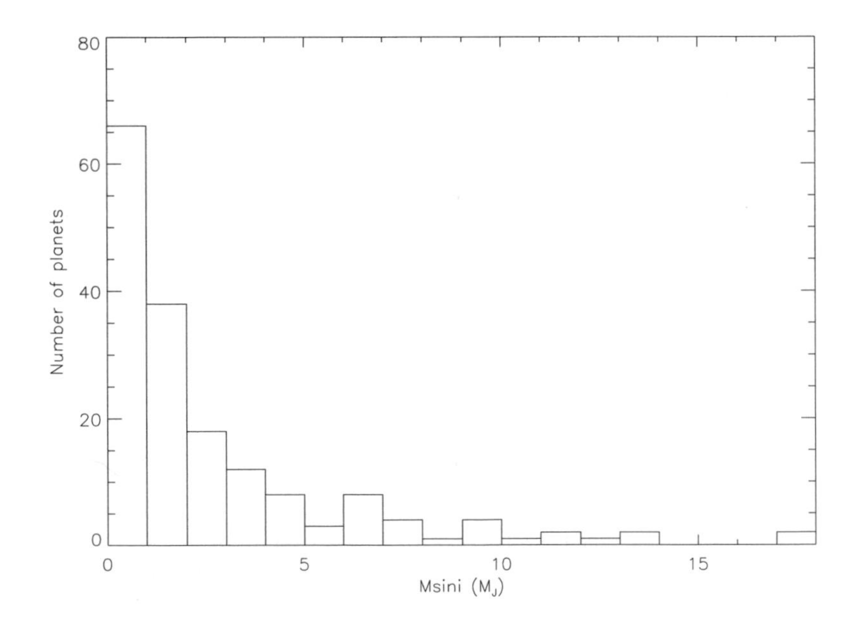

Fig. 7.1. Distribution of $M_p \sin i$ of currently known exoplanets.

heavy planets might be expected. However, most of the planets discovered so far have $M_p \sin i < 10M_J$, and the distribution of planets rises rapidly for smaller masses. The smallest exoplanet discovered to date is GJ 876 d (Rivera et al., 2005) which has an estimated mass of only $\sim 7.5M_{\text{Earth}}$. In contrast, there is an apparent absence of heavy exoplanets planets with mass above the deuterium-burning limit for brown dwarfs of $\sim 13.6M_J$ (Lewis, 2004). This apparent absence of very large mass planets has become known as the 'Brown Dwarf Desert'. It has been suggested that brown dwarfs might be formed by a different process from planets, leading to them orbiting at much greater distances than is currently detectable with the radial velocity technique. However, very recently a few heavy mass exoplanets have been discovered, the heaviest being HD 202206 b, which has an estimated $M_p \sin i$ of $17.4M_J$ (Correia et al., 2005). Hence, the 'Brown Dwarf Desert' may prove not to be quite so barren as has been previously thought, supporting the suggestion of Jorissen et al. (2001) that there is no reason to to ascribe the transition between giant planets and brown dwarfs to the threshold mass of deuterium ignition.

The distribution of exoplanet orbit radii is shown in Fig. 7.2 and it is found that a large fraction of known EGPs orbit within 1 AU. However, given that planets with larger orbital distances take longer to orbit and current observation programmes have only been running for 10 years and are becoming more precise all the time, there is good reason to suspect that there is a large population of planets orbiting beyond 3 AU (Marcy et al., 2003) which will soon be detected.

Fig. 7.3 shows $M_p \sin i$ for known exoplanets plotted against their orbital distance and there is a clear lack of massive planets ($M_p > 4M_J$) orbiting within 0.3 AU.

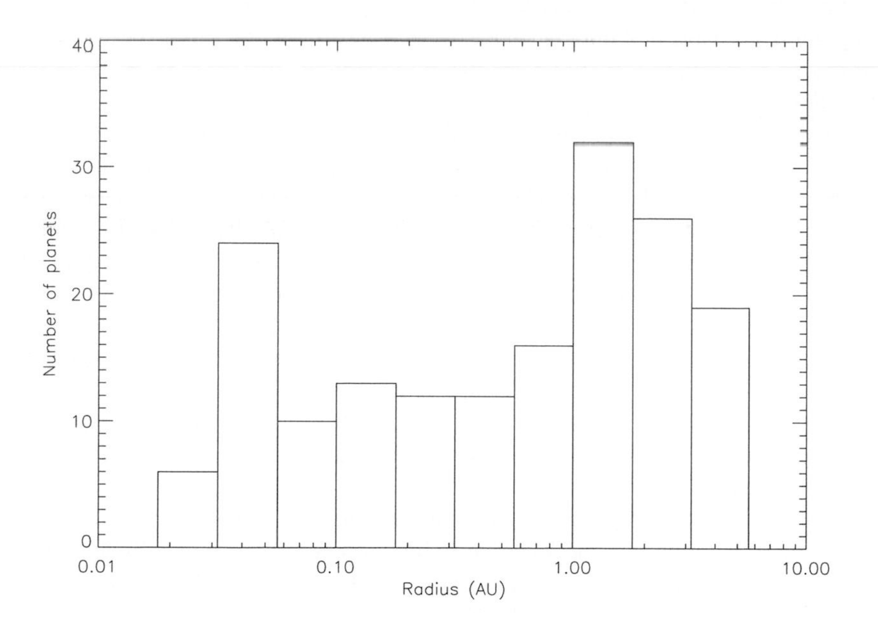

Fig. 7.2. Orbital radius distribution of known exoplanets.

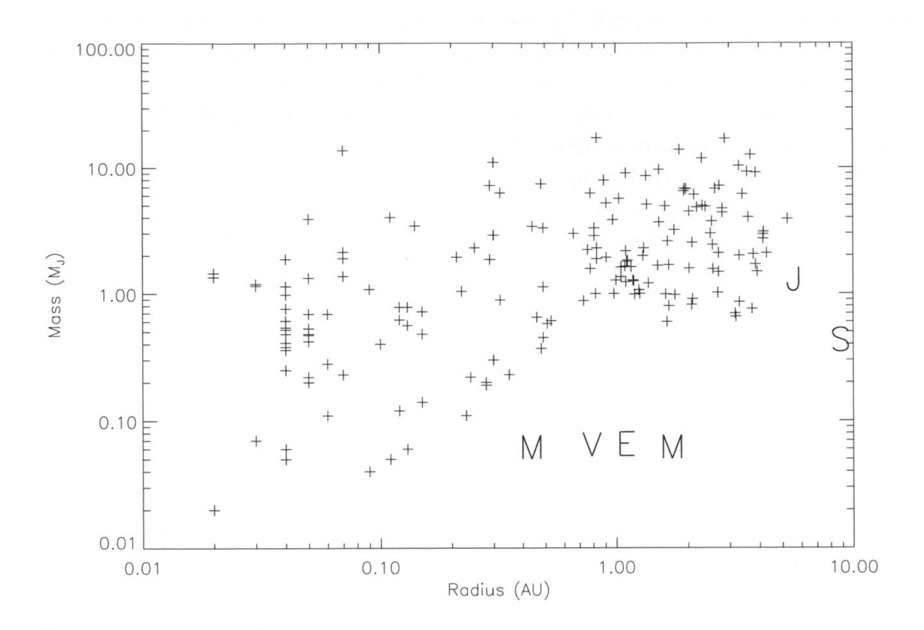

Fig. 7.3. Distribution of mass and radius for known exoplanets. Solar System planets are indicated by letter.

Such planets would be eminently detectable using the radial velocity method so we can be confident that they really aren't there. A possible explanation for this is that the migration mechanism of massive planets (Sect. 7.2.4) is either inefficient within 0.3 AU or too efficient and thus that massive planets straying with in 1 AU fall all the way into the star (Marcy et al., 2003).

There is a massive and uniform spread in the eccentricities of EGPs between 0 and 0.9 (Fig. 7.4), which suggests that there is a common mechanism for pumping the eccentricity of extrasolar planets. It can also be seen from Fig. 7.4 that the eccentricity distribution for planets in multiple-planet systems is indistinguishable from that for single planet systems. For the multiple planet systems known, eccentricity pumping may result from planets migrating in their circumstellar disc, leading to occasional mutual capture and resonance. Subsequent close encounters may lead to scattering and ejection of planets. This scenario explains the orbital resonances commonly seen in multiple-planet systems and also the occurrence of 'hierarchical' systems (ones with only a few, widely separated planets), where some of the planets have presumably been ejected. Single planet systems may be the end result of such interactions, where all other giant planets have been lost through ejection.

The analysis of the metallicity of stars which have planetary companions is very revealing. It is found that most known EGPs orbit stars with a metallicity equal to, or greater than that of our Sun (Sudarsky et al., 2003). Valenti (2003) have extended this to explore the distribution of the [Fe/H] ratio of host stars and find that the distribution rises rapidly at the high metallicity end. A similar conclusion is reached by Santos et

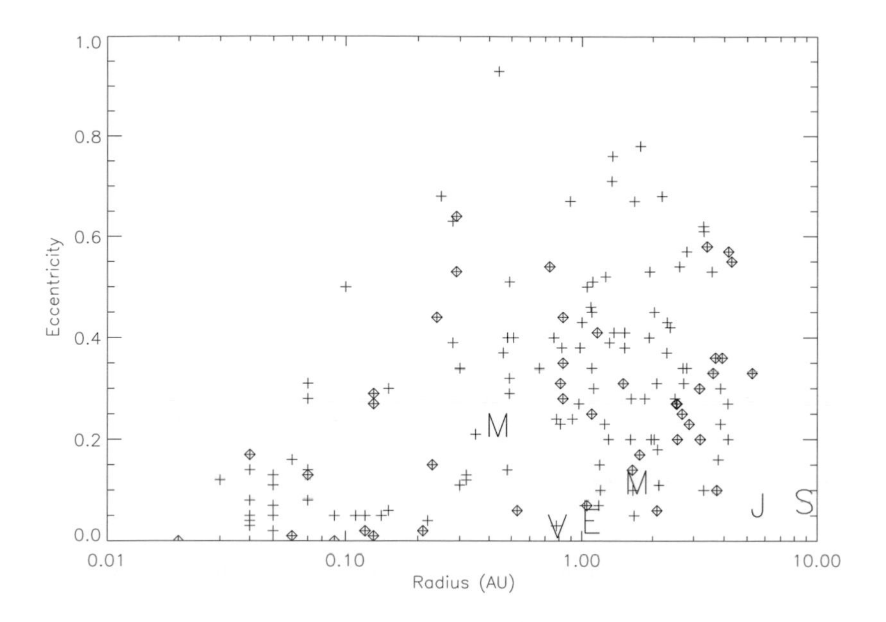

Fig. 7.4. Distribution of eccentricity and radius for known exoplanets. In this plot Solar System planets are indicated by letter and planets in multi-planet systems are indicated by diamonds.

al. (2004). These observations strongly suggest that the presence of dust in proto-stellar nebulas is very important for the formation of planets.

7.2.2 Transit detections

For extrasolar planets, there is a small, but finite chance that the orbital inclination i will be very close to 90° and thus that a planet will periodically pass between the planet's star and the Earth. If the planet is sufficiently large, then the drop of intensity of the starlight can be detected and used to determine both i and also the radius of the planet.

The first published detection of a planetary transit (using the STARE transit camera, Charbonneau et al. (2000), Sect. 7.3.2) was of the planet HD 209458 b, which orbits its star at a distance of 0.046 AU in a period of 3.5 days (Henry et al., 2000). The transit was observed the next year with the Hubble Space Telescope (HST) (Fig. 7.5) and Brown et al. (2001) concluded, from the transit depth, that the planet had a radius of 1.35 R_J (where R_J is the radius of Jupiter).

Assuming HD 209458 b to be typical, and until more transits of this type are observed there is no reason to think otherwise, these observations showed that the massive, close-orbiting planets discovered by the radial velocity survey were not just rocky cores, but large Jupiter-sized objects. The radius observed is considerably larger than that expected from a planet cooling in isolation and Burrows et al. (2000) proposed that irradiation from the star inhibits convection and thus cooling/contraction. This idea was developed by Bodenheimer et al. (2001) and Showman (2002).

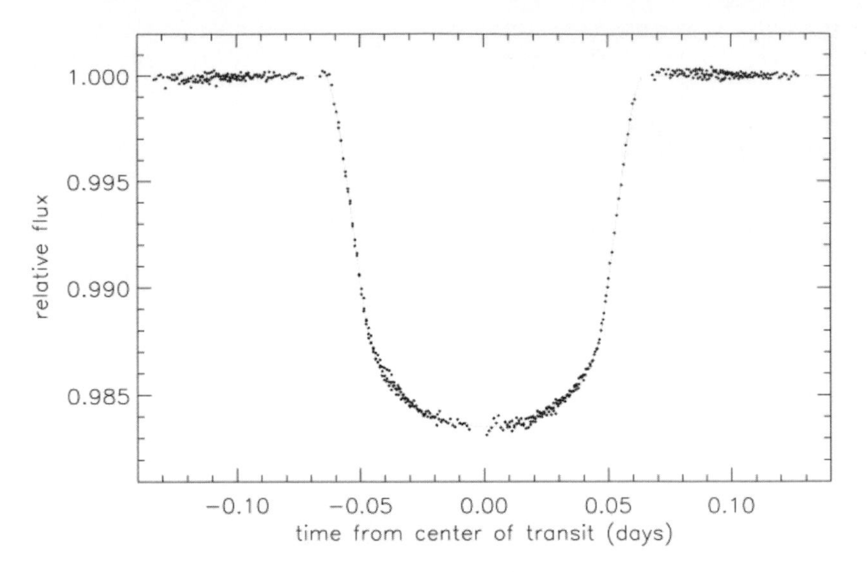

Fig. 7.5. HST observation of transit of HD 209458 b (Brown et al., 2001)

A number of other extrasolar planetary transits have been observed since 1999, using projects such as OGLE (Sect. 7.3.5). Most lead to a dip in intensity of the order of 1%, and at these levels care must be taken to ensure that phenomena such as sunspot variations are not mistaken for planet detections.

Transit Spectroscopy

Soon after the first transit of HD 209458 b was observed, it was realised that observations at a number of different wavelengths might be used to infer the atmospheric transmission of the planet's atmosphere, since a planet's effective cross-sectional area will be larger at wavelengths where its atmosphere is more strongly absorbing than at others. Just such a study is reported by Charbonneau et al. (2002) who used HST observations near 600 nm to search for the atmospheric sodium absorption lines predicted for 'Hot Jupiters' by radiative transfer models such as Sudarsky et al. (2003). The absorption line was duly detected, the first ever detection of an exoplanetary atmosphere, although the magnitude of the absorption was found to be less than predicted by cloud-free radiative transfer models suggesting that clouds high in the atmosphere of this planet reduce the absorption band depth. Brown et al. (2002), Richardson et al. (2003a) and Richardson et al. (2003b) extended this campaign to the infrared, searching for CO, H_2O and CH_4 absorption, and recently Deming et al. (2005) detected a weak absorption due to CO at 4325 cm^{-1} and also suggested the presence of a high level cloud at, or above, 3.3 mbar.

In addition to direct detection of atmospheric absorption during transits, a gas giant orbiting as close to its star as HD 209458 b will get very hot in its upper atmosphere leading possibly to exospheric loss. Vidal-Madjar et al. (2003) report HST observations of atomic hydrogen absorption of starlight during several transits of HD 209458 b. They interpret this observation as being due to absorption by hydrogen atoms that have exospherically escaped the planet's atmosphere and are now beyond the Hill radius[1] of the planet. They further conclude that if the timescale for this evaporation is comparable to the lifetime of the stellar system then it may explain why so few 'Hot Jupiters' are found orbiting with periods less than $\sim$ 3 days. More recent HST observations by Vidal-Madjar et al. (2004) have also detected exospherically escaping carbon and oxygen atoms. Such atoms should be too heavy to escape by the Jean's mechanism, responsible for the hydrogen escape, and instead Vidal-Madjar et al. (2004) suggest that hydrodynamic escape (or 'blow-off') is responsible, whereby the outward flow of exospherically escaping hydrogen atoms carry with them heavier atoms such as carbon and oxygen.

7.2.3 Atmospheres and predicted spectra of EGPs

The temperatures in the upper atmospheres of planets are governed by the balance between incoming stellar flux and outgoing thermal radiation. Hence, planets orbiting close to their star would be expected to have elevated atmospheric temperatures. Furthermore, heating of an atmosphere from above tends to lead to an isothermalisation of the temperature profile, which instead of falling quasi-linearly with height, following an adiabatic lapse rate, would instead decrease much more slowly, leading to a stable radiative layer. In this high temperature layer, the scale height will be large and thus the atmosphere will be vertically distended, an effect only amplified by the reduction of g with altitude.

Modelling the temperature profile in these circumstances is difficult, but a number of studies have been conducted, such as Sudarsky et al. (2003), who conclude that EGPs may be naturally split into five categories depending on their effective mean upper atmospheric temperatures, which are predicted to have very different visible/infrared spectra both due to mean atmospheric temperatures and also due to different upper atmospheric compositions and clouds. The classes range from the Class I to Class V, with Class I being the 'Jovian' class of EGPs, with effective radiating temperatures $T_{\rm eq} \leq 150$K and orbital distances of a few AU, to Class V 'hot roasters' (or 'Hot Jupiters') such as 51 Peg b with effective radiating temperatures $T_{\rm eq} \geq 1400$K and orbital distances $\sim$ 0.05 AU. For Class I EGPs, the heavy elements are predicted to be in their low temperature, fully-hydrogenated form and thus strong methane and ammonia molecular absorption is expected, together with ammonia clouds. For Class V EGPs, strong absorption in the visible spectrum due to sodium and potassium is expected, and clouds of silicate or iron may also condense, greatly affecting the calculated spectra.

Guillot and Showman (2002) consider the state of 'Hot Jupiter' atmospheres and conclude that the observed large radius of the planet HD 209458 b (previous section) requires much higher temperatures in the deep atmosphere than can be accounted for from formation energy. They suggest that a small fraction ($\sim$ 1%) of the stellar radiation is transformed into kinetic energy in the atmosphere and subsequently converted to thermal energy by processes at deep pressures. Showman and Guillot (2002) then go on to consider atmospheric circulations for these 'Hot Jupiters', which are likely to be synchronised by tidal interactions leading to day-night temperature differences of

[1] The Hill radius gives the limit of the gravitational sphere of influence of a body in orbit about another heavier body, in this case the central star.

$\sim$ 500K, generating winds with speeds up to $\sim$ 2km/s. They conclude that significant departure from chemical equilibrium is expected and thus that radiative transfer models assuming homogeneous conditions, such as Sudarsky et al. (2003), are oversimplified.

7.2.4 Implications for Solar System formation models

Core Accretion Theory

The core accretion theory is the most commonly accepted theory of how all planets, including the giant planets, came to form in our Solar System (e.g. Irwin, 2003). In this model, planet growth starts with the concentration of the solid material in a protosolar nebula into a sheet in the disc plane about the protostar. These dust grains slowly clump together, eventually forming small bodies of size 0.1–10 km called planetesimals. Once planetesimals reach a size of the order of 10 km, their gravitation starts to become significant leading to the runaway growth of larger planetesimals, called embryos, at the expense of smaller ones in a time frame of perhaps 500,000 years after the formation of the star. Typical modelled embryo masses at 1 AU are of the order of 0.1 M_{Earth}, while at 5 AU they are of the order of 10 M_{Earth}. Once the outer Solar System embryos reach a mass of approximately 10 M_{Earth}, they start being able to trap the nebula gas itself, plus remaining planetesimals, in a very slow second stage of accretion lasting between 1 and 10 million years. Eventually the mass of some of the planets is predicted to become so large that the remaining nebula gas becomes gravitationally unstable and a third phase of accretion takes place with the local nebula rapidly collapsing onto the planet.

The timing of these phases explains the differences between the giant planets of our Solar System. Jupiter is predicted to have reached the final gravitational collapse stage about 1.5 million years after the formation of the Sun (Hersant et al., 2001) and Saturn after something like 11 million years, when the nebula was less dense. Jupiter is thus more massive than Saturn and has lower abundances of heavy elements because it was able to collapse more nebula gas. As for Uranus and Neptune, it would appear that before they could reach their critical mass, the remnants of the circumsolar disc had been dissipated, either by photo-evaporation by other stars, or blown away by the T-Tauri phase of the Sun, leading to these planets' relatively low abundances of hydrogen and helium. Hence the bulk differences between the giant outer planets, both in terms of overall mass, and proportion of heavy elements, are all elegantly explained.

Gravitational Instability

Although the core accretion model fits the observed characteristics of our Solar System very well, there are concerns that it may be too slow. Haisch et al. (2001) have estimated the ages of circumstellar discs about stars in nearby clusters by analysing the mean colour of the stars in these clusters. They find that the circumstellar discs appear to evaporate long before the T-Tauri phase is reached and that half of the stars in these clusters lose their discs within 3 Myrs, with a mean overall lifetime of 6 Myrs. Similarly, Briceño et al. (2001) find that stars older than 10 Myrs do not have massive, optically thick discs.

A competing view of giant planet formation is that if protostellar discs are dense enough they may become gravitationally unstable and thus giant planets may collapse directly from the disc in the early period of the circumstellar disc evolution (Boss, 1997, 2002, 2004; Mayer et al., 2002). The most widely used criterion for assessing the gravitational stability is the Toomre stability parameter Q (Toomre, 1964; Pickett and Lim, 2004), which depends on a number of factors such as the disc surface density. If the Toomre Q parameter is less than approximately unity, a disc can become unstable to the rapid growth of a spiral structure which can then clump together to form giant planets. However, although this process is fast, the models have considerable difficulty in actually condensing a planet because the disc does not appear to cool rapidly enough (Rice et al., 2003; Pickett and Lim, 2004), although Boss (2004) suggests that convective overturning may assist this. Another drawback of the model is that the condensed planets would all have a composition similar to that of the central star, which does not fit at all well with observations of the Solar System giant planets. The final drawback of these models is that they are not sensitive to the abundance of dust in the nebula, and thus the process should be equally likely around metal-poor stars as metal-rich. This goes against the observation discussed earlier that the exoplanets found so far are discovered preferentially around metal-rich stars.

Migration

The giant planets of our Solar System and of other stars are unlikely to be at the same distance from the Sun as when they initially formed due to gravitational interaction with other growing planetesimals (leading to their ejection from the Solar System) and frictional interactions with the accretion disc itself. Conditions in the early circumstellar disc were probably very turbulent, perhaps including MHD effects (Papaloizou et al., 2004) and this turbulence would have led to a net transfer of mass outward in the outer part of the disc, and inwards toward the Sun nearer the centre, with a dividing radius at around 10 AU (Ida et al., 2000).

Until recently, migration of planets in these turbulent discs was thought to be restricted to two main types. Type I migration is expected when the planet mass is small, such that its Hill radius is much smaller than the disc thickness. In such cases a protoplanet will tidally interact with the disc via Lindblad resonances (Ward, 1986, 1997) and is modelled to migrate rapidly. Larger planets, however, appear to be controlled by Type II migration, where the planetary mass is large enough to open up a gap in the disc, splitting it up into an inner and outer part. The planet is then locked in with long term evolution of the disc (Lin and Papaloizou, 1986a,b) and slowly migrates inwards toward the star (assuming it is within the critical orbital distance described above). In this model, if the planet does not migrate all the way inwards before the disc dissipates, it is more likely to have an orbital distance greater than that of 'Hot Jupiters' (0.05–0.2 AU), which is consistent with the observed lack of 'Hot Jupiters' with mass $> 4M_J$ orbiting within 0.3 AU of their stars. However, objects that have undergone this type of migration would be expected to be heavier for smaller orbital radii, since they would have had more time to accrete the disc material, which is not observed amongst known EGPs. Hence 'Hot Jupiters' do not appear to be explained by either Type I or Type II migration. Instead, 'Hot Jupiters' may be explained by an intermediate, or 'runaway' mode of migration (Masset and Papaloizou, 2003) where, for certain combinations of planet and disc mass, a planet very rapidly moves either toward or away from the star, changing its orbital radius by a factor of 2 or more. The runaway does not carry on indefinitely since eventually the planet becomes so large that it clears out its 'feeding zone' and is then governed by slow Type II migration.

Finally, one solution to the apparent slowness of the core accretion model outlined above is proposed by Rice and Armitage (2003) who suggest that turbulent fluctuations in a protoplanetary disc cause migrating giant planets to perform more of a random walk, with amplitude of a few tenths of an AU, than a steady drift towards the star, resulting in an acceleration of the accretion rate by almost an order of magnitude.

7.3 Further Methods of Detecting Extrasolar Planets

Now that Jupiter-like planets are being regularly discovered about other stars, how long might it be before more earth-like planets are found? Ida and Lin (2004a) and Ida and Lin (2004b) have modelled the expected mass-radius distributions of planetary systems using Monte Carlo techniques and suggest that considerable numbers of both terrestrial and ice giant planets remain to be discovered, although they predict a 'Planet Desert' in the distribution for planet masses of between 10–100 M_{Earth} at a distance < 3 AU, similar to the 'Brown Dwarf Desert' mentioned earlier.

There are a number of ways of trying to detect the existence of an extrasolar planet, most indirect. All the techniques have their own advantages and disadvantages although to date the radial velocity technique has proven to be most effective. The different selection effects of these detection methods are summarised in Fig. 7.6, on which are

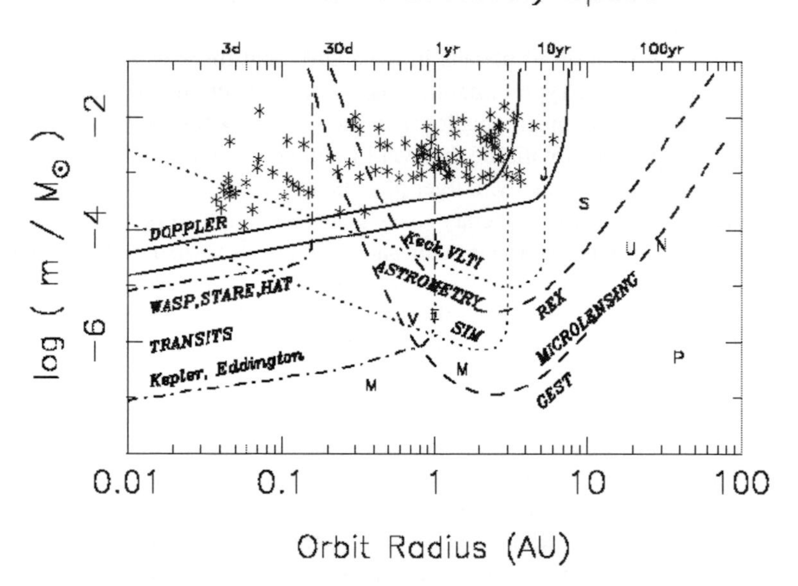

Fig. 7.6. Selection effects of different exoplanet detection programmes. Solar System planets are indicated by letter. (Courtesy of K. Horne)

plotted the mass and orbital radii of known EGPs, together with characteristics of the Solar System planets.

As can be seen, the radial velocity (or Doppler) technique is best for heavy, close orbiting planets, and thus the planets found so far are clustered in the top left corner of the figure. The limit of detectability of existing measurements is shown, together with the expected improvement due to ever increasing sensitivity and longer observation runs. Transit observations also favour shorter periods, but can detect lighter planets, especially the planned space-based missions (Sect. 7.3.2). The selection limits of astrometric observations (Sect. 7.3.3), both for ground-based programmes such as at the Keck Observatory and the VLT, and for space-based missions such as the forthcoming Space Interferometry Mission (SIM), are expected to start probing into the terrestrial planet region of the diagram, especially the space-based missions. This technique is complemented by the microlensing technique (Sect. 7.3.5) which similarly is more sensitive for space-based proposals (such as GEST), than terrestrial-based ones such as REX indicated here. We will now consider future observations using these methods and what may reasonably be expected of them.

7.3.1 Radial velocity programmes

The radial velocity method has reaped rich rewards in the last ten years and with improving sensitivities, and data sets covering ever longer periods, we may expect the discovery of many more EGPs. There are a number of programmes currently under way, including the Anglo-Australian Planet Search (e.g. Carter et al., 2003), the California and Carnegie Planet Search, ELODIE (Naef et al., 2004) and CORALIE (Mayor et al., 2004).

7.3.2 Transit programmes

A number of transit surveys have been planned for the next few years - both ground- and space-based. They may be conveniently split into two categories: 1) Deep surveys, which have small pixel size, can see faint stars, but do not cover a wide area of sky; and 2) Wide surveys which cover a wide area of the sky with large pixel size, but cannot see fainter stars. The *a priori* probability of transit detection is the ratio of the diameter of a star to the diameter of a planet's orbit (Lewis, 2004). For the Sun-Jupiter system this is $1.4 \times 10^{11}/5.2 \times 10^{13} = 1.8 \times 10^{-3}$. Hence, transit surveys need to observe lots of star systems to have sufficient probability of planetary detection and so both survey methods attempt to view many stars simultaneously, taking special care not to confuse real planetary transits with other phenomena (e.g. Mandushev et al., 2005).

One ground-based wide survey is STARE which uses a 99-mm aperture field-flattened Schmidt telescope. STARE made the first observation of a planetary transit (HD 209458 b), and recently discovered its first planet, TrES-1. Another ground-based wide-survey is SuperWASP, which has two facilities: one based in the Canary Islands and another soon to be operating in South Africa. Both instruments are comprised of five 11-cm aperture wide-angle cameras and are developments of the WASP0 prototype instrument (Kane et al., 2004). Both SuperWASPs will be able to monitor 10,000 stars simultaneously over a $15° \times 15°$ field of view.

Space-based wide-survey observations are predicted to be both more sensitive and less prone to false identifications of planetary transits. One such mission is COROT, which is a French-led project to place a small 0.27-m telescope into orbit to study astroseismology and also detect extrasolar planet transits. COROT is due for launch in early 2006 and will observe an area of the sky of size $2.8° \times 2.8°$ for $2\frac{1}{2}$ years. The US-led Kepler mission is a Schmidt telescope with 1.4-m primary mirror and a 0.95-m aperture, due for launch in June 2008. Kepler will continuously and simultaneously monitor the brightness of approximately 100,000 A–K dwarf (main-sequence) stars brighter than 14th magnitude in the Cygnus-Lyra region along the Orion arm, for a period of 4 years.

Deep transit surveys will be conducted by microlensing programmes, which are outlined in Sect. 7.3.5.

7.3.3 Astrometry

Given a sequence of observations of a star's position of sufficiently high accuracy relative to the celestial sphere, the reflex motion[2] of the star caused by the orbit of a planet around it can be detected. This can be used to determine both the absolute mass and orbital inclination of a planet. Considering the motion of the star and planet about their common centre of mass we can see that the reflex amplitude of the star is $a_* = a_p M_p / M_*$, where a_* and a_p are the distances from the centre-of-mass to the star and planet respectively. Thus, this method is most effective for large mass planets orbiting at some distance from their parent stars. In addition, since what is actually measured is the angular position of the star, the method is clearly best for planetary systems within a few parsecs of the Earth.

The accurate measurement of a star's position over a number of years is a challenging task. Current optical systems have an absolute accuracy of a few milliarcseconds. However this precision can be improved through the use of long-baseline interferometry. The VLT and Keck currently have programmes to do this and are expected to achieve accuracies of 30 μas (microarcseconds), and 10 μas respectively, which should be sufficient to observe the reflex motion of the stars of several EGPs already discovered. In addition, there are two space missions planned to exploit this technique. The NASA SIM (Space Interferometry Mission) is due for launch in 2011 and will be able to achieve 1 μas accuracy, while the ESA GAIA spacecraft, which is a follow-up to ESA's Hipparcos mission, is also due to launch in 2011. Although not an interferometric instrument, GAIA aims to observe 1 billion stars with magnitude greater than 20, with accuracy of 10–20 μas at magnitude 15.

7.3.4 Direct optical detection

The methods of detection outlined so far, with the exception of transits, detect the reaction of the star to a planet's motion, rather than the planet itself. However, there are a number of techniques currently being employed or developed to directly detect either the reflected starlight or thermal emission of EGPs about nearby stars.

[2] The reflex motion of the star is caused by both it and the planet orbiting their common centre of mass.

Doppler Spectral Separation

As a planet orbits a star, part of the starlight will be reflected by the planet towards the observer. This component will be Doppler-shifted by an amount depending on the planet's orbital velocity ($\sim$ 100 km/s), rather than the star's ($\sim$ 10 m/s) and can be extracted using very high-resolution ground-based spectroscopy and correlation techniques. Collier Cameron and Leigh (2004) review the current status of a number of direct planetary detections achieved by this technique. Assuming the planetary radius is known, the visible planetary albedo can be estimated. For example, the albedo of τ Bootis b (Collier Cameron et al., 1999; Charbonneau et al., 1999) is estimated by Leigh et al., (2003a) to be less than 0.39. Similarly, the albedoes of υ Andromeda b and HD 75289 b are estimated to be less than 0.3 (Collier Cameron et al., 2002) and 0.14 (Leigh et al., 2003b) respectively. These albedoes are much less than Jupiter's (0.5).

Differential Direct Detection

Models of the expected spectra of EGPs (e.g. Sudarsky et al., 2003) show that the reflected sunlight from such planets will be significantly affected by absorption of atmospheric constituents such as sodium and carbon monoxide, whereas the stellar spectrum is expected to be smoothly varying. Hence, these absorption features provide a possible means of discriminating between the light reflected by a planet and the direct stellar light. Wiedemann et al. (2001) report just such a detection of the 3-μm methane absorption of τ Bootes b. There are other programmes in development, such as ARIES (Arizona Infrared Imager and Echelle Spectrograph) on the 6.5m Multi-Mirror-Telescope (MMT), and TRIDENT on the 3.6m Canada-France-Hawaii-Telescope (CFHT). Both instruments observe the edge of a methane absorption band between 1.5 and 1.8 μm.

Interferometric Imaging

Using two telescopes, separated by a long baseline of precisely controlled optical length D, the beams may be combined with a phase difference of π to completely eliminate the light from the central star. Constructive interference will then occur at a number of angles θ where $D \sin\theta = (2n + 1)\lambda/2$ and n is an integer. By varying the baseline D (assuming fixed wavelength λ), a range of constructive interference angles can be examined to attempt to detect either the weak stellar reflection or thermal emission of an extrasolar planet. The principle was tested successfully on MMT prior to its decommissioning and upgrade in 2000. Both the Keck Observatory and VLT have long baseline interferometric observation programmes. One particularly interesting project is the Large Binocular Telescope Interferometer (LBTI), currently being built in Arizona, which will use adaptive optics and a beam combiner including a dielectric material to correct for colour dependence of light interference. LBTI will operate in the infrared (3-5 μm) and should be able to detect planets further than 0.03" from their stars. Nulling interferometry is also the planned mode of the proposed ESA/Darwin space mission, and a possible mode of the proposed NASA Terrestrial Planet Finder (TPF-IR). In these mission plans, a fleet of large telescopes would fly in formation and the light combined in a central hub using precisely controlled phase delays. Due to their very long baselines, low temperatures and no atmospheric absorption, these missions will be able to not only directly detect extrasolar planets in the infrared, but also measure their emission spectra, allowing the composition of their atmospheres to be determined.

Coronagraphic Imaging

This is a direct imaging technique, suitable only for space, where the light from the central star is eliminated using a mask in the focal plane. The method is used in solar studies to study the corona and prominences of the Sun's atmosphere, from which its name is derived. The technique may be used by a version of the proposed NASA Terrestrial Planet Finder (TPF-C). In addition, the new James Webb Space Telescope will house the NIRCam instrument, which includes a coronagraphic module, operating from 2–5 μm. This system will be capable of 10^8–10^9 high contrast imaging for separations > 0.1". In addition, tunable narrow-band filters will allow the measurement of spectra from 2.5–4.5 μm at low resolution.

7.3.5 Microlensing

For several years now there have been campaigns to observe galactic bulge microlensing events, with a view to searching for dark matter and extrasolar planets. In this technique, light from a distant (source) star is observed as another star at intermediate distance (the lens star) passes close to, or in front of it. Light from the source star is gravitationally bent around the lens star and thus its apparent magnitude changes during the event. Two such campaigns are OGLE (Udalski, 2003) and MOA (Bond et al., 2001). In addition to lensing events, such programmes are also sensitive to planetary transits and to date, OGLE has detected the transits of five previously unknown extrasolar planets.

In 2003, both observatories observed a remarkable microlensing event shown in Fig. 7.7 where, in addition to the central peak in source star brightness due to the gravitational lensing of the lens star, two additional sharp peaks were observed which are interpreted as being due to the microlensing of a planetary companion to the lens star. Bond et al., (2004) conclude, assuming the lens star to be a main sequence M dwarf, that the planet has a mass of 1.5 M_J, and orbits the lens star at a distance of approximately 3 AU.

For future observations it can be seen from Fig. 7.6 that gravitational lensing is the only detection method that is capable of sensing terrestrial planets orbiting some distance from their stars (dubbed 'Cool Earths'). In addition to the continuation of the OGLE and MOA campaigns, other ground-based campaigns include PLANET, which is a collaboration of telescopes in the southern hemisphere observing since 1995, and REX, which is a proposal to link a series of 2 m robotically controlled telescopes around the world to search for exoplanet microlensing events. The sensitivity of microlensing campaigns to 'Cool Earths' would be further advanced by placing the telescope in space and GEST (Galactic Exoplanet Survey Telescope) is just such a proposed mission.

7.3.6 Conclusions

The discovery of extrasolar giant planets has been one of the most exciting astronomical discoveries in the last ten years. It has proved beyond doubt that planetary formation is

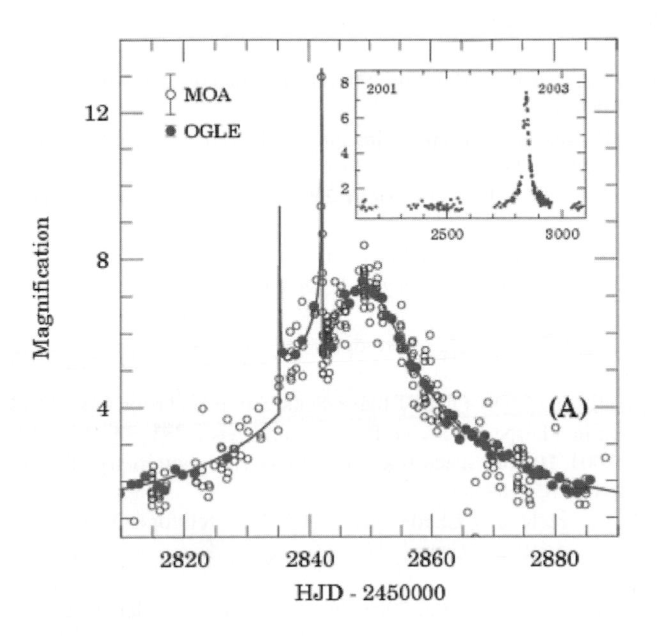

Fig. 7.7. Observation of gravitational microlensing by a planet by OGLE (Bond et al., 2004). Inset panel shows all OGLE data from 2001 to 2003, while the main figure shows a close-up of the data for 2003 for both OGLE and MOA.

a common by-product of star formation, although some of the systems discovered so far appear peculiarly exotic compared to our own. Analysing the characteristics of these systems is already giving us a new insight into how our own Solar System formed and has placed very important constraints on planetary formation theories. With ~ 170 giant extrasolar planets discovered, attention is starting to switch to the detection of more earth-like planets. It may thus come as a surprise to discover that three earth-mass extrasolar planets have already been discovered. However, they have not been discovered about a main sequence star but instead have been observed orbiting the pulsar PSR 1257+12 (Wolszczan and Frail, 1992; Wolszczan, 1994).

If giant planets are apparently so common about other stars, it is natural to ask whether terrestrial planets might be also. One is then naturally led to wonder if terrestrial planets are commonplace, how many 'habitable earths' might exist and if so, how many of these might be inhabited? A number of programmes planned for the next 10–20 years will be able to address these questions. Of particular interest are the NASA TPF-IR and ESA Darwin missions, which through their infrared nulling interferometry approach will be able to measure the thermal emission spectra of any planets discovered. The Earth's atmosphere has been substantially modified by the presence of life, which produces high levels of oxygen. This is photolysed to form ozone in the upper atmosphere which has a very clear signature in the thermal infrared. It is believed that the simultaneous detection of both ozone and methane in an extrasolar planetary atmosphere would be a strong indicator of the development of life similar to our own. Such a discovery would be significant indeed.

References

Bodenheimer, P., D.N. Lin & R.A. Mardling 2001, On the Tidal Inflation of Short-Period Extrasolar Planets, *ApJ*, **548**, 466.

Bond, I.A. et al. 2001, Real-time difference imaging analysis of MOA galactic bulge observations during 2001, *MNRAS*, **327**, 868.

Bond, I.A. et al. 2004, OGLE-BLG-235/MOA 2003-BLG-53: A planetary microlensing event, *ApJL*, **606**, 155.

Boss, A.P. 1997, Giant planet formation by gravitational instability, *Science*, **276**, 1836.

Boss, A.P. 2002, Evolution of the solar nebula. V. Disk instabilities with varied thermodynamics, *ApJ*, **576**, 462.

Boss, A.P. 2004, Convective cooling of protoplanetary disks and rapid giant planet formation, *ApJ*, **610**, 456.

Briceño, C. et al. 2001, The CIDA-QUEST Large-Scale Survey of Orion OB1: Evidence for Rapid Disk Dissipation in a Dispersed Stellar Population, *Science* **291**, 93.

Brown, T.M. et al. 2001, Hubble Space Telescope time-series photometry of the transiting planet of HD 209458, *ApJ*, **552**, 699.

Brown, T.M., K.G. Libbrecht & D. Charbonneau 2002, A Search for CO Absorption in the Transmission Spectrum of HD 209458b, *Publications of the Astronomical Society of the Pacific*, **114**, 826.

Burleigh, M., F. Clarke & S. Hodgkin 2003, Imaging extrasolar planets around nearby white dwarves. In: *Scientific Frontiers in Research on Extrasolar planets*, vol 294, ed by D. Deming & S. Seager (ASP Conference Series) pp. 111–115.

Burrows, A. et al. 2000, On the Radii of Close-in Giant Planets, *ApJL*, **534**, 97.

Carter, B.D. et al. 2003, A planet in a circular orbit with a 6 year period, *ApJL*, **593**, 43.

Charbonneau, D. et al. 1999, An upper limit on the reflected light from the planet orbiting the star tau Bootis, *ApJL*, **522**, 145.

Charbonneau, D. et al. 2000, Detection of Planetary Transits Across a Sun-like Star, *ApJL*, **529**, 45.

Charbonneau, D. et al. 2002, Detection of an extrasolar planet atmosphere, *ApJ*, **568**, 377.

Chauvin, G. et al. 2005, Giant planet companion to 2MASSW J1207334-393254, *Astron. & Astrophys.*, **438**, L25.

Collier Cameron, A. et al. 1999, Probable detection of starlight reflected from the giant planet orbiting tau Bootis, *Nature*, **402**, 751.

Collier Cameron, A. et al. 2002, A search for starlight reflected from nu And's innermost planet, *MNRAS*, **330**, 187.

Collier Cameron, A. 2002, What are hot Jupiters made of? *Astron. & Geophys.*, **43**, 4.21.

Collier Cameron, A. & C. Leigh 2004, Tomographic studies of exoplanet atmospheres, *Astron. Nachr.*, **325**, 252.

Correia, A.C.M. et al. 2005, The CORALIE survey for southern extra-solar planets. XIII. A pair of planets around HD 202206 or a circumbinary planet?, *Astron. & Astrophys.*, **440**, 751.

Deming, D. et al. 2005, A new search for carbon monoxide absorption in the transmission spectrum of the extrasolar planet HD 209458b, *ApJ*, **622**, 1149.

Fischer, D.A. & J.A. Valenti 2003, Metallicities of stars with extrasolar planets. In: *Scientific Frontiers in Research on Extrasolar planets*, vol 294, ed by D. Deming & S. Seager (ASP Conference Series), pp. 117–128.

Grevesse, N. & A.J. Sauval 1998, Standard Solar Composition, *Space Sci. Rev.*, **85**, 161.

Guillot, T. & A.P. Showman 2002, Evolution of "51 Pegasus b-like" planets, *A&A*, **385**, 156.

Haisch, K.E., E.A. Lada & C.J. Lada 2001, Disk frequencies and lifetimes in young clusters, *ApJL*, **553**, 153.

Henry, G.W. et al. 2000, A transiting “51-Peg-like” planet, *ApJL*, **529**, 41.
Hersant, F., D. Gautier & J.M. Hure 2001, A Two-dimensional Model for the Primordial Nebula Constrained by D/H Measurements in the Solar System: Implications for the Formation of Giant Planets, *ApJ*, **554**, 391.
Hubbard, W.B. 1997, Jupiter: Interior Structure. In: *Encyclopaedia of the Planetary Sciences* (Chapman & Hall).
Ida, S. et al. 2000, Orbital migration of Neptune and orbital distribution of trans-neptunian objects, *ApJ*, **534**, 428.
Ida, S. & D.N.C. Lin 2004a, Toward a deterministic model of planetary formation. I. A desert in the mass and semimajor axis distributions of extrasolar planets, *ApJ*, **604**, 388.
Ida, S. & D.N.C. Lin 2004b, Toward a deterministic model of planetary formation. II. The formation and retention of gas giant planets around stars with a range of metallicities, *ApJ*, **616**, 567.
Irwin, P.G.J. 2003, *Giant planets of our solar system: atmospheres, composition and structure*, Springer-Praxis.
Jorissen, A., M. Mayor & S. Udry 2001, The distribution of exoplanet masses, *A&A*, **379**, 992.
Kane, S.R. et al. 2004, Results from the Wide-Angle Search for Planets prototype (WASP0) – I. Analysis of the Pegasus field, *MNRAS*, **353**, 689.
Leigh, C. et al. 2003a, A new upper limit on the reflected starlight from tau Bootis b, *MNRAS*, **344**, 1271.
Leigh, C. et al. 2003b. A search for starlight reflected from HD 75289b, *MNRAS*, **346**, L16.
Lewis, J.S. 2004, *Physics and Chemistry of the Solar System*, 2nd edition, Elsevier Academic Press.
Lin, D.N.C. & J.C.B. Papaloizou 1986a, On the tidal interaction between protoplanets and the protoplanetary disk. III. Orbital migration of protoplanets, *ApJ*, **309**, 846.
Lin, D.N.C. & J.C.B. Papaloizou 1986b, On the tidal interaction between protoplanets and the primordial solar nebula. II. Self-consistent nonlinear interaction, *ApJ*, **307**, 395.
Mandushev, G. et al. 2005, The challenge of wide-field transit surveys: The case of GSC 09144-02289, *ApJ*, **621**, 1061.
Marcy, G.W. et al. 2003, Properties of extrasolar planets. In: *Scientific Frontiers in Research on Extrasolar planets*, vol 294, ed by D. Deming & S. Seager (ASP Conference Series), pp. 1–16.
Masset, F.S. & J.C.B. Papaloizou 2003, Runaway migration and the formastion of Hot Jupiters, *ApJ*, **588**, 494.
Mayer, L. et al. 2002, Formation of giant planets by fragmentation of protoplanetary disks, *Science*, **298**, 1756.
Mayor, M. & D. Queloz 1995, A Jupiter-Mass Companion to a Solar-Type Star, *Nature*, **378**, 355.
Mayor, M. et al. 2004, The CORALIE survey for southern extra-solar planets. XII. Orbital solutions for 16 extra-solar planets discovered with CORALIE, *A&A*, **415**, 391.
Naef, D. et al. 2004, The ELODIE survey for northern extra-solar planets. III. Three planetary candidates detected with ELODIE, *A&A*, **414**, 351.
Papaloizou, J.C.B., R.P. Nelson & M.D. Snellgrove 2004, The interaction of giant planets with a disc with MHD turbulence – III. Flow morphology and conditions for gap formation in local and global simulations, *MNRAS*, **350**, 829.
Pickett, M.K. & A.J. Lim 2004, The race is not to the swift, *Astron. & Geophys.*, **45**, 1.12.
Pollack, J.B. et al. 1996, Formation of the Giant Planets by concurrent accretion of solids and gas, *Icarus*, **124**, 62.
Rice, W.K.M. et al. 2003, Substellar companions and isolated planetary-mass objects from protostellar disc fragmentation, *MNRAS*, **346**, L36.
Rice, W.K.M. & P.J. Armitage 2003, On the formation timescale and core masses of gas giant planets, *ApJL*, **598**, 55.

Richardson, L.J., D. Deming, S. Seager 2003a, Infrared Observations during the Secondary Eclipse of HD 209458b. II. Strong Limits on the Infrared Spectrum Near 2.2 *mu*m, *ApJ*, **597**, 581.

Richardson, L.J. et al. 2003b, Infrared Observations during the Secondary Eclipse of HD 209458b. I. 3.6 Micron Occultation Spectroscopy Using the Very Large Telescope, *ApJ*, **584**, 1053.

Rivera, E.J. et al. 2005, A $\sim 7.5 M_{\mathrm{Earth}}$ Planet orbiting the nearby star, GJ 876, *ApJ*, **634**, 625.

Santos, N.C., G. Israelian & M. Mayor 2004, Spectroscopic [Fe/H] for 98 extra-solar planet-host stars, *A&A*, **415**, 1153.

Showman, A.P. & T. Guillot 2002, Atmospheric circulation and tides on "51 Pegasus b-like" planets, *A&A*, **385**, 166.

Sudarsky, D., A. Burrows & I. Hubeny 2003, Theoretical Spectra and Atmospheres of Extrasolar Giant Planets, *ApJ*, **588**, 1121.

Toomre, A. 1964, On the gravitational stability of a disk of stars, *ApJ*, **139**, 1217.

Udalski, A. 2003, The Optical Gravitational Lensing Experiment. Real time data analysis systems in the OGLE-III survey, *Acta Astron.*, **53**, 291.

Vidal-Madjar, A. et al. 2003, An extended upper atmosphere around the extrasolar planet HD 209458 b, *Nature*, **422**, 143.

Vidal-Madjar, A. et al. 2004, Detection of oxygen and carbon in the hydrodynamically escaping atmosphere of the extrasolar planet HD 209458 b, *ApJL*, **604**, 69.

Ward, W.R. 1986, Density waves in the solar nebula: Differential Lindblad torque, *Icarus*, **67**, 164.

Ward, W.R. 1997, Protoplanet migration by nebula tides, *Icarus*, **126**, 261.

Wiedemann, G., D. Deming & G. Bjoraker 2001, A Sensitive Search for Methane in the Infrared Spectrum of tau Bootis, *ApJ*, **546**, 1068.

Wolszczan, A. & D.A. Frail 1992, A planetary system around the millisecond pulsar PSR1257+12, *Nature*, **355**, 145.

Wolszczan, A. 1994, Confirmation of earth mass planets orbiting the millisecond pulsar PSR B1257+12, *Science*, **264**, 538.

8 The Icy Moons of Jupiter

Richard Greenberg

Abstract. The Galilean satellites formed in a nebula of dust and gas that surrounded Jupiter toward the end of the formation of the giant planet itself. Their diverse initial compositions were determined by conditions in the circum-jovian nebula, just as the planets' initial properties were governed by their formation within the circum-solar nebula. The Galilean satellites subsequently evolved under the complex interplay of orbital and geophysical processes, which included the effects of orbital resonances, tides, internal differentiation, and heat. The history and character of the satellites can be inferred from consideration of the formation of planets and the satellites, from studies of their plausible orbital evolution, from measurements of geophysical properties, especially gravitational and magnetic fields, from observations of the compositions and geological structure of their surfaces, and from geophysical modeling of the processes that can relate these lines of evidence. The three satellites with large water-ice components, Europa, Ganymede, and Callisto are very different from one another as a result of the ways that these processes have played out in each case. Europa has a deep liquid-water ocean with a thin layer of surface ice, Ganymede and Callisto likely have relatively thin liquid water layers deep below their surfaces, and Callisto remains only partially differentiated, with rock and ice mixed through much of its interior. A tiny inner satellite, Amalthea, also appears to be largely composed of ice. Each of these moons is fascinating in its own right, and the ensemble provides a powerful set of constraints on the processes that led to their formation and evolution.

8.1 Introduction

Of Jupiter's four large Galilean satellites, three of them, Europa, Ganymede, and Callisto, have compositions that include substantial fractions of H_2O (Fig. 8.1a, Table 8.1). Each of the three can be considered icy, in that an outer layer of frozen water dominates the satellite's superficial appearance and much of its observable character. At the same time, each also has a major complement of rock and iron, which plays an important role: For example, despite its icy exterior, Europa is a predominantly rocky body; Ganymede's magnetic field requires a molten iron core; and Callisto probably contains a large fraction of rock even near its surface. Moreover, while the H_2O on each surface is frozen, there is evidence that below the surface of each of these three "icy" moons lies a layer of liquid water.

The evidence for such liquid water is strongest for Europa, where a deep ocean lies just below the surface (Cassen et al. 1979, Khurana et al. 1998, Hoppa et al. 1999, Kivelson et al. 2000) and plays a major role in continually renewing the satellite's appearance (Greenberg & Geissler 2002, Greenberg 2005). Consequently, of the three icy moons Europa is distinguished by its young, active surface, with a wide variety of tectonic and thermally driven features. The paucity of craters shows the surface to have been entirely replaced during the past 50 million years (Zahnle et al. 2003).

Table 8.1. The Icy Satellites of Jupiter.

Satellite	discovery year	semi-major axis a (1000 km)	radius (km)	density (water=1)	Moment of inertia C/MR^2	H_2O (ice + liquid)	Liquid ocean	Rock	Iron (+FeS)
Callisto	1610	1883	2410 ± 2	1.834 ± 0.003	0.355 ± 0.004	large fraction		Mixed with ice?	
Europa	1610	671	1565 ± 8	2.99 ± 0.05	0.346 ± 0.005	outer $\sim$150 km	Most of the H_2O	Mantle below H_2O	Core radius is 200–700 km
Ganymede	1610	1070	2631 ± 2	1.942 ± 0.005	0.312 ± 0.003	outer $\sim$900 km	Layer below $\sim$100 km ice	Mantle below H_2O	Core radius is 650–900 km
Amalthea	1882	181	125x73x64	0.86 ± 0.1		Mixed with variable amount of rock to depth of $\sim$1000 km	Layer below $\sim$100 km ice	Mixed with ice in outer $\sim$1000 km, mixed with iron below	Mixed with rock inside radius $\sim$1200 km

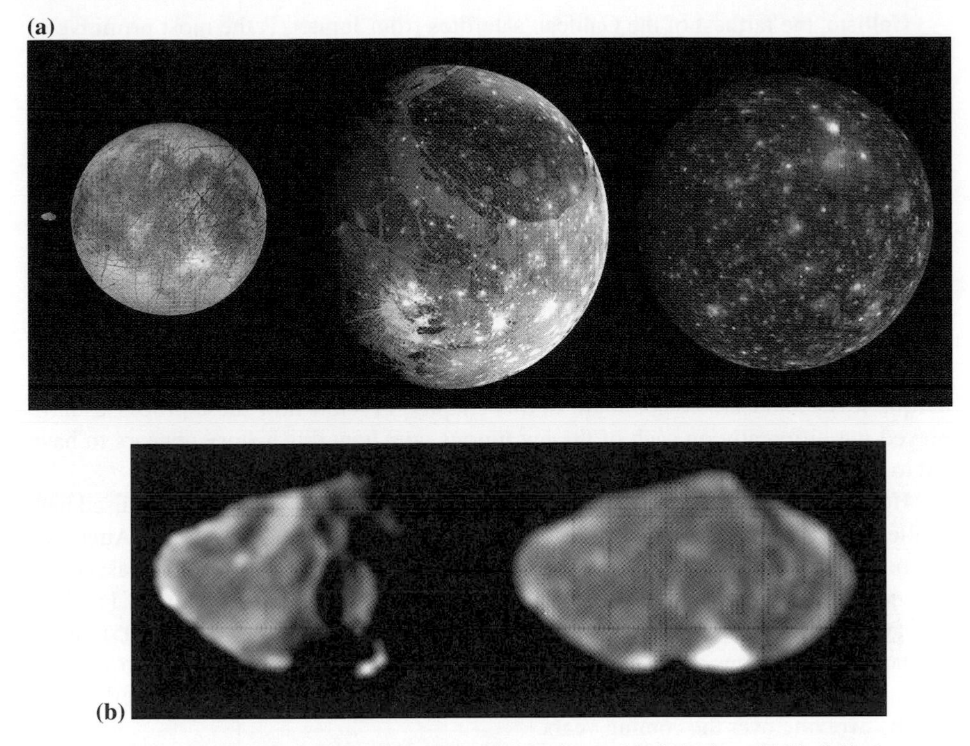

Fig. 8.1. **(a)** A color portrait of the icy satellites of Jupiter, with their sizes (Table 8.1) shown to scale. At the right is Callisto, the farthest from Jupiter, with its dark silicate veneer. With minimal endogenic activity, this surface shows at this scale the record of cratering bombardment. To the left of Callisto, at the center of the montage, is Ganymede, with 1/3 of its surface similar to Callisto's, and 2/3 consisting of younger, bright grooved terrain, strongly modified by extensional tectonic processing. To the left of Ganymede is Europa, with global scale lineaments testifying to a complex and active tectonic history, and dark splotches marking thermally driven chaotic terrain. Craters are few on Europa, indicating that this surface has continually been reprocessed by tectonics and chaos formation until very recently and may still be active. One crater, Pwyll, has a large system of bright ejecta rays, evident in this image. The tiny satellite Amalthea at the far left is the closest to Jupiter and the smallest of the group. Its elongated shape (Table 8.1)is barely visible at this scale. Amalthea was only recently recognized to be an icy body, a discovery that raises new questions about the origin of the Jupiter system. **(b)** Two views of Amalthea, which has recently been found to have a low density indicative of a substantially icy composition. The bright crater at the south (bottom) may represent excavation to purer ice, similar to the bright craters on Callisto or on the dark terrain of Ganymede.

On Ganymede, the next farthest out from Jupiter, much of the surface has been modified by tectonics, with some limited evidence for fluid exposure and flow, while the rest of the surface shows only minimal modification of any kind (Pappalardo et al. 2004). The crater record suggests that the major resurfacing occurred about 2 billion years ago, but uncertainty in that estimate admits the possibility that even this modified portion of the surface dates back to twice that age (Zahnle et al. 2003).

Callisto, the farthest of the Galilean satellites from Jupiter, is the most primitive of the three icy satellites, with its interior only partially differentiated by density (Anderson et al. 2001), and its surface dominated by impact features with little sign of endogenic activity (Moore et al. 2004). Both in its orbital location and its physical properties, Callisto is the opposite extreme from the innermost Galilean satellite, Io, which has no water or ice and is rapidly and continually resurfaced by active volcanism (Lopes and Gregg 2004).

The diversity among the Galilean satellites, especially in terms of the amount of water and its relationship with other constituents, results from the conditions of their formation at various locations and times within the early nebula that surrounded the just-forming Jupiter. Later, the orbital resonance among Io, Europa, and Ganymede drove tides that had major effects on their structure and surfaces, affecting each satellite in unique ways, and Callisto not at all. In this chapter, I review how these processes have played out differently on each of the icy moons, and how this history appears to have led to the distinctive appearance and characteristics of each one.

However, the recent discovery that another jovian satellite is also icy has raised new challenges to models of the formation and evolution of the system. Satellite Amalthea has been known for over a century, but had been assumed to be rocky on the basis of its very regular orbit closer to Jupiter than any of the Galilean satellites (Table 8.1). Now, recognition that it may be the iciest of all jovian satellites (Anderson et al. 2005), raises fundamental questions about the entire Jupiter system. Thus while this chapter reviews the current best understanding of the satellites, continuing research ensures that the story will be evolving over the coming years.

8.2 Internal Structure and Composition

Our knowledge of the bulk structure and composition of these satellites (Table 8.1) comes from observations of their size from images, of their mass and moments of inertia from their gravitational effects on spacecraft and on one another, and from our understanding of the types of materials likely to have condensed from the gaseous nebula surrounding the early Jupiter (Schubert et al. 2004). In addition, from spacecraft-based magnetometer measurements, we can infer the states of iron cores that generate satellites' magnetic fields and of electrically conducting layers (probably salty liquid water) that modify Jupiter's magnetic field.

8.2.1 Europa

Europa, with a radius of 1565 km, has a density of 2.99 ± 0.05 gm/cm^3, much greater than water (density 1 gm/cm^3, by definition) or ice which is a few percent less dense than liquid water (Schubert et al. 2004). From its measured moment of inertia (Table 8.1), we know that internal heating by radioactivity and tidal friction must have differentiated the internal composition by density and the H_2O of Europa is confined to an outer layer of thickness <170 km. The gravity constraints would allow the low-density portion to include a substantial layer of hydrated silicates (i.e. clays) below the water, in which case the H_2O layer could be as thin as 80 km. On the other hand, geochemical considerations

Fig. 8.2. Cycloidal crack patterns (chains of arcs), are ubiquitous on Europa, usually displayed in the form of double ridges lining the sides of the cracks. This image from the Voyager mission in 1979 shows beautiful, but typical, examples of cycloidal ridges, including Cilicia, Delphi, and Sidon Flexus. Arcs are typically about 100 km long, and many cycloids comprise a dozen arcs or more. Older cycloids have been sliced and diced by subsequent tectonic and thermal processing, and thus may only appear as short arcuate features or isolated cusps.

argue against the expectation of much clay. Thus most estimates for the thickness of the H_2O layer are ~150 km.

Substantial tidal heating of Europa probably provides adequate heating to keep most of this water in the liquid state, with only a relatively thin outer shell of ice (Peale et al. 1979, Squyres et al. 1983, O'Brien et al. 2002). The first observational evidence that most of the water is liquid came from interpretation of characteristic cycloidal crack patterns (Fig. 8.2), which could be explained by tidal stress, but only if a thick fluid layer allowed sufficient amplitude for the tides (Hoppa et al. 1999). However, strictly speaking, that evidence really only applied to the time that the cracks were formed, which although very recent in geological terms (less than ~1 million years ago) could not definitively address the presence of the ocean now. Evidence for the existence of the ocean more recently, at least up to the end of the 20th century, came from the Galileo spacecraft's magnetometer, which showed variations in Jupiter's magnetic field in the vicinity of Europa that could be explained by an ocean within 200 km of the surface of Europa (Khurana et al. 1998, Kivelson et al. 2000, Zimmer et al. 2000).

In fact the ocean comes quite close to the surface, as indicated by a variety of geological features (described in Sect. 8.4.1 below). These features are best explained by linkage to the ocean below (Greenberg & Geissler 2002, Greenberg 2005). The ice shell is probably permeable due to ever-changing tectonic and thermal connections, suggesting ice thinner than 10 km, but with thickness varying with time and place. Uncertainties about the amount of internal heating and the heat transport processes through the ice have frustrated geophysical modeling of the ice thickness. In any case, the bulk of the H_2O on Europa probably forms a deep liquid water ocean just below the ice.

Below the ocean lies the rocky "mantle" and the iron core. Estimates of the size of the core depend on assumptions about the amount of iron sulfide included, so the core radius could be as small as 200 km or as large as 700 km (Schubert et al. 2004). Because the Galileo magnetometer found no internally generated magnetic field, the core must be solid or, if liquid (consistent with the tidal heating implied by the liquid water ocean), it must not be convecting (Schilling et al. 2004).

In bulk structure, Europa could hardly be called an icy satellite, but in terms of most of our observational data, which is imaging of the surface, most of what we have seen is ice (e.g. Greenberg 2005).

8.2.2 Ganymede

Ganymede's moment of inertia value (Table 8.1) is low enough that, like Europa, it implies complete interior differentiation into separate layers of water, rock and iron. Of the total radius of 2410 km, the outer layer of H_2O is about 900 km thick, the interior core is 650–900 km in radius, and in between is the rocky mantle (Schubert et al. 2004). Early heating by radioactivity and by dissipation during accretional impacts was probably adequate to explain the complete differentiation, but tidal heating due to the orbital resonance with Europa and Io may have been a factor as discussed in Sect. 8.3.2.

Ganymede is the only Galilean satellite with a magnetic field, which has a strength at the surface of about 1/10 of that of the Earth (Kivelson et al. 2002). To generate this field, the core must be currently molten. In addition to the internally generated magnetic field, the Galileo spacecraft's magnetometer detected variations in the jovian magnetic field in the neighborhood of Ganymede that can plausibly be explained by a electromagnetic induction in a conducting layer, suggesting (as for Europa) a layer of liquid salty water. However, in contrast to Europa, the conducting layer must be well below the surface. Accordingly, observations of the surface geology do not show any signs of interaction with the liquid layer.

Interior models of ice properties at depth are consistent with the possibility of a liquid layer (Spohn & Schubert 2003, Schubert et al. 2004). As is well known, the melting temperature of water decreases slightly with modest pressure, but as the pressure increases further, the melting point rises (Chizov 1993). For Ganymede, the minimum melting temperature would be at 150 km below the surface, assuming pure water. Thus any liquid would be around that depth, with ice layers above and below it. Assuming a reasonable heat flux, if the outer ice is conductive, the equilibrium thickness of ice above the liquid layer would be about 100 km. In that case, the liquid layer would be quite thick, extend from depth 100 km down past the minimum-melting-point depth of 150 km. On the other hand, if the outer layer of ice is convecting, it conveys heat more efficiently so its equilibrium thickness is greater, leaving space for only a few km of liquid at most. Salt in the water changes the story only slightly. Ammonia, which is another plausible substance (Kargel 1992), could make a very big difference: An ammonia-water layer could be 200–300 km thick and as close to the surface as 70 km (Sotin et al. 1998). Without knowledge of the interior heating rate, the thermal history, the actual composition of the solute, or the mode of heat transport (convection vs. conduction), these figures are all speculative, Even so, they do demonstrate the plausibility of a fluid layer within the ice as an explanation for the magnetic field variation (Spohn & Schubert 2003).

Even with the liquid layer buried too deep to have a direct influence, the surface of Ganymede displays the effects of substantial geological activity and variability. Only about 1/3 of the surface is dark and heavily cratered (Fig. 8.1), with relatively little evidence of endogenic activity since formation of the satellite (Shoemaker et al. 1982, Pappalardo et al. 2004). The crater record here indicates a surface age of >4 Gyr. The darkening may result from silicate or other impurities that remained in the cold icy outer crust, even while the interior underwent heat-driven differentiation. Sublimation of ice may have further concentrated dark material in a thin layer near the surface. In considering the low albedo of this terrain, it is important to bear in mind that even a relatively small concentration of dark material dispersed through ice can absorb a surprisingly large amount of light.

The remaining 2/3 of the surface has been converted into relatively bright terrain characterized by complex sets of roughly parallel grooves (Golombek 1982, Shoemaker et al. 1982). The surface here seems to have been stretched, and thus heavily faulted, exposing the brighter ice from below (Fig. 8.3). In addition, this modified terrain shows evidence of volcanic resurfacing, a process often called cryovolcanism, because on Ganymede liquid water flows out and over ice, rather than molten lava flowing over rock.

The crater record on the bright two-thirds of the surface confirms that it is younger than the old, dark, heavily cratered one-third. The nominal age of the reprocessed portion is about 2 billion years (Zahnle et al. 2003), a figure that suggests that global-scale endogenic processes occurred long after formation of the Jovian satellites system, but then turned off very long ago. This timing appears to be a major constraint on the

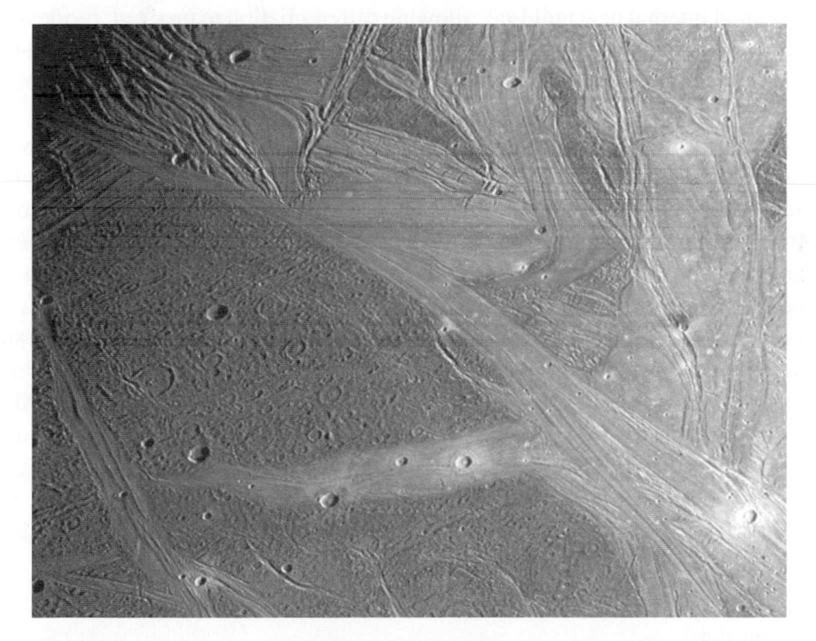

Fig. 8.3. The bright grooved terrain on Ganymede contrasts with the dark ancient terrain in this much higher resolution view than that shown in Fig. 8.1. The greater crater density on the dark terrain is evident, as is the predominantly tectonic character of the brighter terrain. This region is about 660 by 520 km in size with resolution about 1 km per pixel (JPL Image PIA 1617).

evolution of the Jovian system. However, the crater-based age is so uncertain, that the resurfacing could have been associated with the initial formation of the satellite, rather than a subsequent heating event. The range of uncertainty also admits the possibility that the surface is relatively recent. However, in contrast to Europa's surface which may still be undergoing continual renewal, Ganymede's has probably been relatively unchanged for at least 1/2 billion years and perhaps for as long as 4 billion years.

8.2.3 Callisto

Callisto is of similar size and density to Ganymede (Table 8.1, Fig. 8.1) but its moment of inertia is much larger (Anderson et al. 2001). Thus, while its composition of ice, rock and iron is probably similar, there has been only partial differentiation into layers. (Note that fully-differentiated Europa has a moment of inertia closer to Callisto's than to Ganymede's because it has a relatively thin H_2O layer.) If Callisto were completely uniform, its moment of inertia would have coefficient 0.4; gravitational compression without differentiation would reduce that value to about 0.38. The actual value of 0.355 indicates that there has been at least some differentiation, but not much.

Most likely the partial differentiation has resulted in a distinctly ice-rich layer, because the alternative, a fairly continuous increase in the fraction of rock and metal with depth, would be unstable (Schubert et al. 2004): Without the distinct ice-rich layer, convection would be suppressed. In that case, radiogenic and primordial heat could only escape via conduction, which is much less efficient than convection and would thus allow substantial internal heat build-up, allowing differentiation to proceed. Evidently, this scenario did not occur, so there must be some layering. However, there is a wide range of uncertainty about the thickness of the ice-rich layer, its purity, and the compositional profile below it.

While iron must be concentrated somewhat toward the center, there is no evidence for a core from the geophysical constraints. The lack of an observed magnetic field is consistent with that structure. Similar to Ganymede however, magnetometer data does suggest a deep conducting layer, again most plausibly explained by a liquid water layer (Khurana et al. 1998). The depth, thickness, and composition of this liquid layer would be subject to the same constraints as discussed above for Ganymede.

Like Europa and Ganymede, Callisto's surface is dominated by water ice. The surface is darkened by the presence of hydrated silicates (clays) and organic chemicals that may have rained onto the surface from comets and asteroids (Calvin & Clark 1991, McCord et al. 1998). Sublimation of the ice has concentrated a thin layer of these darker materials on the surface so that many craters, which dominate the geology of Callisto, appear as bright spots on the generally darker surface (Figs. 8.1 & 8.4). The numbers of craters show the surface to be at least 4 billion years old (Zahnle et al. 2003), but with a few craters having formed somewhat more recently.

8.2.4 Amalthea

The trend of compositions among the Galilean satellites, from rocky Io closest to Jupiter out to the icy Ganymede and Callisto is strong and is broadly consistent with expectations for formation in a nebula around the hot giant planet. Hence, Amalthea, which was

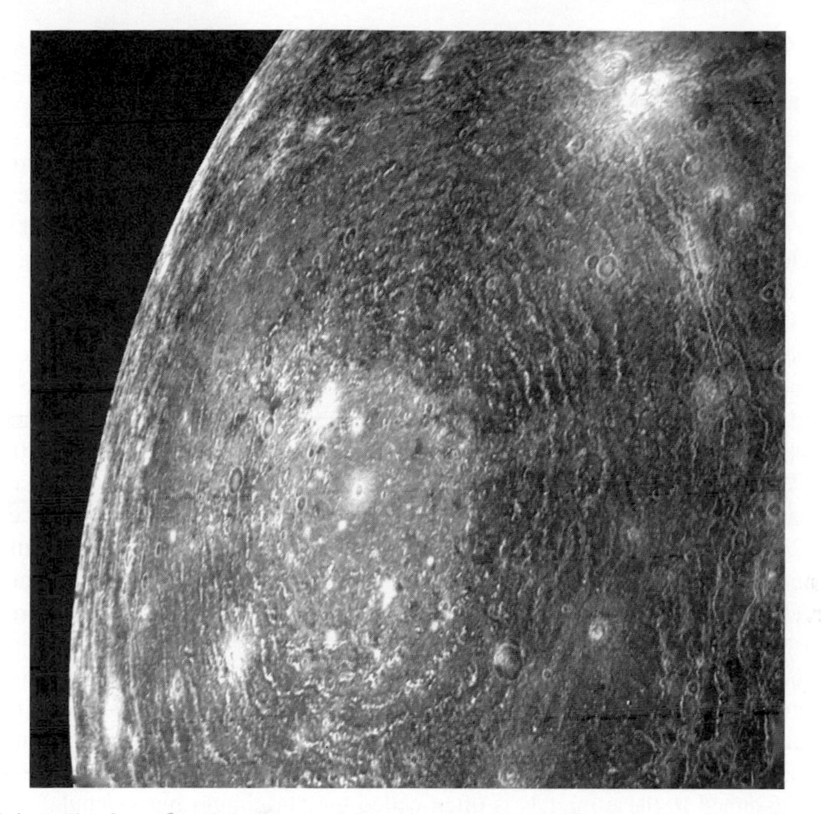

Fig. 8.4. Callisto's surface is ancient and heavily cratered (c.f. Fig. 8.1). This image shows the huge multi-ringed impact feature Valhalla, with its 300-km-wide bright center, where underlying ice has been exposed (NASA/JPL Image PIA 2277, from the Voyager mission). To the right, a string of small craters (a *catena*) is probably the result of impact with a comet that had been pulled apart by Jupiter's tidal effect, before hitting Callisto. Several such catenae are found on Ganymede and Callisto.

discovered by Barnard in 1892 on a regular, circular, co-planar orbit interior to the Galilean satellites was long assumed to have formed along with them and to be rocky. Its fairly dark surface was consistent with this assumption.

The Galileo spacecraft passed close enough to Amalthea (Fig. 8.1b) that a measurable effect on the spacecraft's orbit allowed determination of the satellite's mass. Combining the mass value with determination of its volume from images yielded a surprisingly low density of 1.0 ± 0.5 gm/cm^3 (Anderson et al. 2002). Because the density of most rock is closer to 3 gm/cm^3, this result was taken as evidence that the body must be quite porous, as a rubble pile of gravitationally bound debris. Recently, however, refinement of the Galileo radio-tracking results has reduced the density value to 0.86 ± 0.1 gm/cm^3, implying a substantial, probably dominant, ice component, and still requiring considerable porosity (Anderson et al. 2005).

The presence of this icy satellite, so close to Jupiter and in a very regular orbit, means that models of the formation of the Jupiter system will need to be reevaluated carefully.

8.3 Formation and Evolution of Icy Satellites

8.3.1 Formation of the jovian system

The bulk characteristics of the icy satellites described above result from the processes of formation of the satellites, followed by tidal effects, especially heating, due to the orbital resonance that has affected Europa and Ganymede. Indeed, it is the observed characteristics of the satellites that place essential constraints on the modeling of the formation and evolution of the jovian system.

Jupiter itself formed from the gravitational collapse of primarily gas in the circumsolar nebula, probably nucleated by a large core of solid material that had formed from collisionally accreted planetesimals (as reviewed by Lunine et al. 2004). The newly formed planet was likely surrounded by a flattened, spinning nebula of dust and gas. The nebula may have been left there during planet formation because it was rotating too fast to collapse (e.g. Korycansky et al. 1991), or else it formed as an accretion disk of material entering the region around the planet from the circumsolar nebula (Canup & Ward 2002). Formation of the regular satellites of Jupiter, including the Galilean satellites and Amalthea, has long been assumed to have progressed within the circumjovian nebula by a process of condensation and accretion of solids analogous to the formation of most of the planets of the solar system in the circumsolar nebula.

Models of the process of satellite formation have demonstrated the plausibility of a pressure and temperature profile that would place the condensation of water ice outside the orbit of Europa, explaining why Ganymede and Callisto accreted largely from H_2O, while Europa contained relatively little (e.g. Lunine & Stevenson 1982). The jovian nebula assumed in these models is often called the "minimum-mass nebula" because it assumes the lowest possible mass (assuming solar-nebula composition) that would contain enough condensable materials to produce the composition of the Galilean satellites. Although the minimum-mass nebula remained a standard model for many years, a fundamental problem has been the rapid orbital migration of satellites in such a nebula: While accretion and migration would occur on a timescale of thousands of years, the nebula would endure for millions of years. Satellites would have rapidly spiraled in toward Jupiter long before the nebula dissipated (Schubert et al. 2004).

Another problem was introduced with the recognition that Callisto is only partially differentiated. The rapid accretion in the minimum-mass nebula model would have resulted in enough accretional heating to differentiate all of the Galilean satellites. In order for Callisto to have avoided complete internal differentiation, accretion would have had to have been slow enough that the heat generated during each impact could largely radiate away before it is buried by the accumulation of subsequently accreted material (Canup & Ward 2002).

Two models have been suggested to account for the slow accretion of Callisto and to avoid the rapid orbital shrinking of the other Galilean satellites. In one model (Mosqueira & Estrada 2004), Callisto is assumed to have accreted much further out in the Jovian system where lower density would yield longer accretion times, and then spiraled in to near its current orbit. In that model, the other satellites exert torques that open gaps in the disk, slowing their orbital evolution. The other model is based on the assumption of a "gas-starved nebula" (Stevenson 2001, Canup & Ward 2002), which at no given time contained even near the minimum mass required to build the satellites. Instead, this nebula is a thin accretion disk continually fed by solar nebula material on its way to accretion by Jupiter. The slow feeding is consistent with the expected gap in the solar nebula opened by the gravitational influence of Jupiter, and it continues until the solar nebula is finally removed. The lower mass of the nebula at any given time accounts for the required slowing of Callisto's accretion and of the orbital drift of the other satellites. An open question is how Amalthea, on an orbit interior to Io's, was able to form with an icy composition.

8.3.2 Tidal effects of the Laplace resonance

While heating due to radioactivity and accretion probably played a significant role during the formation period, Io, Europa, and to some extent Ganymede have been affected by substantial tidal heating over a much longer period of time. Tides periodically distort the figure of any satellite if it rotates non-synchronously or if its orbit is eccentric. Such continual deformation results in heating and tectonic stress. But for most satellites it is usually short-lived, because the tides tend to turn themselves off by exerting torques that quickly (a) circularize a satellite's orbit and (b) change the rotation to be synchronous with the orbital period (e.g. Peale 1977, Burns 1977). With a circular orbit and synchronous rotation, there is no continual tidal distortion.

For Io, Europa, and Ganymede, an orbital resonance known as the Laplace relation maintains the eccentricities (Greenberg 1982). The resonance results from a commensurability of the orbital periods near a ratio of 1:2:4. Their mutual gravitational effects are enhanced, with the effects of forcing eccentricities and maintaining the resonance. Thus, tidal heating is maintained as well.

Tides are dependent on the mass of the tide-raising body, in this case massive Jupiter, and even more strongly dependent on the distance from that body. Therefore tidal heating is greatest in Io, smaller but significant in Europa, and only slight at the distance of Ganymede. This trend explains the basic differences among the satellites: the great volcanic activity and depletion of any water or volatiles that may have once been on Io; the thick layer of liquid water and evidence for long-term, probably currently active, surface reprocessing on Europa; and the relatively little thermally driven processing on Ganymede.

The extent to which the differences between Ganymede and Callisto have been enhanced by the Laplace resonance is not clear. Currently, given Ganymede's distance from Jupiter, the resonance is not strong enough to drive significant tidal heating. Thus the differences in internal differentiation and past geological activity between these two satellites might be due only to different formation circumstances, as discussed above. However, at some time in the past the resonance may have played a significant role in modifying Ganymede. A key issue is the still unknown history of the Laplace resonance and how its strength has varied over time.

One of the effects of tides is to exert torques that change orbital periods. Tides raised on the satellites by Jupiter tend to increase the orbital periods (and correspondingly increase the sizes of the orbits). Tides raised on Jupiter by satellites tend to have the opposite effect. These torques affect the ratios of the periods and thus the long-term evolution of the resonance.

A more precise description of the current state of the Laplace resonance takes into account the slight deviation from the commensurability of orbital periods (Greenberg 1982). In terms of the mean motion n, (defined as the orbital angular velocity averaged over an orbit), the relationship is $n_1 - 2n_2 = n_2 - 2n_3 = \nu$, where subscripts 1, 2, and 3 refer to Io, Europa, and Ganymede, respectively, and the quantity ν is much smaller than any n. If the periods were in an exact commensurability, ν would be zero. Currently, ν is about 0.7°/day, while the mean motions are about 200, 100, and 50°/day. If ν were smaller, we would describe it as *deeper* in resonance, and in that case the eccentricities forced into the orbits would be greater (generally $e \propto 1/\nu$).

For many years, a canonical assumption was that the Laplace resonance is in a steady state with a fixed value of ν (Yoder 1979), and hence constant orbital eccentricities and uniform tidal heating rates. The difficulty with that assumption was that the strong tides on Io, evidenced by measured heat and observed volcanism, needed to be balanced by torques due to tides raised on Jupiter by Io. Yet models of the structure of Jupiter and its tidal response did not produce adequate torque. With tides on Io dominating, n_1 would increase, with a corresponding increase in ν. Rather than remaining in a steady state, the system should be evolving away from deeper resonance, and eccentricities would have been greater in the past (Greenberg 1982, 1989).

Nevertheless, the canonical model was that somehow tides in Jupiter prevailed and even dominated, so that the current system evolved from an initial condition not in resonance (Yoder 1979). In that case the forced eccentricities would not have developed until long after the formation of the system. Such a delayed onset of the resonance might explain how a burst of geological activity on Ganymede could have occurred a couple of billion years after the satellite formed. Models of hypothetical evolutionary histories were developed that passed through various resonance conditions which might have included periods of large forced orbital eccentricity for Ganymede (Showman & Malhotra 1997). Those evolutionary models included various ad hoc assumptions about feedback loops in which changes in tidal heating modify Ganymede's interior structure, which in turn changes tidal dissipation rates. While ingenious, the application to observations is not clear: The crater-count-based age for the thermally-driven, bright, younger surface could be nearly as old as the satellite, not necessarily billions of years later.

Unless the processed grooved terrain of Ganymede can be shown to significantly post-date the satellite's formation, a more straightforward history of the orbits would have the system evolving under the dominant effect of tides on Io from deep resonance (very small ν) toward the current moderate value of ν. It is possible that such evolution has been episodic (Greenberg 1982), with ν increasing and decreasing as feedback from changes in the tidal heating of Io alternately allow Jupiter tides and Io tides to dominate, with a ~100 million year periodicity. Heating of the icy satellites Europa and Ganymede would oscillate accordingly, with much more heating 100 million years ago than now (Ojakangas and Stevenson 1986).

Another plausible scenario is that the system simply formed in a deep resonance and has gradually been evolving (under the tidal influence of Io) to larger values of ν ever since (Greenberg 1982, Peale & Lee 2002). Orbital change during the satellite formation process could have initiated the original resonance. In this scenario, the partial resurfacing of Ganymede would have occurred as a result of some combination of radiogenic, accretional, and tidal heating, all of which substantially diminished early in its history;

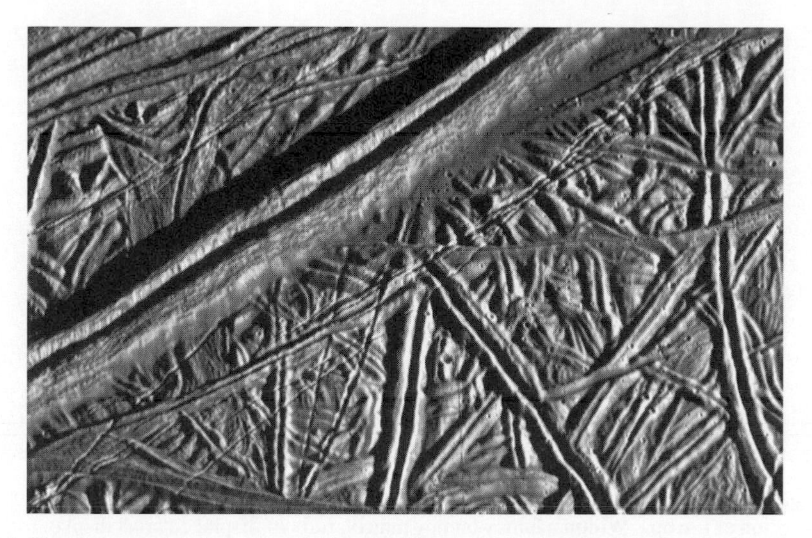

Fig. 8.5. The common denominator of ridges on Europa is the double ridge. In this densely ridged area (about 12 km across at 21 m/pixel), double ridges of various sizes have formed on top of one another at various orientations. Ridge formation is a major surface renewal process on Europa.

Europa's geological activity would have been driven by the resonance and associated tides ever since its formation, and has been even greater in the past than in the relatively recent epoch recorded on its surface.

8.4 Surface Appearance and Geology

8.4.1 Europa

The dramatic continual resurfacing of the ice shell of Europa has produced two main types of geological terrain: tectonic and "chaotic", representing the tidal effects of stress and of thermal processing, respectively (e.g. Greenberg & Geissler 2002, Greenberg 2005). On a global scale (Fig. 8.1), these features are recognizable as lineaments and splotches marked by slight darkening, probably representing deposition of oceanic substances through cracks or melt zones. Few craters are visible because of the recent resurfacing.

The tectonic terrain is dominated by double ridges, whose central grooves mark cracks in the crust (Fig. 8.5). Global-scale lineament patterns correlate roughly with tidal tension patterns, and often consist of multiple sets of roughly parallel double ridges. Elsewhere, terrain is densely covered by criss-crossing double ridges. The ridges may form as cracks initiated by tidal stress are continually worked by periodic tides that squeeze up crushed ice and slush. Many crack features follow the cycloidal patterns predicted for tidal stress (Sect. 8.2.1, Fig. 8.2), with chains of as much as a dozen arcs, each arc typically $\sim$100 km long.

Large plates of the ice shell, often hundreds of kilometers across have undergone major displacement, including dilation of cracks, which has allowed infilling from below, and strike-slip (shear) displacement. The strike-slip is probably driven by diurnal tides,

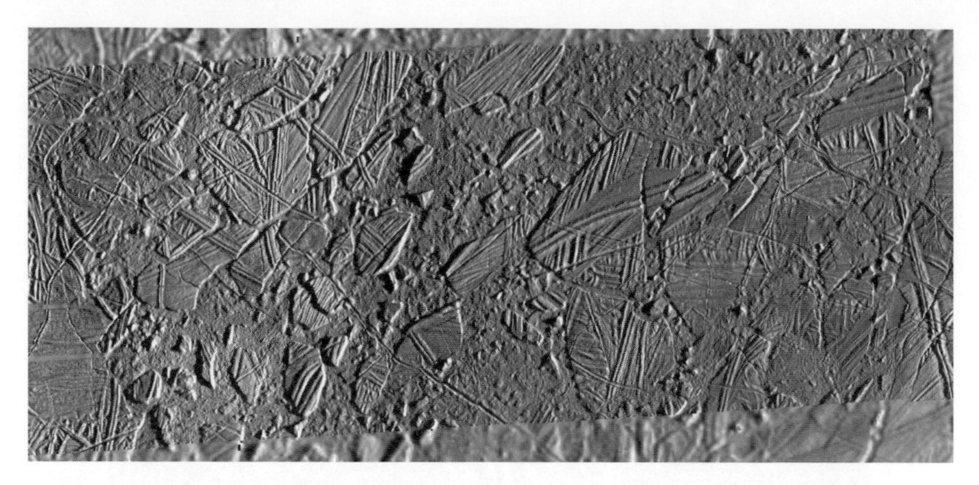

Fig. 8.6. A mosaic of high-resolution images (54m/pixel) of chaotic terrain within the Conamara Chaos region of Europa. Within a lumpy bumpy matrix, rafts of displaced crust display fragments of the previous tectonic terrain. Subsequent to the formation of the chaotic terrain, probably by melt-through from below, the fluid matrix refroze and new double ridges have formed across the area and begun the process of tectonic resurfacing.

and the ubiquitous displacement is facilitated by the low-viscosity of the underlying liquid ocean. Because the tectonic patterns can be fit to the predictions of tidal-stress theory, evidence has been developed for non-synchronous rotation (which further enhances tidal stress), and for polar wander in which the icy shell slips around relative to the spin orientation of the satellite.

Nearly half of the surface of Europa comprises chaotic terrain, where the surface has been disrupted, with rafts of older terrain displaced within a lumpy matrix (Fig. 8.6). Chaos likely represents melt-through, perhaps from the ocean below, followed shortly by refreezing. Only modest concentrations of tidal heat are needed to melt through the ice. While creation of chaotic terrain destroys older surfaces, chaotic terrain in turn can be destroyed by subsequent tectonics. The history of Europa has been an on-going interplay of resurfacing, by tectonics and by chaos formation, with each destroying what was there before, and with each seemingly involving breakthrough of the ocean to the surface (Greenberg 2005).

8.4.2 Ganymede

The darker 1/3 of Ganymede's surface is enriched in impurities, probably predominantly hydrated silicates, which avoided the internal differentiation and may have been accreted late in the formation of the satellite (Pappalardo et al. 2004). This older terrain is relatively heavily cratered and also includes other impact features: palimpsests and concentric furrows. Palimpsests are round, subtly brightened patches where the topography of an impact has evidently relaxed away. Furrows $\sim$10 km wide and typically spaced $\sim$50 km apart are generally part of large circular systems that record large early impacts. Other groups of furrows in the dark terrain may be extensional features, like graben, and sometimes are oriented parallel to the borders with the younger brighter terrain (Fig. 8.7).

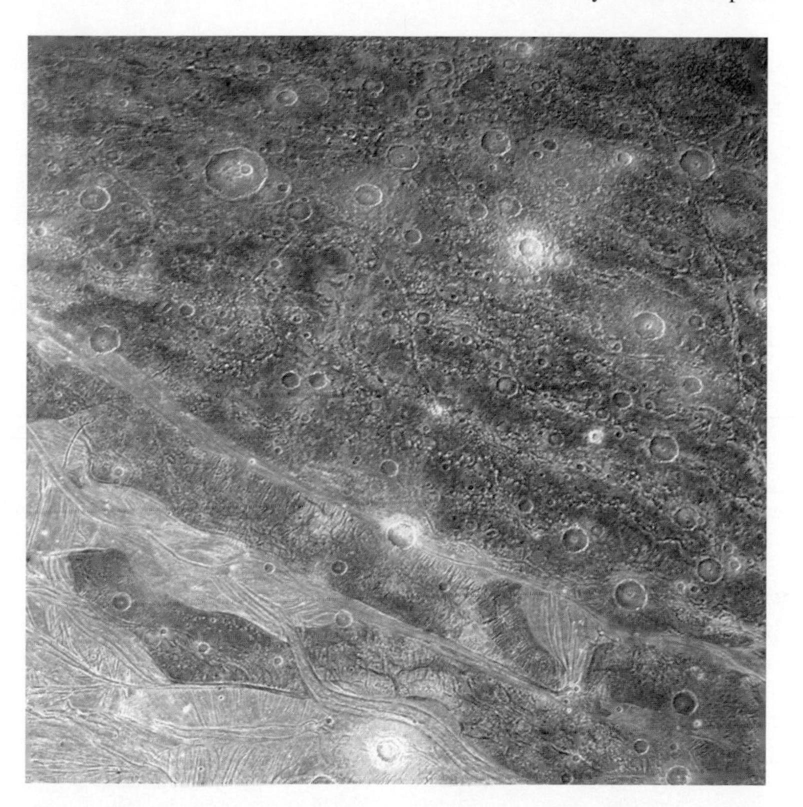

Fig. 8.7. The dark regions on Ganymede contain sets of furrows, which are often parallel to the borders with the bright grooved terrain as seem here. These dark-region furrows may have formed by surface extension, an incipient version of the kind of tectonic processing that produced the bright grooved terrain. Furrows in the dark regions also often display concentric patterns suggesting that they formed as rings around impact basins, similar to those of Valhalla on Callisto (Fig. 8.4). In the foreground (lower left) is typical bright grooved terrain. Craters of various morphologies are displayed in this area. The area shown is about 450 km wide in a Voyager image (NASA/JPL PIA 2281).

The brighter terrain comprises bands of roughly parallel grooves that have cut across the darker terrain, and cover 2/3 of the surface. This terrain (Fig. 8.8) appears to be the result of surface extension, probably more mature versions of the extensional furrows on the dark terrain (Golombek 1982, Pappalardo et al. 2004). The extension of Ganymede's icy surface is very different from that on Europa. On Europa, dilational bands formed where cracks have opened tens of km wide, allowing infilling by new material, all due to the mobility provided by the near-surface ocean. Terrains on opposite sides match and can be reconstructed by removing the band material. On Ganymede, the old terrain on opposite sides of the bright grooved terrain cannot be matched. The material in between is generally not newly emplaced, but rather old surface that has been stretched and modified. The grooves and furrows that characterize the brighter terrain result from the extensional tectonic processes, starting with graben-like furrowing, like what is seen on the dark terrain, followed by finer scale tilt-block faulting and extensional necking

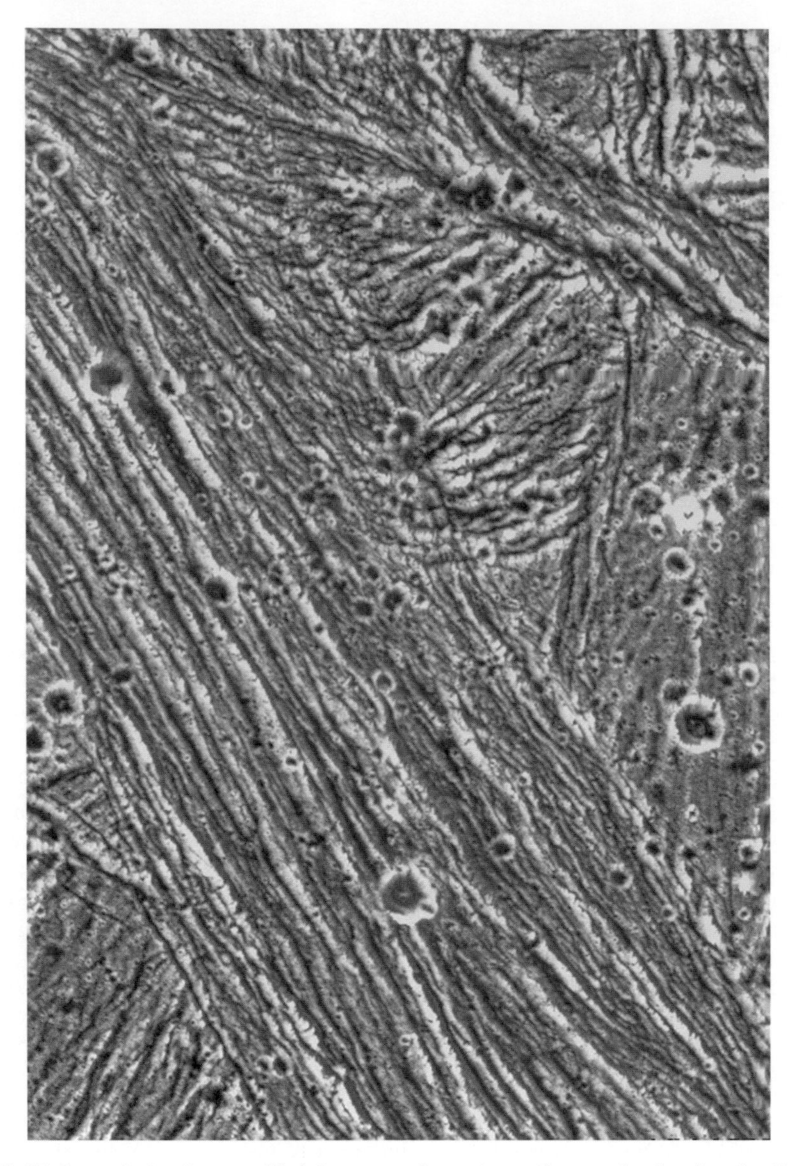

Fig. 8.8. High resolution image of bright, grooved terrain on Ganymede showing details of the extensional tectonics. This image, taken by the Galileo spacecraft, shows an area 35 km wide at resolution 74m/pixel (NASA/JPL PIA 0277).

of the lithosphere (the cold, brittle-elastic upper portion of the ice). The brightening of this extensional terrain relative to the older dark terrain is in part due to exposure of the purer ice just below the surface by the faulting. In addition, liquid water has oozed to the surface ("cryovolcanism") to some degree (Schenk et al. 2001).

The dominance of extensional geology on a global scale seems to require that Ganymede has expanded significantly since the formation of the old, dark terrain. The most plausible explanation for such global expansion follows from Ganymede's icy composition and the internal differentiation: When the lower density ice rose up from the deep interior, the release of pressure allowed substantial expansion. Thus, whenever the surface modification occurred, the first major internal heating must have occurred at about the same time, an important constraint on models of formation and orbital evolution. These events may have been either long after Ganymede's formation as indicated by the nominal crater age, or early in the lifetime of the satellite as permitted by crater-age uncertainty,

Ganymede's dark terrain has also undergone "mass wasting" (downslope movement of surface materials) as evidenced by landslide morphologies (Moore et al. 1999), exposure of brighter material on slopes and accumulation of darker material in topographic lows (Oberst et al. 1999). Sublimation has also modified the dark terrain, seeming to have concentrated darker material as ice sublimates off sun-facing slopes with frost deposits forming preferentially on the opposite slopes. Diffuse polar brightening can be explained by such frost deposition near the poles.

8.4.3 Callisto

Callisto's surface is quite similar to the old, dark terrain of Ganymede, unmodified by endogenic geological processes and consistent with the failure of the interior to have been heated enough to differentiate fully (Moore et al. 2004). The surface is heavily cratered by impacts, with younger craters usually appearing relatively bright where they have exposed purer ice below the relatively concentrated silicates at the surface. As on Ganymede, pallmpsests record the locations of ancient craters with circles of subtle brightness. Larger impacts have left multi-ring structures of concentric scarps and troughs, including the 3800 km wide Valhalla system (Figs. 8.4 & 8.9).

The terrain has a strange softened topography with knobby protrusions, usually the modified rims of craters. This character has been formed by considerable mass wasting, which has concentrated dark material in a fairly smooth mantle spread over the lower elevations. The bright, erosion-modified knobs are products of erosion of crater rims, or of other ancient topography, that has been reformed by sublimation as well as mass wasting. Sublimation and deposition of frosts is controlled by local orientation toward the sun and by enhanced heating due to the dark impurities at the bases of scarps and knobs. These processes steepen slopes, creating the knobby topography and smoothing the dark terrain at the lower elevations (Moore et al. 2004).

8.5 Conclusions and Implications

Planetary geology and geophysics have largely consisted of "comparative planetology", a discipline that applies lessons learned on some planets to new information about others. This approach has been reasonably successful when dealing with rocky bodies for which we have a closely studied archetype, the Earth. Icy satellites pose greater challenges. Some terrestrial analogs are useful: Some rock processes may be analogous to solid state processes in ice; Arctic and Antarctic ice geology may have features similar to those on icy satellites; Glacier studies provide constraints on material behavior. However, the

Fig. 8.9. A high resolution Galileo-spacecraft view (46 m/pixel, and 33 km across) of one of the concentric scarps of the Valhalla multi-ring impact structure (Fig. 8.4). The image also shows the softened lower terrain between knobby remnants of sublimation-eroded hills and crater rims. Dark material has moved downslope leaving the relatively bright icy knobs (NASA/JPL PIA 0561).

conditions on icy satellites are so alien that such comparisons require great caution. We do not understand how ice behaves in the deep interiors of planetary bodies, or even what the conditions are there. Similarly, surface morphologies have formed at temperatures unknown on Earth and in materials whose properties and detailed compositions are uncertain. Clearly we have a great deal more to learn.

At the same time, there is an unusually strong motivation to continue to pursue studies of the icy satellites. We have seen that all three icy Galilean satellites probably have liquid water layers, and one, Europa, almost certainly has an ocean just below the surface. Naturally, liquid water raises the possibility of extraterrestrial life. On Ganymede and Callisto, and on Europa as well if the ice is so thick that the ocean is isolated below, life would face an inhospitable setting. The ecosystem would be isolated from both oxygen and from sunlight. The possibility of life would require alternative biochemistries, based on assumed volcanism and exotic hypothetical metabolisms. Even with the freedom to model deep-sea conditions unconstrained by observations, theoretical considerations indicate that life would be very limited if it could exist at all.

On Europa, however, the surface geology suggests linkages between the ocean and the surface, with dynamic flow of liquid to (or close to) the surface. The surface of the ice is bombarded by energetic charged particles from Jupiter's magnetosphere, which has released oxidants into the ice. Cometary material lands on the surface, depositing a suite of organic and other substances. Organisms within a few centimeters of the surface would be killed by the radiation, but enough sunlight could penetrate a few meters below to drive photosynthesis. Relatively warm sea water periodically reaching the surface could conceivably support a rich ecology, both in the crust and in the ocean. A biosphere on Europa, if any, probably extends from deep in the ocean up to within a few centimeters of the surface.

Such possibilities make further investigation of the icy satellites especially exciting and urgent. On Europa, life, or its remains, may well be accessible close to the surface. In preparation for such a venture, comparative studies of all icy satellites will be essential. But even if extraterrestrial life were out of the question, the remarkable properties and processes on these complex worlds make them worthy of continuing exploration.

Acknowledgments

Preparation of this review was supported by NASA's Outer Planet Research program.

References

Anderson, J.D., et al., 2001. Shape, mean radius, gravity field and interior structure of Callisto, *Icarus* **153**, 157–161.

Anderson, J.D., et al., 2005. Amalthea's density is less than that of water. *Science* **308**, 1291–1293.

Burns, J.A., 1977. Orbital evolution. In *Planetary Satellites*, ed. J.A. Burns, University of Arizona Press, Tucson, Arizona, pp. 113–156.

Calvin, W.N., and R.N. Clark, 1991. Modeling the reflectance spectra of Callisto. *Icarus* **89**, 305–317.

Canup, R.M., and W.R. Ward, 2002. Formation of the Galilean satellites: Conditions of accretion. *Astron. J.* **124**, 3404–3423.

Cassen, P., R.T. Reynolds, and S.J. Peale, 1979. Is there liquid water on Europa? *Geophys. Res. Lett.*, **6**, 731–734.

Chizov, V.E., 1993. Thermodynamic properties and equation of state of high pressure ice phases. *Prikladnay Mekhanika I Tekhnichskaya Fizika* **2**, 113–123.

Golombek, M.P., 1982. Constraints on the expansion of Ganymede and the thickness of the lithosphere. *J. Geophys. Res.* **87**, 77.

Greenberg, R., 1982. Orbital evolution of the Galilean satellites. In *The Satellites of Jupiter*, ed. D. Morrison, University of Arizona Press, Tucson, Arizona, pp. 65–92.

Greenberg, R. 1989. Time-varying orbits and tidal heating of the Galilean satellites. In *NASA SP-494, Time-Variable Phenomena in the Jovian System*, eds. M.J.S. Belton, R.A. West, and J. Rahe, NASA, Washington D.C., pp 100–115

Greenberg, R., 2005. *Europa, the Ocean Moon*, Springer-Praxis, Chichester, UK.

Greenberg, R., & P. Geissler, 2002. Europa's dynamic icy crust, An invited review. *Meteoritics and Planetary Science* **37**, 1685–1711.

Hoppa, G.V., et al., 1999. Formation of cycloidal features on Europa. *Science* **285**, 1899–1902.

Kargel, J.S., 1992. Ammonia-water volcanism on icy satellites: Phase relations at 1 atm. *Icarus* **100**, 556–574.

Khurana, K.K., et al., 1998. Induced magnetic fields as evidence for sub-surface oceans in Europa and Callisto. *Nature, 395*, 777–780.

Kivelson, M.G., et al., 2000. Galileo magnetometer measurements: A stronger case for a subsurface ocean at Europa. *Science* **289**, 1340–1343.

Kivelson, M.G., et al., 2002. The permanent and inductive magnetic moments of Ganymede. *Icarus* **157**, 507–522.

Korycansky, D.G., et al., 1991. Numerical models of giant planet formation with rotation. *Icarus* **92**, 234–251.

Lopes, R.M.C., and T.K.B. Gregg, 2004. *Volcanic Worlds*, Springer-Praxis, Chichester, UK.

Lunine, J.I., & D. Stevenson, 1982. Formation of the galilean satellites in a gaseous nebula. *Icarus* **52**, 14–29.

Lunine, J.I., et al., 2004. The origin of Jupiter. In *Jupiter: The Planet, Satellites, and Magnetosphere*, eds. F. Bagenal, T.E. Dowling and W.B. McKinnon, University of Arizona Press, Tucson, Arizona, pp. 19–34.
McCord, T.B., et al., 1998. Non-water-ice constituents in the surface material of the icy galilean satellites from the Galileo near-infrared mapping spectrometer investigation. *J. Geophys. Res.* **103**, 8603–8623.
Moore, J.M., et al., 1999. Mass movement and landform degradation on the icy galilean satellites. *Icarus* **140**, 294–312.
Moore, J.M., et al., 2004. Callisto. In *Jupiter: The Planet, Satellites, and Magnetosphere*, eds. F. Bagenal, T.E. Dowling and W.B. McKinnon, University of Arizona Press, Tucson, Arizona, pp. 397–426.
Mosqueira, I., and P.R. Estrada, 2004. Formation of the regular satellites of giant planets in an extended gaseous nebula. *Icarus* **163**, 198–255.
Oberst, J., et al., 1999. The distribution of bright and dark material on Ganymede in relationship to surface elevation and slopes. *Icarus* **140**, 283–293.
O'Brien, D.P., et al., 2002. A melt-through model for chaos formation on Europa. *Icarus* **156**, 152–161
Ojakangas, G.W., and D.J. Stevenson, 1986. Episodic volcanism of tidally heated satellites with application to Io. *Icarus,* **66**, 341–358.
Peale, S.J., 1977. Rotational histories of the natural satellites. In *Planetary Satellites*, ed. J.A. Burns, University of Arizona Press, Tucson, Arizona, pp. 87–112.
Peale, S.J., et al., 1979. Melting of Io by tidal dissipation. *Science* **203**, 892–894.
Pappalardo, R., et al., 2004. Geology of Ganymede. In *Jupiter: The Planet, Satellites, and Magnetosphere*, eds. F. Bagenal, T.E. Dowling and W.B. McKinnon, University of Arizona Press, Tucson, Arizona, pp. 363–396.
Peale, S.J., and M.H. Lee, 2002. A primordial origin of the Laplace relation among the Galilean satellites. *Science* **298**, 593–597.
Schenk, P.M., et al., Flooding of Ganymede's bright terrains by low-viscosity water-ice lavas. *Nature* **410**, 57–60.
Squyres, S.W., et al., 1983. Liquid water and active resurfacing on Europa. *Nature* **301**, 225–226.
Schilling, N., et al., 2004. Limits on the intrinsic dipole moment in Europa. *J. Geophys. Res.* **109**, E05006.
Schubert et al., 2004. Interior composition, structure, and dynamics of the galilean satellites. In *Jupiter: The Planet, Satellites, and Magnetosphere*, eds. F. Bagenal, T.E. Dowling and W.B. McKinnon, University of Arizona Press, Tucson, Arizona, pp. 281–306.
Shoemaker, E.M., The geology of Ganymede. In *The Satellites of Jupiter*, ed. D. Morrison, University of Arizona Press, Tucson, Arizona, pp. 65–92.
Showman, A.P., and R. Malhotra, 1997. Tidal evolution into the Laplace resonance and the resurfacing of Ganymede. *Icarus* **127**, 93–111.
Sotin, C., et al., 1998. Thermodynamical properties of high pressure ices. Implications for the dynamics and internal structure of large icy satellites. In *Solar System Ices*, eds. B. Schmitt, C. de Bergh and M. Festou, Kluwer Academic Publishers, The Netherlands, pp. 79–96.
Spohn, T., & G. Schubert, 2003. Oceans in the icy galilean satellites of Jupiter? *Icarus* **161**, 456–467.
Stevenson, D., 2001. Jupiter and its moons. *Science* **294**, 71–72.
Yoder, C.F., 1979. How tidal heating in Io drives the Galilean orbital resonance locks. *Nature* **279**, 767–769.
Zahnle, K.L., et al., 2003. Cratering rates in the outer solar system. *Icarus* **163**, 263–289.
Zimmer, C., 2000. Subsurface oceans on Europa and Callisto. *Icarus* **147**, 329–347.

9 Cassini at Saturn: The First Results

Ellis D. Miner, Dennis L. Matson and Linda J. Spilker

Abstract. The international Cassini-Huygens Mission is a joint mission of the National Aeronautics and Space Administration and the European Space Agency, with the Italian Space Agency as a major partner. The spacecraft was launched from Cape Canaveral, Florida, on October 15, 1997, and was inserted into orbit around the planet Saturn on July 1, 2004. The Huygens Probe separated from the Cassini Orbiter on December 26, 2004 and coasted to a descent by parachute through the atmosphere of Titan to the surface on January 14, 2005. This chapter summarizes the primary science results to date from both the orbiter and the probe. Many of the analyses are preliminary in nature, and some may be revised as additional data enables better characterization of Saturn and its system of rings, magnetosphere and satellites. The initial results summarized in this chapter include images and other data from a close encounter of Phoebe prior to Saturn orbit insertion, from ring studies and magnetospheric studies during orbit insertion and the first year in orbit, from studies of Titan during the first several flybys and from the Huygens Probe mission, and from close flybys of several of Saturn's icy satellites. The results also include discovery of several new satellites and a new radiation belt between the rings and Saturn's atmosphere.

9.1 Introduction

The Voyager 1 spacecraft made a close flyby of Saturn and its largest satellite, Titan, in November 1980. The images of Titan returned by Voyager 1 showed no discernible surface features, but instead revealed a moon enshrouded in a thick orange haze. While the images were somewhat disappointing, data returned by other scientific investigations on both Voyager 1 and 2 revealed Titan as an intriguing world, one unique among the myriads of solar system satellites observed prior to or since the Voyager flybys in 1980 and 1981. The second-largest satellite (behind Jupiter's Ganymede), Titan is the sole example in the solar system of a satellite with a thick atmosphere. That atmosphere extends roughly ten times as far above the surface as does Earth's atmosphere and has a surface pressure roughly 60% greater than Earth's. Both have nitrogen as their primary gaseous constituent. Titan may also be the only body in the solar system other than Earth to have gas, liquid and solid all present at its surface. Small wonder that a proposal to return a properly equipped orbiting spacecraft to the Saturn system was contemplated even before the Voyager flybys of Saturn were complete (Ip, Gautier & Owen, 2004).

The international Cassini-Huygens Mission, a joint mission of the National Aeronautics and Space Administration (NASA) and the European Space Agency (ESA), with the Italian Space Agency (ASI) as a major partner, is the fulfillment of that desire to return to the ringed world of Saturn and to study it in detail. The mission takes its name from Italian-French astronomer Jean Dominique Cassini and Dutch astronomer

Christiaan Huygens. Cassini was the first director of the Paris Observatory, discoverer of four of Saturn's satellites (Tethys, Dione, Rhea and Iapetus) and the first to observe the major division between in the main rings of Saturn which now bears his name. Huygens was the discoverer of Titan and the first to recognize the nature of Saturn's rings as a thin, detached disk of particles that orbit Saturn. The Cassini-Huygens Mission had its official start in 1990, although both ESA (for the Huygens Titan probe) and NASA (for the Cassini Saturn orbiter) had earlier issued separate Announcements of Opportunity for participation by qualified scientists in a Titan-probe / Saturn-orbiter mission (ESA, 1989; NASA, 1989). Scientist selection for nine Huygens probe science investigations occurred in 1989; selection of seventeen science investigations for the Cassini orbiter occurred in 1990–1. One of the Cassini interdisciplinary investigations was later dropped with the untimely death of James Pollack of NASA Ames Research Center. The current Cassini-Huygens scientific investigations are listed in Table 9.1.

Table 9.1. Cassini-Huygens Scientific Investigations

	Scientific Investigation	**Lead Investigator**
Cassini Orbiter	Cassini Plasma Spectrometer	David Young
	Cosmic Dust Analyzer	Eberhard Gruen
	Dual Technique Magnetometer	David Southwood
	Ion and Neutral Mass Spectrometer	Hunter Waite
	Magnetospheric Imaging Instrument	Stamatios Krimigis
	Radio and Plasma Wave Science	Donald Gurnett
	Cassini Radar	Charles Elachi
	Composite Infrared Spectrometer	Virgil Kunde
	Imaging Science	Carolyn Porco
	Radio Science	Arvidas Kliore
	Ultraviolet Imaging Spectrograph	Larry Esposito
	Visible and Infrared Mapping Spectrometer	Robert Brown
	Interdisciplinary Investigation	Michel Blanc
	Interdisciplinary Investigation	Jeffrey Cuzzi
	Interdisciplinary Investigation	Tamas Gombosi
	Interdisciplinary Investigation	Tobias Owen
	Interdisciplinary Investigation	Laurence Soderblom
	Interdisciplinary Investigation	Darrell Strobel
Huygens Probe	Aerosol Collector and Pyrolyser	Guy Israel
	Descent Imager and Spectral Radiometer	Martin Tomasko
	Doppler Wind Experiment	Michael Bird
	Gas Chromatograph and Mass Spectrometer	Hasso Niemann
	Huygens Atmospheric Structure Instrument	Marcello Fulchignoni
	Surface Science Package	John Zarnecki
	Interdisciplinary Investigation	Daniel Gautier
	Interdisciplinary Investigation	Jonathan Lunine
	Interdisciplinary Investigation	Francois Raulin

Limits on mass, power and cost were all major considerations in the design of the spacecraft. Electrical power for the Cassini orbiter is provided by three radioisotope thermoelectric generators (RTGs), which also provided power for the Huygens probe prior to its separation from the orbiter. After separation, the Huygens probe was powered by five lithium sulfur dioxide batteries. The RTGs provide insufficient power to operate all Cassini orbiter subsystems simultaneously; timesharing of the power resources is necessary. The spacecraft at launch (including the launch vehicle adaptor) had a mass of 5722 kg, more than half of which was liquid propellant and pressurant. To keep mass and cost within available limits, all instruments were affixed to the body of the spacecraft; hence pointing them independently is not feasible. This reduced significantly the pre-launch costs of the mission, but has resulted in more complex and costly post-launch mission operations. The launch vehicle was a Titan IV-B / Centaur multistage rocket, the largest launch vehicle available in the United States at the time. Launch occurred in the pre-dawn hours of October 15, 1997, from Cape Canaveral, Florida. The launch vehicle performed flawlessly, but the massive Cassini-Huygens spacecraft still required four gravity-assist planetary flybys, two at Venus and one each at Earth and Jupiter, to reach Saturn on July 1, 2004. More detailed descriptions of the spacecraft and its mission can be found in NASA SP-533, *Passage to a Ringed World* (Spilker, 1997) and in Matson, Spilker and Lebreton (2002). Fig. 9.1 depicts the Cassini spacecraft during the main engine burn at Saturn orbit insertion.

Fig. 9.1. Artist's depiction of the Cassini spacecraft during its insertion into Saturn orbit. Note that the Huygens Probe is still attached at the bottom. The 10-meter-long magnetometer boom protrudes to the left. Also visible are the three antennas of the Radio and Plasma Wave instrument, the 4-meter high gain antenna (at the left), the optical remote sensing instruments (at the top), the three RTG power sources, and the two main engine nozzles, all seen against the background of Saturn and its rings.

9.2 Science En Route to Saturn

The gravity-assist flybys of Venus, Earth and Jupiter and the passage through interplanetary space on the way to Saturn would normally have provided an abundance of opportunities to utilize the sophisticated scientific payload to collect new and useful scientific data about each of these targets. A number of factors made such data collection impractical or impossible. Several of the instruments were not designed to function well in the warm environment of the inner solar system. To protect them from damage, the spacecraft was oriented to keep its high-gain antenna (HGA) pointed toward the Sun, thereby providing shade for most of the instruments. Until after outbound passage of Earth, communication was generally possible only via the spacecraft Low-Gain Antenna (LGA), which in turn required a roll orientation that was seldom ideal for science data collection. Consequently, relatively little science data was garnered during the Venus and Earth flybys.

Thermal limitations on spacecraft orientation eased somewhat after Earth flyby, particularly since communications could generally use the HGA instead of the LGA. However, HGA-to-Sun (or Earth) pointing continued to be the norm. Interplanetary dust stream, magnetic field and charged-particle monitoring became somewhat routine, but optical remote sensing observations were seldom possible. During passage through the asteroid belt, images of Asteroid 2685 Masursky were obtained from a distance of about 1.6 million km. 2685 Masursky, named for the late Harold Masursky, planetary astronomer and friend and colleague of the authors, was found to have a diameter of 15 to 20 km. Rough estimates of its rotation period, composition and reflectivity were also possible.

The cost of additional interplanetary observations was high, but a Saturn-like data collection sequence at Jupiter was needed to overcome operational complexities and to provide valuable experience for the flight team and science investigators at Saturn. Although the Jupiter observation sequences were routine and repetitive, valuable data on Jupiter and its environment were obtained during a period spanning 180 days surrounding the December 30, 2000, closest approach date. The data collection opportunity was further enhanced by the continued presence of the Jupiter-orbiting Galileo spacecraft, which enabled both cross-correlation and cross-calibration of data from the two spacecraft. All Cassini orbiter investigations except the Ion and Neutral Mass Spectrometer and the Radio Science investigations were active during the Jupiter flyby. The former was not functional until after the Saturn orbit insertion maneuver; the latter was too far from Jupiter and its satellites to provide useful data.

A collection of four papers in *Nature* (Gurnett et al., 2002, Bolton et al., 2002, Kurth et al., 2002, and Krimigis et al., 2002) describe some of the complex interactions between Jupiter's magnetosphere, the solar wind, three of the four Galilean satellites (Io, Europa, and Ganymede) and the planet. During its flyby of Jupiter, the Cassini orbiter skirted along the sunset flank of the magnetosphere, spending several days within the magnetopause, but occasionally dipping into the magnetosphere itself.

Cassini imaging results at Jupiter (Fig. 9.2) are published in a paper in *Science* (Porco et al., 2003). The imaging data show that upwelling motions within Jupiter's atmosphere appear to occur exclusively in Jupiter's dark belts, implying that downwelling must occur primarily in the brighter zones; these conclusions are directly opposite what had been

Fig. 9.2. Jupiter as seen by the Cassini imaging narrow-angle camera on December 29, 2000. The Great Red Spot is seen just below center. Dark belts, bright zones, and discrete storm systems are visible over much of the planet.

supposed with lower-resolution images. The characteristics of light-scattering from the rings of Jupiter imply that they are irregular in shape and are therefore likely to be the result of bombardment of the surfaces of one or more satellites; the data seem to imply that Metis and Adrastea are the primary sources of ring particles. Images of the north polar region of Jupiter's atmosphere revealed a dark, swirling oval about the size of the well-known Great Red Spot and showed that persistent bands of globe-circling winds extend northward of the conspicuous dark bands and bright zones. Images of Jupiter's satellites show that the tenuous atmospheres of Io and Europa glow during passage of those satellites through Jupiter's shadow, that a volcanic plume exists over Io's north polar region, and that the small outer satellite Himalia has an irregular shape, with a long dimension of about 120 km.

Extensive infrared coverage of Jupiter's atmosphere by the Composite Infrared Spectrometer has been analyzed to provide detailed composition information. This information is published in the September 2004 issue of *Science* (Kunde et al., 2004).

Other interesting findings of Cassini and Huygens can be found in news releases on the Cassini website at URL: http://saturn.jpl.nasa.gov/news/press-releases.cfm

9.3 Phoebe Science Results

Timing of the Saturn orbit insertion was adjusted to permit a close encounter with Phoebe on June 11, 2004, just nineteen days before the Cassini-Huygens spacecraft arrived at Saturn. Voyager 2 imaging of Phoebe (Fig. 9.3) was from a distance of about 2.2 million km, which served to provide an approximate diameter and rotation period, but little else. By contrast, Cassini's flyby was at a distance of about 2,068 km, and the tiny moon was observed at this close range by a number of different investigations. They revealed a small irregularly shaped satellite, composed primarily of ice, rock and carbon-bearing compounds, very unlike asteroids or the two asteroid-like satellites of Mars observed to date by other spacecraft. (Porco et al. 2005a; Esposito et al., 2005; Clark et al., 2005a) Instead, Phoebe appears to be a prototype for one of the icy bodies in the region beyond Neptune's orbit, a region known as the Kuiper Belt. (Johnson & Lunine, 2005) It is a heavily cratered body and may be the parent body for a number of recently discovered smaller satellites of Saturn with orbits similar to that of Phoebe. Precise radio tracking of the spacecraft during the flyby led to a determination of the mass and density (about $1.6\,\mathrm{g\,cm^{-3}}$) of the strange satellite. (Johnson & Lunine, 2005) Radar observations are also consistent with the dirty, rocky, icy surface suggested by other observations. (Ostro et al., 2004)

9.4 Early Magnetospheric Findings

The Cassini-Huygens spacecraft breached Saturn's bow shock and magnetopause about two days before insertion into orbit around the planet. The magnetopause is the outer edge of Saturn's magnetic field (see Fig. 9.4) and marks the transition from the solar wind to Saturn's magnetosphere. It is akin to the hull of a ship, which separates the interior of the ship from the water through which the ship is traveling. As in the case of the water through which a ship travels, the solar wind is slowed from supersonic to subsonic speeds as it approaches the magnetopause and flows around the magnetosphere, forming a bow shock wave in front of the magnetosphere. The region of affected solar wind flow exterior to the magnetopause is known as the magnetosheath.

Once inside the magnetosphere, the magnetic field of Saturn can be sensed with proper instrumentation. Cassini's Dual Technique Magnetometer is the prime instrument for such measurements, but half a dozen Cassini instrument packages contribute to our understanding of the magnetosphere and its interaction with the solar wind, the rings, the satellites, and the planet's atmosphere.

One of the prime objectives of magnetospheric studies was to determine the angular offset between the magnetic pole direction and Saturn's spin axis. Unlike any of the other

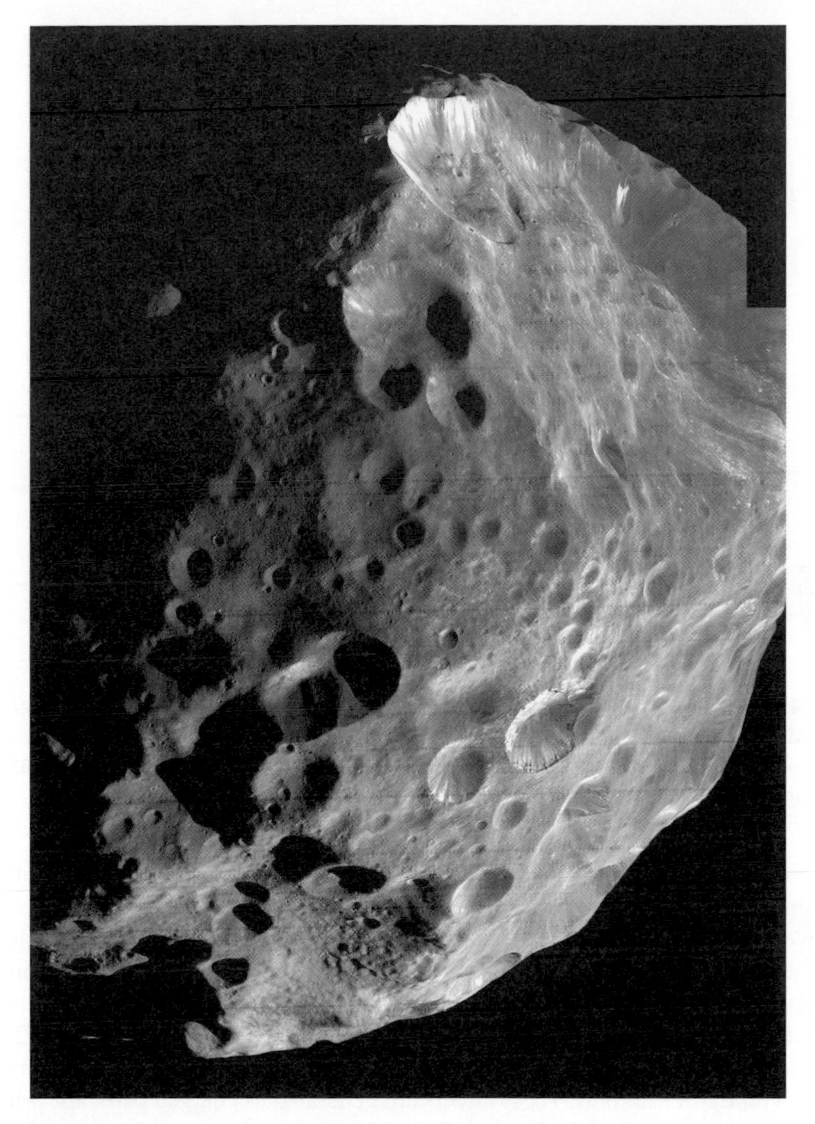

Fig. 9.3. Mosaic of six images showing Phoebe shortly before the time of closest approach on June 11, 2004. Close examination of some of the craters reveals that Phoebe, with one of the darkest surfaces in the solar system, has bright material (presumably water ice) beneath a thin layer of darker material.

planets, Saturn's substantial magnetic poles appear to be almost perfectly aligned with its rotation axis. Dynamo theory is fairly well developed, but perfect alignment between rotation and magnetic axes poses a potential problem for magnetospheric scientists, whose simple dynamo theories require at least a small misalignment for the planet to maintain its magnetic field. (See, for example, Cowling, 1981; or Parker, 1987) More complex (or multiple) dynamos may result in alignment between magnetic and rotation

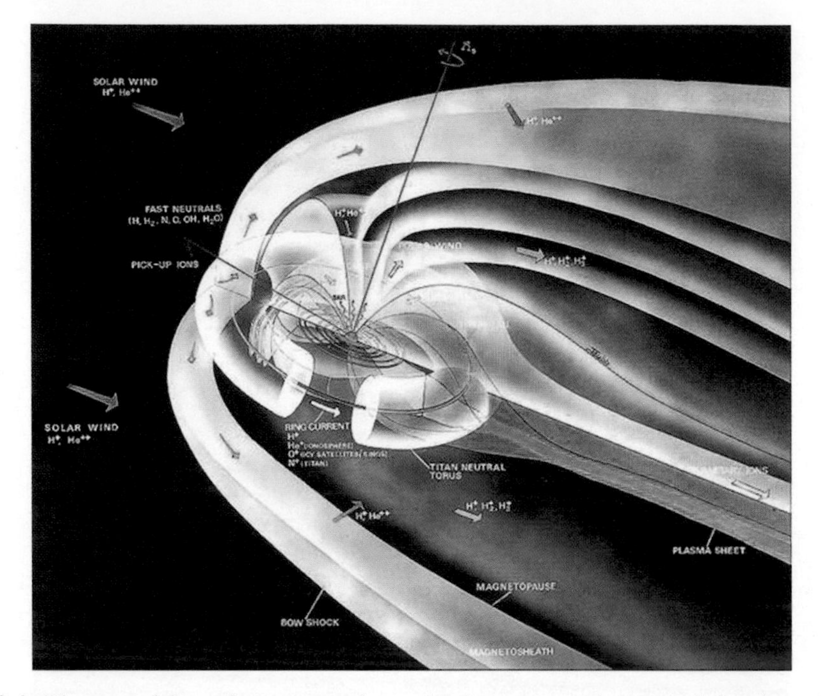

Fig. 9.4. Diagram of Saturn's magnetosphere, showing compression toward the planet in the sunward direction through interaction with the solar wind, which also creates a downstream "windsock" tail in the anti-sun direction. A doughnut-shaped torus of neutral particles surrounds the orbit of Titan.

axes, but determination of their precise parameters is correspondingly more complicated. Measurements during the Saturn orbit insertion, when the spacecraft was as close as it will ever be to the planet, were to have resolved the issue of the magnitude of the misalignment, if any exists. Unfortunately, the magnetometer was not configured properly during that period, and some of the anticipated data were not collected. Subsequent close passes to Saturn should eventually provide the relevant data. (Dougherty et al., 2005)

Observations of Saturn's "northern lights" (aurora borealis) have shown many similarities with Earth's auroras. These effects are generated as charged particles, primarily from the solar wind, spiral down magnetic field lines into the atmosphere near the magnetic poles. On Earth, dramatic brightening in such auroral displays characteristically last a few minutes; on Saturn such auroral brightening seems to be stable for days at a time. Saturn's auroras seem to be affected more strongly by the Sun's magnetic field than by the magnetic fields generated by the flow of charged particles in the solar wind; for Earth, the opposite is true. As a consequence, the auroras on Saturn seem to be relatively simple, remaining relatively fixed with respect to the Sun's direction, whereas those on Earth and Jupiter tend to be tied more strongly to the rotation of the planet. However, observations to date have revealed very little azimuthal variation with the Saturn auroras, and these conclusions may need to be revised as more data are collected. (Clarke et al., 2005; Crary et al., 2005; Kurth et al., 2005)

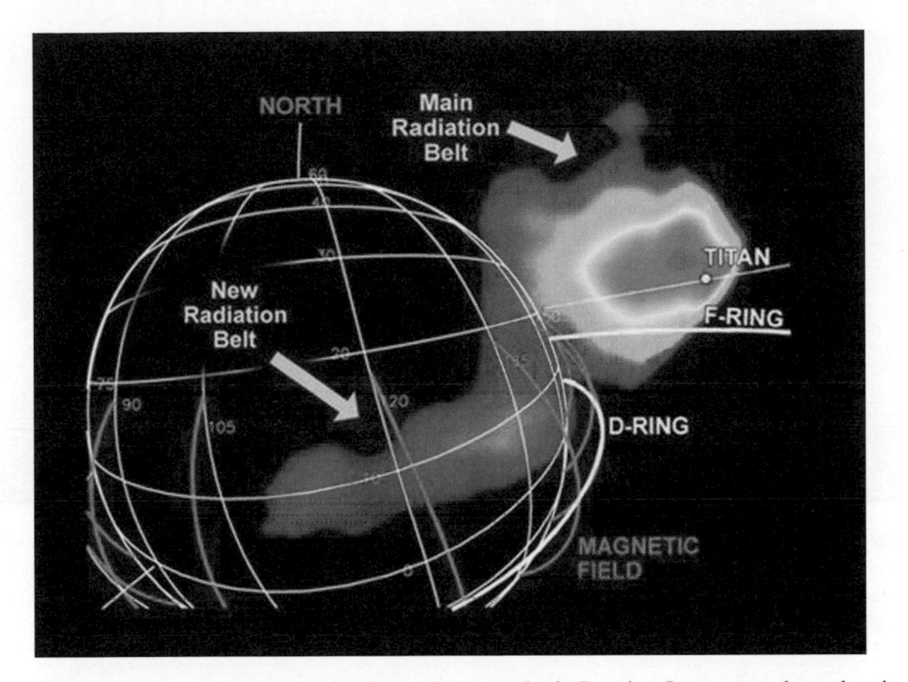

Fig. 9.5. Interpretive image produced from Magnetospheric Imaging Instrument data, showing both a portion of the main radiation belt (upper right) and part of the newly discovered radiation belt (near the bottom center) between the inner edge of the rings and the top of Saturn's atmosphere.

The Magnetospheric Imaging Instrument (MIMI) has detected a new radiation belt (Fig. 9.5) between the inner edge of Saturn's D Ring and the top of the atmosphere; this detection is only possible with an instrument like MIMI that can detect charged particle interactions from a distance rather than having to fly through the region. (Krimigis et al., 2005)

The Radio and Plasma Wave Science instrumentation also discovered that pulsed radiation from Saturn (similar to that used by Voyager to determine the rotation period of the magnetosphere and hence the rotation rate of Saturn's interior) has a period six minutes longer than the Voyager-determined rotation rate of Saturn. It is highly unlikely that Saturn's rotation period has lengthened from 10 h 39 m 22.4 s $\pm$ 7s in 1980 to 10 h 45 m 45 s $\pm$ 36 s in 2005, so other explanations for this surprising finding are being sought. It is possible that the difference is caused by mass loading within the magnetosphere, which might cause the external magnetic field to "slip" with respect to the internally generated magnetism. (Gurnett et al., 2005; Sánchez-Lavega and Pérez-Hoyos, 2005)

Voyagers 1 and 2 made estimates of the composition of the charged particle population within Saturn's magnetosphere, but could not distinguish, for example, between water and nitrogen. Titan's predominantly nitrogen atmosphere was expected to contribute nitrogen ions to the magnetosphere, but Cassini, which can easily distinguish between the two constituents, found surprisingly small concentrations of nitrogen and larger than expected water ion concentrations in the inner magnetosphere. (Young et al.,

Fig. 9.6. Saturn and its ring system, seen at high resolution in its visible-light appearance a few days before the Cassini spacecraft went into orbit around the planet. Note the detail in the radial structure of the rings, the shadow of the planet on the rings (at the left), and the shadows of the rings on the planet (at the top).

2005; Gombosi & Hansen, 2005). Overall, the magnetosphere was observed to be much more variable and complex than had been expected.

9.5 Early Atmospheric Observations

Cassini possesses much more capable imaging systems than were available on the two Voyager spacecraft. Some of the observed higher level of atmospheric activity may be due to that fact, but Saturn's atmosphere seems to have more storms than were seen twenty-five years ago. As summarized in Porco et al. (2005b), equatorial wind speeds have slowed from about 500 m/s during the Voyager flybys to less than 400 m/s, although there is a spread of velocities, perhaps associated with different altitudes. Oval spot production seems associated with bright storms, which may be associated with atmospheric lightning (see also Gurnett et al., 2005). Thermal infrared data (Flasar et al., 2005a) also indicate that equatorial winds decrease sharply with altitude. If the lower equatorial wind speeds recently noted by Hubble Space Telescope imaging are associated with higher altitude clouds rather than changing wind structure of the atmosphere, the clouds observed by HST must have been about 130 km higher in altitude than the equatorial clouds presently being measured by Cassini.

It is often instructive to view near infrared images of the atmosphere. Figure 9.7a, for example, is a composite of three near-infrared images, presented in the original false-color image as blue, yellow, and red. In particular, it shows the difference in altitudes of the clouds associated with Saturn's belts and zones. Figure 9.7a emphasizes the distribution of methane gas in the atmosphere and shows a complex storm structure referred to as the Dragon Storm by Cassini's imaging scientists.

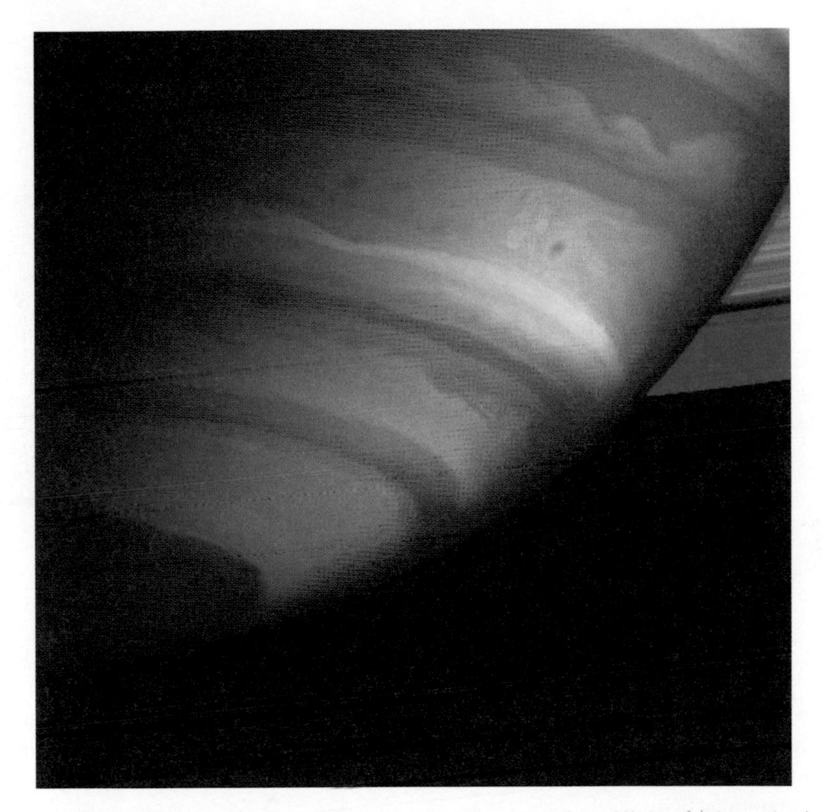

Fig. 9.7. (a) The southern hemisphere of Saturn as seen in near infrared light which emphasizes the methane gas in the atmosphere. The bright bands are higher in the atmosphere; the darker (redder) bands are deeper. Above and to the right of center (just above the bright band) is a complex storm system called the Dragon Storm. Note the obscuration near the limb caused by methane absorption.

A little further into the infrared, the contrast is even more startling. As reported in September 2005 (Baines et al., 2005), data from the Visible and Infrared Mapping Spectrometer show discrete cloud and storm activity about 30 km beneath the upper clouds usually seen on Saturn (Fig. 9.7b). These lower clouds are likely composed of either water or ammonium hydrosulfide, not the ammonia ice of the upper cloud layer. They are actually seen by their blockage of 5-micrometer radiation emanating from Saturn's interior, and would appear like negative images in the infrared data. In Fig. 9.7b, the image is presented as bright clouds to give the clouds a more normal appearance.

A huge cloud of atomic oxygen was detected by the Ultraviolet Imaging Spectrograph early in 2005; the cloud developed very suddenly and then dissipated over a period of about two months. (Esposito et al., 2005) The Composite Infrared Spectrometer has detected an atmospheric methane abundance that is about twice as large as that from Voyager and Earth-based measurements. This new methane value yields a carbon-to-hydrogen ratio for Saturn that is about seven times the solar ratio. Carbon is enriched in Saturn's atmosphere by twice the amount found on Jupiter, a finding more consistent with models of giant planets in the solar system. (Flasar et al., 2005a)

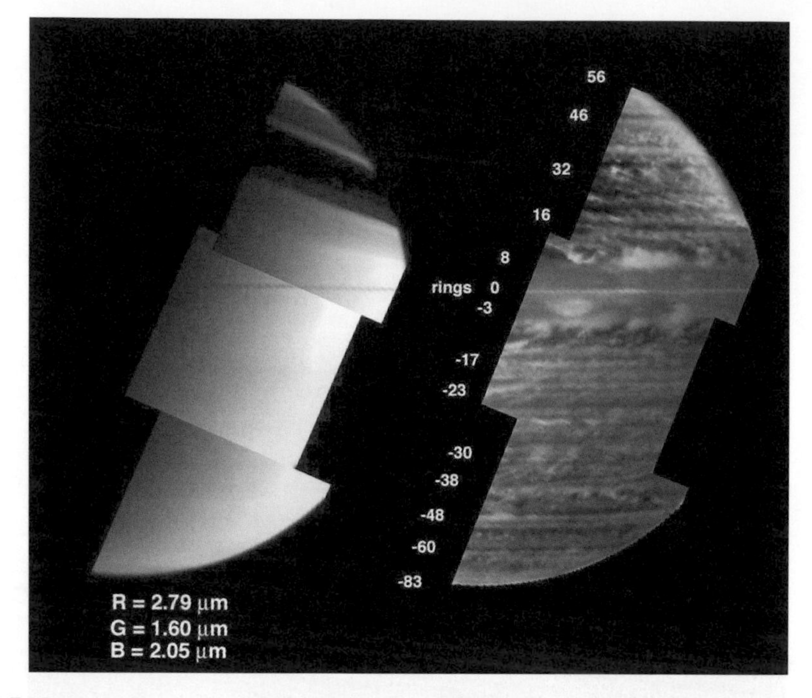

Fig. 9.7. (b) An infrared view of clouds located about 30 km beneath the visible ammonia ice clouds of Saturn. At the left is the appearance of Saturn in the 1–3 micrometer wavelength range; at the right is (at 5 micrometers) the lower cloud layer (with reversed contrast) displays major cloud and storm activity.

9.6 Early Findings on Saturn's Rings

Saturn's most unique characteristic are its bright rings, so long an enigma to astronomers following Galileo's 1610 sighting of Saturn's mysterious triple sphere, later confirmed by Christiaan Huygens to be the "cup handles" (ansae) of a continuous thin ring, circling but nowhere touching Saturn. (That observation was made in 1655, the same year Huygens also discovered Titan.) Because of its rings, Saturn has often been called the "Jewel of the Solar System." The Cassini orbiter has now observed those rings in more detail than ever before, and those details have done nothing to diminish their intrigue.

During Saturn Orbit Insertion, images and other remote and in-situ sensing of the rings provided dramatic proof of their beauty, complexity, and characteristics. In addition to their appearance in high-resolution images (Porco et al., 2005c), an example of which is shown in Fig. 9.8. Infrared data have displayed temperature variations across the rings, showing that temperatures of the ring particles are slightly lower in the thicker A and B rings, where mutual shadowing by other ring particles occurs; temperatures range from about 70K in the optically thick portions of the A and B rings to about 110K in the optically thin C Ring and Cassini Division. (Flasar et al., 2005a) Interestingly, the infrared data also show a decrease in temperature with increasing phase angle for all of the main rings, something that could only occur if a substantial fraction of the

Fig. 9.8. Both a density wave (left, due to Prometheus) and a bending wave (right, due to Mimas) in Saturn's A Ring were captured in a single high-resolution frame near the time of Saturn orbit insertion on 1 July 2004. The view is of the unilluminated northern face of the rings and spans a radial distance of about 300 km in the inner A Ring.

ring particles have rotational periods that are long compared with the time required for radiative cooling. (Spilker et al., 2005) Composition measurements confirm that the rings are mainly composed of water ice, but they show a tendency for there to be more silicate materials near ring edges, near gaps and in the C Ring and Cassini Division. (Esposito et al., 2005; Brown et al., 2005) Much of this variation can be explained by continued meteoritic bombardment over the age of the rings. Particle size variation with radial distance from Saturn were determined from 3-frequency (0.94, 3.6, and 13 cm wavelength) radio signals generated at the spacecraft during its passage behind the rings as viewed from ground receivers. The data indicate that the abundance of ring particles with diameters greater than 5 cm increases outwardly in the A ring, with such particles dominating the population of the area between the Encke Division and the outer edge of the A Ring. 5-cm ring particles are almost absent in the inner portions of the A Ring. Small grains deposited on the ring particles have also been studied by the Visible and Infrared Mapping Spectrometer, which indicates a range of grain sizes from fine snow to snow pellets. The VIMS experiment also confirms the predominantly water ice composition of the rings. The radio and VIMS results were discussed at the annual meeting of the American Astronomical Society's Division for Planetary Sciences in September 2005 in Cambridge, England (Marouf et al., 2005; Clark et al., 2005b).

Several new rings have been found in the images, and careful measurements of the spacing of weak density waves in the rings have yielded accurate masses for the moons whose gravitational forces generate those density waves. When one compares the derived masses with the measured sizes of Atlas (orbiting just exterior to the A Ring) and Pan (orbiting in the A-Ring Encke Gap), the implication is that these moons are very porous and are perhaps loose rubble piles. (Porco et al., 2005c

Images of the F Ring (Fig. 9.9) reveal multiple strands and substantial azimuthal variation. Some of the new rings are also kinked, most likely as a result of nearby satellites not yet seen in the images. A denser ring arc has been found within the otherwise tenuous G Ring, which lies outside the F Ring. (Caption with the September 5, 2005, release of image; see http://photojournal.jpl.nasa.gov/catalog/PIA07718.) Imaging scientists were also surprised to spot narrow finger-like structures one or two km in length in the troughs of the strongest density waves, apparently a clumping of ring particles into long, narrow transient structures which appear to fade as the density wave damps. Another unexpected and intriguing discovery is a series of regularly spaced, ropey features near the outer edge of the Encke Gap (Fig. 9.10). Computer models show that straw-like and ropey structures can form when ring particles are forced close together in the crests of density waves or wakes, forming loosely bound elongated clumps of particles. The ultraviolet instrument succeeded in measuring the length, width, and height for these fingerlike structures (Colwell et al., 2005), which may be the cause of azimuthal brightness variations in the A Ring reported by many ground-based observers and by Voyagers 1 and 2. (see, for example, Smith et al., 1981) The ghostly radial spokes observed by

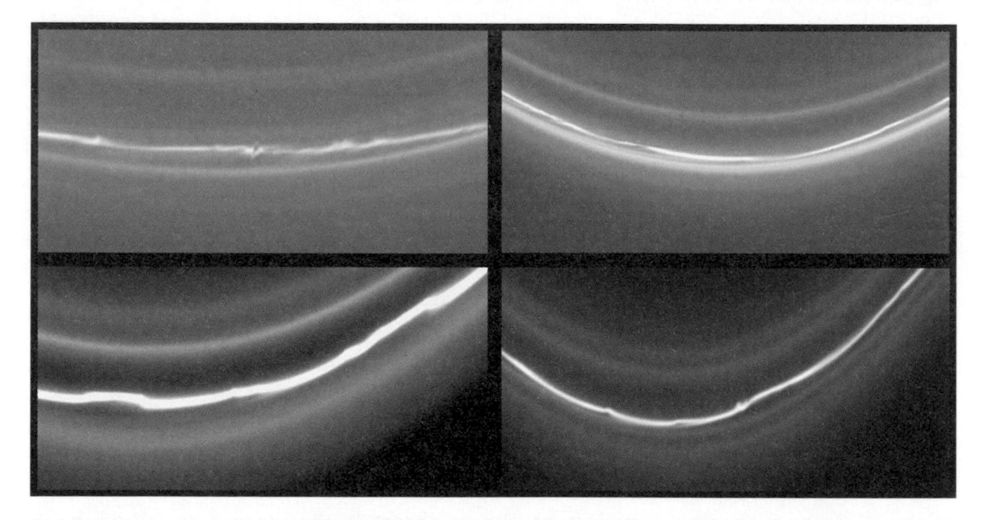

Fig. 9.9. Four views of the F Ring, showing kinks and other non-axisymmetric features, assumed to be due primarily to gravitational interactions with the two shepherding moons, Prometheus and Pandora. These images were taken at different azimuths, all within a few hours. Note that even the number of strands is not the same in all four views. These images have widths of about 400 to 600 km.

Fig. 9.10. Three views of non-axisymmetric features within the A Ring of Saturn. (A) A mottled region near the outer edge of the ring. (B) Same mottled region contrast-enhanced and reconstructed into an azimuth vs. radial distance image, covering about 1% of the ring's circumference. (C) Outer edge of the Encke Gap, where the structure appears to be rope-like in nature; individual strands are 10 to 20 km in length.

Voyagers 1 and 2 in the outer half of the B Ring eluded detection by Cassini instruments for the early part of the tour, but they have now been seen, with essentially the same characteristics as observed by the two Voyager spacecraft in the early 1980s. (See NASA press release of September 16, 2005; full text at http://saturn.jpl.nasa.gov/news/press-release-details.cfm?newsID=602)

An interesting result from the Ion and Neutral Mass Spectrometer is the discovery of molecular oxygen above the ring plane. (Waite et al., 2005a) Molecular oxygen in Earth's atmosphere is uniquely tied to the presence of life, but the process which created the molecular oxygen above Saturn's rings must one that is abiotic, such as the molecular decomposition of water in the ice.

During Cassini's approach to Saturn, the Cosmic Dust Analyzer discovered streams of tiny (less than 20 nanometers) high-velocity (~100 km/s) "dust" particles escaping from the Saturn system. (Kempf et al., 2005) Some of these dust particles were determined to have originated from the outer regions of the A Ring. Composition of these particles was primarily silicates, implying that the particles are impurities from the icy ring material rather than the ice particles themselves.

The complexities of the ring system make it difficult to estimate with any degree of certainty the age of the rings. Cassini data may well constrain that age, but it will take considerably more analysis before we know just how long these magnificent rings have existed or will continue to exist. It seems clear thus far that the rings which presently exist have not been around for a large fraction of the age of Saturn itself. Northrup & Connerney (1987) used arguments based on estimated micrometeoroid impact rates to estimate an age for the rings of between 4.4 and 76 million years, orders of magnitude shorter than the estimated 4.5 billion years that Saturn has existed.

9.7 Observations of Titan

9.7.1 Huygens probe observations of Titan

The Huygens Titan probe was built and operated by ESA and carried piggyback into orbit around Saturn by the Cassini orbiter. Initially, the probe was to descend to Titan's surface during the first close flyby of Saturn's largest moon, but the Doppler shift created by the 6 km/sec relative velocity shifted the frequency of the signal transmitted by the probe out of the orbiter's probe receiver frequency range. There was no solution available to mission designers other than to reduce the relative velocity of the two craft. The redesigned mission added an additional Titan flyby and targeted the probe descent for the third flyby instead of the first. Accordingly, the Huygens probe was released from the orbiter on December 25, 2004, and, following a three-week ballistic cruise, descended by parachute through Titan's atmosphere on January 14, 2005. All six instrument packages on the Huygens probe operated as designed, and almost all the anticipated data was collected. The Doppler Wind Experiment involved several radio telescopes (including the large Green Bank telescope) and the Parkes Radio Telescope in Australia, even beyond cessation of the data recording on the orbiter. The resulting data reduction revealed high-altitude eastward winds with velocities of up to 430 km/hr near 120 km altitude; below 60 km wind speeds decline monotonically to a gentle breeze near Titan's surface.

The data collection period was to be approximately 2.5 hours for the descent and, if the probe survived landing, up to an additional hour on the surface of Titan. As it turned out, the surface of Titan at the landing site had the consistency of wet mud, and Huygens easily survived the landing. The battery pack, designed to have a minimum operating lifetime of three hours, continued to power the probe instruments and other electronics until after the Cassini orbiter disappeared below the Titan horizon as viewed from the landing site, more than 100 minutes after landing. Doppler Wind data were received at Earth well after data collection by the Cassini Orbiter ended.

Detailed results from the landing are published in a series of seven articles in the December 8, 2005, issue of *Nature*. A press conference covering very early findings was held January 21, 2005, one week after the successful landing. More than 474 megabits of data were received from Huygens by the Probe Support Equipment in a period of 3 hours 44 minutes. Titan's atmosphere was probed and sampled at altitudes from 160 km to the surface. The mix of nitrogen and methane gases was uniform in the upper atmosphere, but the relative abundance of methane increased with decreasing altitude, reaching a maximum of about 4% (by number) near the surface, much like water vapor on Earth. Surprisingly, no argon-36 or argon-38 was detected in the atmosphere, although small amounts of argon-40, a byproduct of cryovolcanism, were detected. The probe passed through a methane cloud layer at an altitude of about 20 km, and a fog of methane was detected near the surface of Titan. Near the surface the temperature was about −180° Celsius. Images from the Descent Imager and Spectral Radiometer revealed a

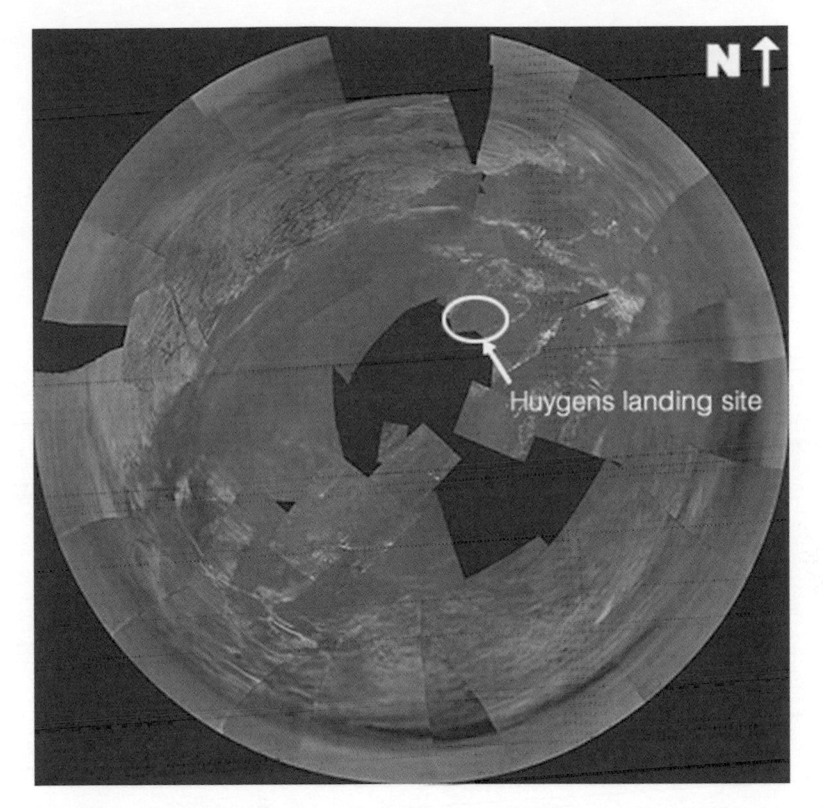

Fig. 9.11. Fisheye view of Titan as seen from the descending Huygens Probe. The approximate landing site is indicated. The data comes from reconstruction of a large number of images from the on-board Descent Imager and Spectral Radiometer.

surface that has been greatly altered by what appears to be liquid erosion (see Figs. 9.11 and 9.12). There were drainage channels, shoreline-like features, erosional rounding of surface rocks, evidence of soil erosion beneath rock edges, and some evidence of size sorting of pebbles on the surface, but no direct evidence of present surface liquid was detected. There is some speculation that Titan's surface may be somewhat akin to the arid southwestern United States, where river channels carry liquid only for brief episodes during the wet season of the year. On Titan, of course, the liquid is not water; temperatures are far too cold for liquid water to exist. Instead, the liquid is most likely methane. Heat emitted by the Huygens probe warmed the surface beneath the probe, and the Gas Chromatograph and Mass Spectrometer measured a sudden increase of methane gas boiling out of the surface, reinforcing the idea that methane forms clouds and produces rain that erodes the surface. However, no ethane gas was detected near the surface, although the vapor pressure of ethane would be about three orders of magnitude smaller than that of methane. The probe imaging system continued to operate after landing, providing a view and spectra of the surface materials near the landing site (Fig. 9.13).

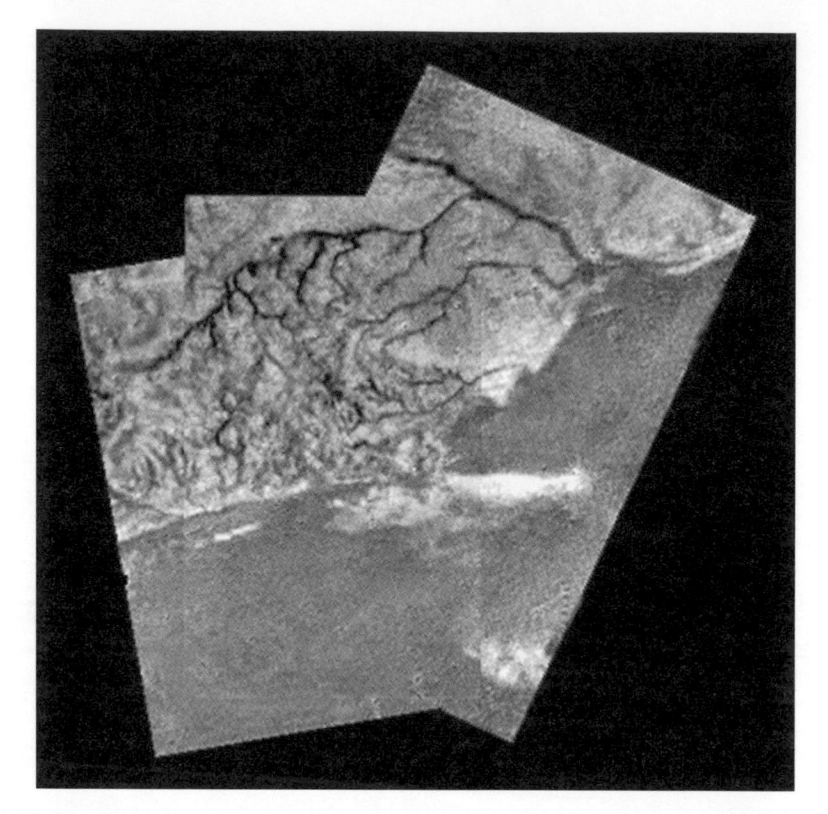

Fig. 9.12. Dendritic channels seen by the Huygens Probe descent imager lead from a brighter, higher region to the darker, nearly featureless lowlands. While they are indicative of past erosion, there does not appear to be liquid in the channels at present. The dark areas are not liquid at present, although they may have held liquid at times in the past.

There are suggestions (although no direct detection) that ammonia is relatively abundant in the surface and subsurface of Titan. That would be particularly significant, if true, because of the potential for a subsurface layer water-ammonia liquid, raising the possibility of ongoing geologic activity on Titan. Titan has an orbital eccentricity of about 0.03, and if that in turn gives rise to solid surface tides and tidal dissipation, the heat might be provided to drive a small amount of surface modification. The detection of argon-40, normally produced at depth on Earth, may further indicate a connection of interior to surface, perhaps giving additional credence to such a process.

Analysis of the Huygens data will continue for years. The first task is the reconstruction of the descent trajectory. That process was complicated by wide swings of the probe in the upper atmosphere, sometimes exceeding 60 degrees from vertical. Lower in the atmosphere, the swings were only a few degrees in extent.

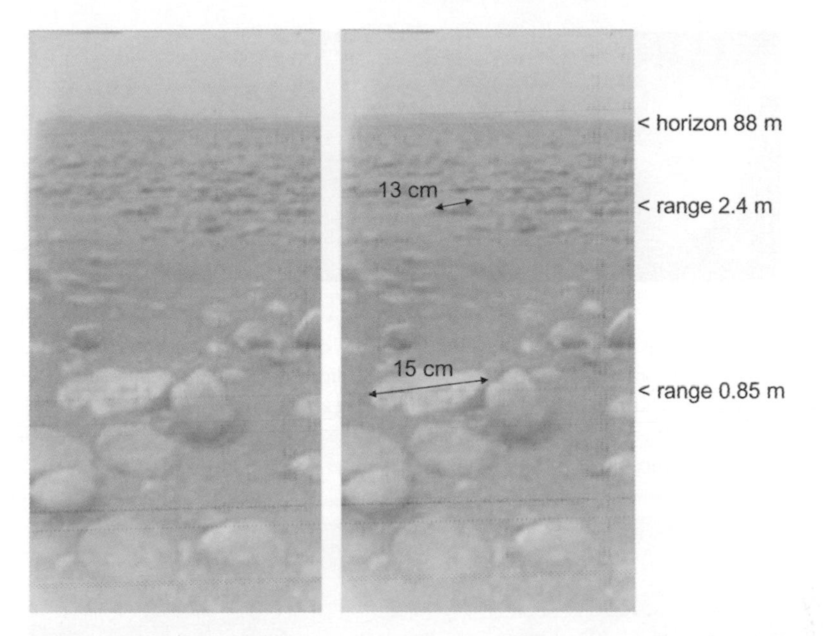

Fig. 9.13. Following its relatively soft landing, the Huygens Probe continued to collect data, including this hazy image of the surface at the landing site. The rocks may be silicate, but are more likely chunks of water or methane ice.

9.7.2 Cassini orbiter observations of Titan

When the Cassini orbiter spacecraft was being planned and designed, great attention was paid to the inclusion of scientific instruments that would be optimized for studies of Titan. Two of the orbiter instruments were specifically included and designed for Titan observations, the Titan Radar and the Ion and Neutral Mass Spectrometer. Many of the other investigations included explicit accommodations for the in-depth study of Titan. Indeed, when one considers the inclusion of the Huygens probe as a part of the science payload, it is clear that understanding Titan is the prime (but not sole) objective of the Cassini-Huygens Mission.

High-resolution surface coverage by the Titan Radar was a prime objective of the mission. About half of the 45 close Titan flybys were designed to include such coverage. Each of these passes covers about one percent of the surface, so during the four-year primary mission, about 20% of the surface is covered by high-resolution radar imaging. Fortunately, other optical remote-sensing instruments also have designs which permit surface observations. Both the Imaging Science instrumentation, with its high spatial resolution, and the Visible and Infrared Mapping Spectrometer, with lower spatial resolution but good spectral resolution, have been successful in detecting Titan surface features.

Cassini orbiter images, including the radar images, (Porco et al., 2005d; Elachi et al., 2005) of the surface show evidence of a variety of processes that have occurred on Titan's surface. These include fluid and wind-driven erosion and transport of surface materials, surface folding and faulting, and possibly volcanic activity. There are also

Fig. 9.14. Titan is seen here at different wavelengths. At the left is a view of Titan as it would appear to the human eye, with the surface completely obscured by the overlying haze. At the center is a view of Titan in the near infrared (984 nanometers), where the haze becomes almost transparent. The view at the right is a false-color representation, which combines two infrared images with a blue image, creating an image in which the surface dominates the view in the center of the disk and the atmosphere and haze layers dominate near the limb of Titan.

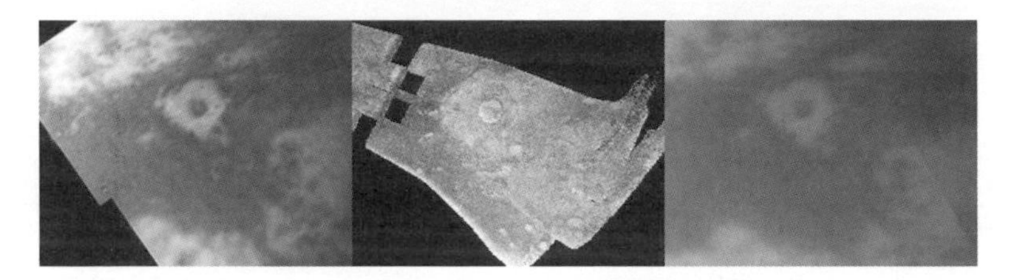

Fig. 9.15. Although early in the mission there was not a lot of overlap between high-resolution radar views and visible or infrared imaging of Titan's surface, these views of the same crater, obtained by the Visible and Infrared Mapping Spectrometer (left and right) and the Cassini Radar (center), are representative of the kinds of synergism that will facilitate better understanding of Titan's surface processes.

circular features which may be impact craters (Fig. 9.15). Among the more surprising features is a river channel extending for 1500 km across the surface of Titan. Bright and dark regions with sharp boundaries are plentiful, but their precise nature is still unknown. Occasional clouds at mid-latitudes show that winds in the lower atmosphere are prograde (eastward – the same direction as Titan's rotation) and have velocities of up to 120 km/h. Persistent but changing clouds hover over Titan's south pole; it is possible that methane rain and enhanced erosion may be taking place in the polar region. In fact, one feature seen in the south polar region (Fig. 9.16) may be the only lake yet detected, either at Titan's pole or elsewhere on the surface. Infrared spectral images show marked lateral differences in surface composition (Brown et al., 2005).

A detached haze layer exists at an altitude of 500 km, which is 150 to 200 km higher than a similar haze layer observed by Voyager 25 years ago. (Porco et al., 2005d) During the first Titan flyby, the Ion and Neutral Mass Spectrometer obtained the first *in situ* measurements of the upper atmosphere. (Waite et al., 2005b) Among the more abundant species in the upper atmosphere were molecular nitrogen, methane, molecular hydrogen, argon, and a host of stable carbon-nitrile compounds. A stellar occultation of Titan also provided information on the upper atmosphere composition, identifying six species:

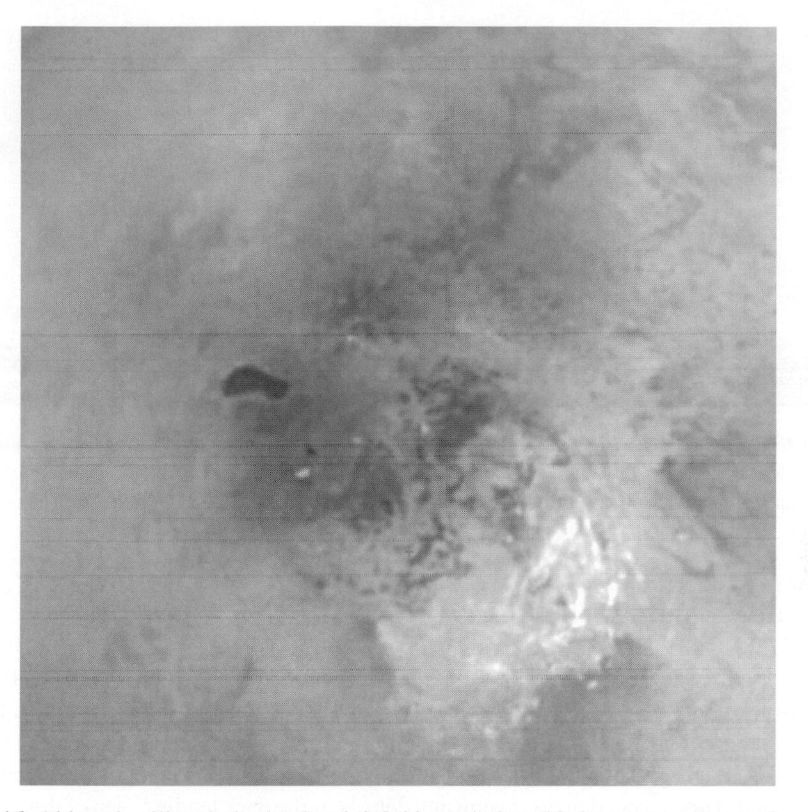

Fig. 9.16. Although still early in the Cassini Orbiter mission, this image near the south pole of Titan may be the first to show the possible presence of a lake on Titan, the dark spot left of center in the image. The south pole is marked by a small "+" just below center. Close to the pole in the lower right are the persistent (methane?) clouds seen in that region.

methane (CH_4), acetylene (C_2H_2), ethylene(C_2H_4), ethane (C_2H_6), diacetylene (C_4H_2), and hydrogen cyanide (HCN). (Shemansky et al., 2005) Deeper in the atmosphere, infrared spectral measurements yielded mole fractions of methane and carbon monoxide in the stratosphere as $(1.6 \pm 0.5) \times 10^{-2}$ and $(4.5 \pm 1.5) \times 10^{-5}$, respectively. (Flasar et al., 2005b) The fields and particle instruments also measured Titan's interaction with the Saturn magnetosphere (Mitchell et al., 2005), detected no evidence of an internal magnetic field at Titan (Backes et al., 2005), and measured the ionospheric electron density ($3800\,\mathrm{cm}^{-3}$) and escape flux (1025 ions/s) (Wahlund et al., 2005).

Because Titan is so diverse with its complex atmosphere and surface, it is the richest target in the Saturn system and is therefore the primary objective for both the Cassini Orbiter and the Huygens Probe; it is clear that it will continue to hold that distinction for the foreseeable future. The 45 close encounters of Titan included in the primary orbital mission may have seemed somewhat excessive to some, but only 20% of the surface will have been covered by high-resolution radar. In retrospect, many more flybys are needed to provide a reasonably clear definition of Titan's relatively static characteristics; even more are needed to characterize weather and possible seasonal changes.

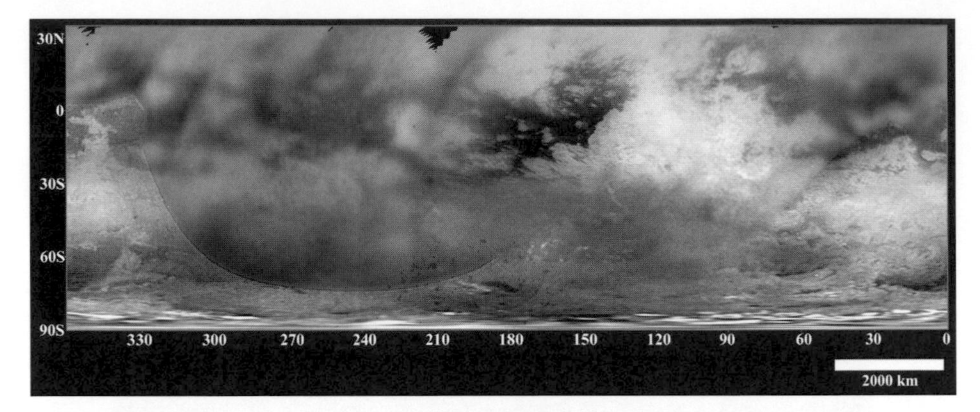

Fig. 9.17. Mercator map of Titan's surface constructed from orbiter imaging obtained in the first few passes of Titan. The bright region centered near a longitude of 120 has been unofficially designated as Xanadu. Resolution varies greatly from area to area because of the uneven coverage by mid-2005.

9.8 Early Findings on Iapetus

When Iapetus was first sighted through a telescope, its discoverer (Cassini) could only spot it on one side of Saturn. He correctly concluded that the satellite was in locked rotation, keeping its same face toward the planet and that the satellite's leading face in its orbit is much darker than its trailing side. Voyager imaged Iapetus from moderate range, but did not succeed in disclosing any details of the dark leading face of the satellite.

The Cassini orbiter has succeeded not only in bringing out features within the dark face of Iapetus, but has revealed one of the most amazing features seen on any body in the solar system: a linear ridge 1300 km in length which approximately bisects the dark face of Iapetus (Fig. 9.18). (Porco et al., 2005a) The ridge also very nearly marks the satellite's equator. Along parts of that ridge there are mountains that rival Olympus Mons on Mars in height, almost three times the height of Mount Everest on Earth. The diameter of Iapetus is only 21% that of Mars and 8% that of Earth.

Another very interesting finding of Cassini is its remarkably large oblateness (equatorial bulge). Iapetus has a shape equivalent to that of a fluid body with a 17-hour rotation period, not that of its present 79.33-day rotation period. By modeling its thermal and tidal evolution, Castillo et al. (2005) have shown that Iapetus is slightly older than Earth's age of 4.5×10^9 years. Theoretical scenarios involving rotation periods even shorter than 17 hours could possibly account for the equatorial ridge, which may extend completely around Iapetus.

The images alone cannot presently differentiate between external and internal sources for the dark material. However, multi-spectral imaging of Iapetus show that the bright regions are rich in water ice and the dark regions are rich in organic materials, which would seem to favor an external source. The reflectivity of the dark area of Iapetus (Cassini Regio, named for the man, not the mission) is about 4%, roughly that of fresh coal. Measurements of temperature variation with time of day indicate that the dark surface of Iapetus is finely divided and doesn't store heat well (Flasar et al., 2005a),

Fig. 9.18. An enormous ridge bisects the dark face of Iapetus and very nearly defines its equator. The relationship of the ridge to the dark deposits is still being debated. Note the slumping of the near wall of the large crater at the right into the floor of a smaller crater in its interior.

again possibly favoring external deposition. A third piece of evidence favoring external deposition is feathery-looking black streaks near the boundary between dark and light sides, similar in form to ejecta from impact craters on the Moon and other bodies in the solar system. While scientists agree that the dark material is most likely deposited from an external source, there is little consensus about the nature of that source. A very close encounter with Iapetus is scheduled to occur in September 2007; image resolution during that encounter is a hundred times better than imaging obtained prior to that time, and Cassini scientists hope therefrom to learn more about the possible source of the material that coats the dark face of this strange satellite.

Another startling feature on Iapetus is a landslide extending halfway across the floor of a 120-km diameter impact crater. The crater is at the interior edge of a much larger 600-km wide crater (see Fig. 9.18). The landslide appears to have originated from the collapse of part of the 15-km high cliff that forms the rim of the larger crater.

9.9 Early Findings on Enceladus

Three close-range targeted flyby of Enceladus are to be executed during the four-year orbital mission of the Cassini orbiter. Enceladus is of high interest because of its bright and geologically young surface (Fig. 9.19a). It is also of interest to Cassini scientists because of its orbital position near the densest part of Saturn's tenuous E Ring, leading to speculation that Enceladus may be the source of E-Ring particles. The first of the targeted flybys occurred March 9, 2005; the second occurred July 14, 2005. The third was not slated to occur until late in the four-year orbital tour of Cassini.

Enceladus is seen to have a very tenuous atmosphere, probably of water vapor. The atmosphere was not detected by conventional means, but by the distortion of Saturn's magnetic field in the vicinity of the small satellite. That is an indication that ionized particles are abundant near Enceladus. Because of its small size and low gravity, the finding

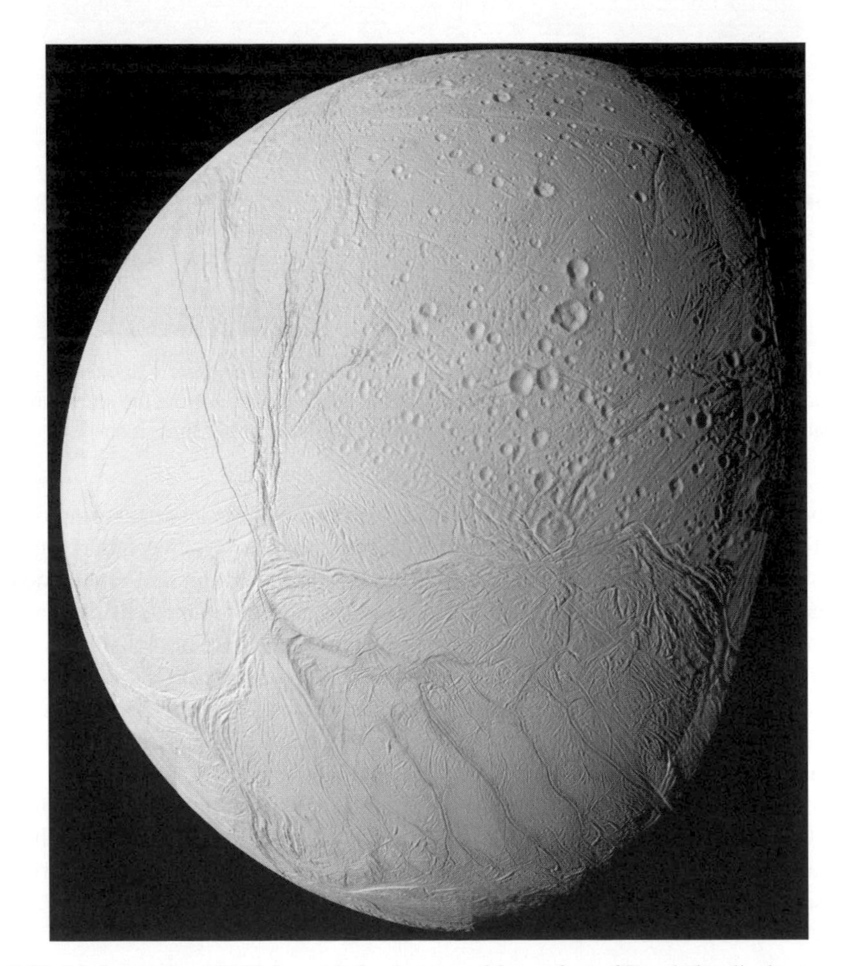

Fig. 9.19. (a) A mosaic of 21 high-resolution images of the surface of Enceladus displays a variety of surface features, including some moderately cratered areas, some lightly cratered areas, and areas devoid of craters but crisscrossed by fractures and other geologically young features.

is particularly significant; it implies a strong ongoing process to feed that atmosphere. There is a suspicion that the source may be geysers or volcanoes on the surface of the satellite, but images and other data have not yet revealed such activity. "Enceladus could be Saturn's more benign counterpart to Jupiter's dramatic Io," quipped Fritz Neubauer, Cassini magnetometer scientist and professor at Germany's University of Cologne. Since the initial discovery, confirmation of a tenuous atmosphere was provided by Ultraviolet Imaging Spectrograph observation of a stellar occultation by Enceladus.

Of particular interest are several long fractures in the south polar region of Enceladus. Their color is distinctively bluer that the surrounding plains and reminiscent of highly compressed water ice seen in glaciers and icebergs on Earth. A temperature scan across one of these fractures revealed that they are at least 20°C warmer than their surroundings. It is natural to conclude that the fractures are related to heat escaping from the interior of Enceladus, although the source of that excess heat has not yet been determined. Late in November 2005, an image (Fig. 9.19b) of a crescent Enceladus with these fractures at the illuminated limb, show that they are closely related to fountain-like sources of fine particles, presumably water ice, that spray material high above the surface of this geologically active moon.

Another intriguing discovery came from the highest resolution images yet obtained for Enceladus (Fig. 9.20). Taken from a distance of 200 km above the surface, the wide-angle image shows that the fracturing of the surface is even more extensive than had

Fig. 9.19. (b) Plume or fountain-like activity is seen on Enceladus near the south polar fracture region, verifying that fine-grained material is escaping from these fractures into space. These may be the source of particles feeding and replenishing Saturn's tenuous E Ring.

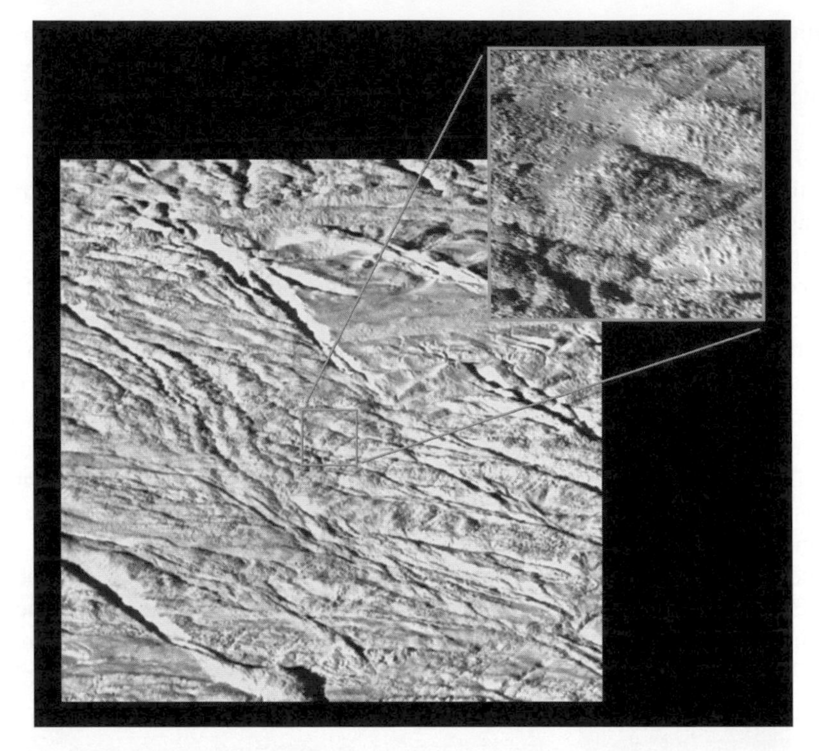

Fig. 9.20. A wide-angle image of Enceladus and an embedded narrow-angle image were obtained from an altitude of 208 km. The high-resolution image shows a scene strewn with 10-m to 100-m blocks of ice which may or may not be typical of much of the surface of this tortured moon.

been supposed. The accompanying narrow-angle image shows a boulder-strewn terrain, with blocks of ice generally in the size range from 10 to 100 m. If this type of terrain is ubiquitous across the surface of Enceladus and not confined to the area imaged in the figure, it is possible that there is explosive venting of subsurface volatiles through surface fractures or as-yet-undiscovered volcanic vents.

9.10 Other Preliminary Icy Satellite Findings

Saturn has at least 47 satellites. Cassini discovered four of these, and many more may be discovered before the mission is completed. The difficulty is sometimes in distinguishing between moons and loose rubble piles, especially for those near Saturn's rings. One of the newly discovered satellites (Fig. 9.21) was discovered in the Keeler Gap near the outer edge of the A Ring. Another (Polydeuces) shares its orbit with Dione.

In Fig. 9.22, Prometheus appears to be capturing material from the multi-stranded F Ring. Another series of images shows Rhea being eclipsed by a closer Dione; another series shows Mimas eclipsing Janus. Tiny Atlas, which orbits just outside the A Ring was seen for the first time since Voyager images of it in 1981. Careful measurements of density waves induced in the rings by Atlas show that its mass is small enough that

Fig. 9.21. A 7-km-wide satellite, first seen in images in May 2005, is responsible for creating the Keeler Gap in the outer A Ring. The satellite, provisionally designated S/2005 S1, is Saturn's 47th known moon. Note the wakes created by the moonlet at the inner and outer edges of the Keeler Gap.

Fig. 9.22. Prometheus, one of the F-ring shepherds, appears to be pilfering some of the ring material in addition to creating (along with the other F-ring shepherd, Pandora) kinks and twists in the F Ring.

Fig. 9.23. Hyperion does not have a long-term fixed rotation period, thanks to its gravitational interactions with Titan. Its "chaotic" rotation and irregular shape are not its only unusual characteristics, as evidenced by this surface image obtained from a range of about 62,000 km on September 26, 2005.

Atlas must be very porous; the same is true of Pan in the Encke Gap of the A Ring. Perhaps they too are rubble piles. Wispy terrain on Dione is seen to be the consequence of bright ice cliffs created by tectonic activity, not from thick ice deposits, as previously supposed. A large multi-ringed crater straddles the long Ithaca Chasma on the surface of Tethys.

Tiny Hyperion, which tumbles chaotically in an orbit just outside that of Titan, has a surface appearance reminiscent of a sea sponge (Fig. 9.23). The floors of many of the craters appear dark, but the dark material may be only tens of meters or less in thickness. (See NASA Press Release of September 29, 2005; the full text of the release is found at http://saturn.jpl.nasa.gov/news/press-release-details.cfm?newsID=605) This bizarre little moon may offer some severe challenges for planetary geologists attempting to interpret the images.

Other discoveries related to the icy satellites are too numerous to mention, and each new view brings more surprises. The observations continue, and as Cassini continues its orbital tour, it is rewriting the books on satellite characteristics (and on all other aspects of Saturn's system).

9.11 Conclusions

Cassini appears on the way to being the most prolific planetary mission in history in terms of scientific return. Even considering its considerable cost relative to more recent, less complex robotic missions, the cost relative to amount of useful data Cassini-Huygens is collecting may make it one of the most fiscally efficient missions ever. The mission has been almost problem-free thus far. There have been a few human errors, but the hardware has been remarkably resilient to the demands placed on it. This testifies to the care in design, maintenance and operation of that hardware provided by Cassini and Huygens personnel, past and present. It also testifies to the robustness of the spacecraft and mission designs. Cassini was built with abundant electrical power, plentiful fuel and other resources, and adequate margin in its navigation. The trouble-free operation has enabled the scientists, and the world in general, to focus on the spectacular results from Saturn. A major factor in the success of the mission is the international teamwork and cooperation that has characterized it. Indeed, the very survival of the mission prior to its launch is largely due to the strong international support in the face of United States Congressional efforts to trim NASA's budget. Let us hope that the Cassini-Huygens Mission continues to be so successful.

Acknowledgement

All figures reproduced with this review are courtesy of NASA Jet Propulsion Laboratory.

References

Backes, H., Neubauer, F.M., Dougherty, M.K., Achilleos, N., André, N., Arridge, C.S., Bertucci, C., Jones, G.H., Khurana, K.K., Russell, C.T., Wennmacher, A., 2005. Titan's Magnetic Field Signature During the First Cassini Encounter. *Science* **308**, 992–995.

Baines, K.H., Momary, T.W., Roos-Serote, M., 2005. The Deep Winds of Saturn: First Measurements of the Zonal Windfield Near the Two-bar Level. *Bull. Amer. Astron. Soc.* **37**, 658 (DPS2005 abstract 20.07).

See also the NASA News Release on September 5, 2005. (http://saturn.jpl.nasa.gov/news/features/feature20050905a.cfm).

Bolton, S.J., Janssen, M., Thorne, R., Levin, S., Klein, M., Gulkis, S., Bastian, T., Sault, R., Elachi, C., Hofstadter, M., Bunker, A., Dulk, G., Gudim, E., Hamilton, G., Johnson, W.T.K., Leblanc, Y., Liepack, O., McLeod, R., Roller, J., Roth, L., West, R., 2002. Ultra-relativistic electrons in Jupiter's radiation belts. *Nature* **415**, 987–991.

Brown, R.H., Baines, K., Bellucci, G., Buratti, B., Capaccioni,F., Cerroni, P., Clark, R. N., Coradini, A., Cruikshank, D., Drossart, P., Formisano, V., Jaumann, R., Matson, D., McCord, T., Mennella, V., Nielson, R., Nicholson, P., Sicardy, B., Sotin, C., 2005. Cassini VIMS at Saturn: The first 6 months. Oral paper delivered at 36th Annual Lunar and Planetary Science, 14–18 March 2005, Houston, Texas, Abstract #1166.

Castillo, J.C., Matson, D.L., Sotin, C., Johnson, T.V., Lunine, J.I., Thomas, P. C., 2005. A Geophysical Study of Iapetus: The Need For and Consequences of Al26. *Bull. Amer. Astron. Soc.* **37**, 705 (DPS2005 abstract 39.04).

Clark, R.N., Brown, R.H., Jaumann, R., Cruikshank, D.P., Nelson, R.M., Buratti, B.J., McCord, T.B., Lunine, J., Baines, K.H., Bellucci, G., Bibring, J.-P., Capaccioni, F., Cerroni, P., Coradini, A., Formisano, V., Langevin, Y., Matson, D.L., Mannella, V., Nicholson, P.D., Sicardy, B., Sotin, C., Hoefen, T.M., Curchin, J.M., Hansen, G., Hibbits, K., Matz, K.-D., 2005a. Compositional maps of Saturn's moon Phoebe from imaging spectroscopy. *Nature* **435**, 66–69.

Clark, R.N., Brown, R., Baines, K., Bellucci, G., Bibring, J.-P., Buratti, B., Capaccioni, F., Cerroni, P., Combes, M., Coradini, A., Cruikshank, D., Drossart, P., Filacchione, G., Formisano, V., Jaumann, R., Langevin, Y., Matson, D., McCord, T., Mennella, V., Nelson, R., Nicholson, P., Sicardy, B., Sotin, C., Curchin, J., Hoefen, T.. Cassini VIMS Compositional Mapping of Surfaces in the Saturn System and the Role of Water, Cyanide Compounds and Carbon Dioxide. *Bull. Amer. Astron. Soc.* **37**, 705 (DPS2005 abstract 39.05).

Clarke, J.T., Gérard, J.-C., Grodent, D., Wannawichian, S., Gustin, J., Connerney, J., Crary, F., Dougherty, M., Kurth, W., Cowley, S.W.H., Bunce, E.J., Hill, T., Kim, J., 2005. Morphological differences between Saturn's ultraviolet aurorae and those of Earth and Jupiter. *Nature* **433**, 717–719.

Colwell, J.E., Esposito, L.W., Larsen, K., Stewart, A.I.F., McClintock, W.E., Sremcevic, M., Shemansky, D.E., Hallett, J.T., Hansen, C.J., Hendrix, A. R., West, R.A., Ajello, J.A., Pryor, W.R., Yung, Y.L., 2005. UV Spectroscopy of the Saturn System. *Bull. Amer. Astron. Soc.* **37**, 631 (DPS2005 abstract 10.05).

Cowling, T.G., 1981. The present state of dynamo theory. *Ann. Rev. Astron. Astrophys.* **19**, 115–135.

Crary, F.J., Clarke, J.T., Dougherty, M.K., Hanlon, P.G., Hansen, K.C., Steinberg; J.T., Barraclough, B.L., Coates, A.J., Gérard, J.-C., Grodent, D., Kurth, W.S., Mitchell, D.G., Rymer, A.M., Young, D.T., 2005. Solar wind dynamic pressure and electric field as the main factors controlling Saturn's aurorae. *Nature* **433**, 720–722.

Dougherty, M.K., Achilleos, N., Andre, N., Arridge, C.S., Balogh, A., Bertucci, C., Burton, M.E., Cowley, S.W.H., Erdos, G., Giampieri, G., Glassmeier, K.-H., Khurana, K.K., Leisner, J., Neubauer, F.M., Russell, C. T., Smith, E.J., Southwood, D.J., Tsurutani, B.T., 2005. Cassini Magnetometer Observations During Saturn Orbit Insertion. *Science* **307**, 1266–1270.

Elachi, C., Wall, S., Allison, M., Anderson, Y., Boehmer, R., Callahan, P., Encrenaz, P., Flamini, E., Franceschetti, G., Gim, Y., Hamilton, G., Hensley, S., Janssen, M., Johnson, W., Kelleher, K., Kirk, R., Lopes, R., Lorenz, R., Lunine, J., Muhleman, D., Ostro, S., Paganelli, F., Picardi, G., Posa, F., Roth, L., Seu, R., Shaffer, S., Soderblom, L., Stiles, B., Stofan, E., Vetrella, S., West, R., Wood, C., Wye, L., Zebker, H., 2005. Cassini Radar Views the Surface of Titan. *Science* **308**, 970–974.

ESA, 1989. Announcement of Opportunity, Cassini Mission: Huygens, ESA SCI(89)2. 61 pp.

Esposito, L.W., Colwell, J.E., Larsen, K., McClintock, W.E., Stewart, A.I.F., Hallett, J. T., Shemansky, D. E., Ajello, J. M., Hansen, C. J., Hendrix, A. R., West, R. A., Keller, H. U., Korth, A., Pryor, W. R., Reulke, R., Yung, Y. L., 2005. Ultraviolet Imaging Spectroscopy Shows an Active Saturnian System. *Science* **307**, 1251–1255.

Flasar, F. M., Achterberg, R. K., Conrath, B. J., Pearl, J. C., Bjoraker, G. L., Jennings, D. E., Romani, P. N., Simon-Miller, A. A., Kunde, V. G., Nixon, C. A., Bézard, B., Orton, G. S., Spilker, L. J., Spencer, J. R., Irwin, P. G. J., Teanby, N. A., Owen, T. C., Brasunas, J., Segura, M. E., Carlson, R. C., Mamoutkine, A., Gierasch, P. J., Schinder, P. J., Showalter, M. R., Ferrari, C., Barucci, A., Courtin, R., Coustenis, A., Fouchet, T., Gautier, D., Lellouch, E., Marten, A., Prangé, R., Strobel, D. F., Calcutt, S. B., Read, P. L., Taylor, F. W., Bowles, N., Samuelson, R. E., Abbas, M. M., Raulin, F., Ade, P., Edgington, S., Pilorz, S., Wallis, B., Wishnow, E. H., 2005a. Temperatures, Winds, and Composition in the Saturnian System. *Science* **307**, 1247–1251.

Flasar, F. M., Achterberg, R. K., Conrath, B. J., Gierasch, P. J., Kunde, V. G., Nixon, C. S., Bjoraker, G. L., Jennings, D. E., Romani, P. N., Simon-Miller, A. A., Bézard, B., Coustenis, A., Irwin, P. G. J., Teanby, N. A., Brasunas, J., Pearl, J. C., Segura, M. E., Carlson, R. C., Mamoutkine, A., Schinder, P. J., Barucci, A., Courtin, R., Fouchet, T., Gautier, D., Lellouch, E., Marten, A., Prangé, R., Vinatier, S., Strobel, D. F., Calcutt, S. B., Read, P. L., Taylor, F. W., Bowles, N., Samuelson, R. E., Orton, G. S., Spilker, L. J., Owen, T. C., Spencer, J. R., Showalter, M. R., Ferrari, C., Abbas, M. M., Raulin, F., Edgington, S., Ade, P., Wishnow, E. H., 2005b. Titan's Atmospheric Temperatures, Winds, and Composition. *Science* **308**, 975–978.

Gombosi, T. I., Hansen, K. C., 2005. Saturn's Variable Magnetosphere. *Science* **307**, 1224–1226.

Gurnett, D. A., Kurth, W. S., Hospodarsky, G. B., Persoon, A. M., Zarka, P., Lecacheux, A., Bolton, S. J., Desch, M. D., Farrell, W. M., Kaiser, M. L., Ladreiter, P., Rucker, H. O., Galopeau, P., Louarn, P., Young, D. T., Pryor, W. R., Dougherty, M. K., 2002. Control of Jupiter's radio emission and aurorae by the solar wind. *Nature* **415**, 985–987.

Gurnett, D., Kurth, W., Hospodarsky, G., Persoon, A., Averkamp, A., Cecconi, B., Lecacheux, A., Zarka, P., Canu, P., Cornilleau-Wehrlin, N., Galopeau, P., Roux, P., Harvey, C., Louarn, P., Bostrom, R., Gustafsson, G., Wahlund, J.-E., Desch, M., Farrell, W., Kaiser, M., Kellogg, P., Goetz, K., Fischer, G., Ladreiter, H.-P., Rucker, H., Alleyne, H., Pedersen, A., 2005. An Overview of Recent Results from the Cassini Radio and Plasma Wave Science Investigation. *Bull. Amer. Astron. Soc.* **37**, 625 (DPS2005 abstract 6.03).

Ip, W., Gautier, D., Owen, T., 2004. The Genesis of Cassini-Huygens. In *ESA SP-1278*, Proceedings of the International Conference *Titan – from discovery to encounter,* 13–17 April 2004, ESTEC, Noordwijk, Netherlands, pp. 211–227.

Johnson, T. V., Lunine, J. I., 2005. Saturn's moon Phoebe as a captured body from the outer Solar System. *Nature* **435**, 69–71.

Kempf, S., Srama, R., Postberg, F., Burton, M., Green, S. F., Helfert, S., Hillier, J. K., McBride, N., McDonnell, J. A. M., Moragas-Klostermeyer, G., Roy, M., Grün, E., 2005. Composition of Saturnian Stream Particles. *Science* **307**, 1274–1276.

Krimigis, S. M., Mitchell, D. G., Hamilton, D. C., Dandouras, J., Armstrong, T. P., Bolton, S. J., Cheng, A. F., Gloeckler, G., Hsieh, K. C., Keath, E. P., Krupp, N., Lagg, A., Lanzerotti, L. J., Livi, S., Mauk, B. H., McEntire, R. W., Roelof, E. C., Wilken, B., Williams, D. J., 2002. A nebula of gases from Io surrounding Jupiter. *Nature* **415**, 994–996.

Krimigis, S. M., Mitchell, D. G., Hamilton, D. C., Krupp, N., Livi, S., Roelof, E. C., Dandouras, J., Armstrong, T. P., Mauk, B. H., Paranicas, C., Brandt, P. C., Bolton, S., Cheng, A. F., Choo, T., Gloeckler, G., Hayes, J., Hsieh, K. C., Ip, W.-H., Jaskulek, S., Keath, E. P., Kirsch, E., Kusterer, M., Lagg, A., Lanzerotti, L. J., LaVallee, D., Manweiler, J., McEntire, R. W., Rasmuss, W., Saur, J., Turner, F. S., Williams, D. J., Woch, J., 2005. Dynamics of Saturn's Magnetosphere from MIMI During Cassini's Orbital Insertion. *Science* **307**, 1270–1273.

Kunde, V. G., Flasar, F. M., Jennings, D. E., Bézard, B., Strobel, D. F., Conrath, B. J., Nixon, C. A., Bjoraker, G. L., Romani, P. N., Achterbert, R. K., Simon-Miller, A. A., Irwin, P., Brasunas, J. C., Pearl, J. C., Smith, M. D., Orton, G. S., Gierasch, P. J., Spilker, L. J., Carlson, R. C., Mamoutkine, A. A., Calcutt, S. B., Read, P. L., Taylor, F. W., Fouchet, T., Parrish, P., Barucci, A., Courtin, R., Coustenis, A., Gautier, D., Lellouch, E., Marten, A., Prangé, R., Biraud, Y., Ferrari, C., Owen, T. C., Abbas, M. M., Samuelson, R. E., Raulin, F., Ade, P., Césarsky, C. J., Grossman, K. U., Coradini, A., 2004. Jupiter's Atmospheric Composition from the Cassini Thermal Infrared Spectroscopy Experiment. *Science* **305**, 1582–1586.

Kurth, W. S., Gurnett, D. A., Hospodarsky, G. B., Farrell, W. M., Roux, A., Dougherty, M. K., Joy, S. P., Kivelson, M. G., Walker, R. J., Crary, F. J., Alexander, C. J., 2002. The dusk flank of Jupiter's magnetosphere. *Nature* **415**, 991–994.

Kurth, W. S., Gurnett, D. A., Clarke, J. T., Zarka, P., Desch, M. D., Kaiser, M. L., Cecconi, B., Lecacheux, A., Farrell, W. M., Galopeau, P., Gérard, J.-C., Grodent, D., Prangé, R., Dougherty,

M. K., Crary, F. J., 2005. An Earth-like correspondence between Saturn's auroral features and radio emission. *Nature* **433**, 722–725.

Marouf, E., French, R., Rappaport, N., Kliore, A., Flasar, M., Nagy, A., McGhee, C., Schinder, P., Anabtawi, A., Asmar, S., Barbinis, E., Fleischman, D., Goltz, G., Johnston, D., Rochblatt, D., Thompson, F., Wong, K.. Structure of Saturn's Rings from Cassini Diametric Radio Occultations. 2005, *Bull. Amer. Astron. Soc.* **37**, 763 (DPS2005 abstract 62.02).

Matson, D. L., Spilker, L. J., Lebreton, J.-P., 2002. The Cassini/Huygens Mission to the Saturnian System. *Space Science Reviews* **104**, 1–58.

Mitchell, D. G., Brandt, P. C., Roelof, E. C., Dandouras, J., Krimigis, S. M., Mauk, B. H., 2005. Energetic Neutral Atom Emissions from Titan Interaction with Saturn's Magnetosphere. *Science* **308**, 989–992.

NASA, 1989. Announcement of Opportunity, Cassini Mission: Saturn Orbiter, A.O. No. OSSA-1–89. 70 pp.

Northrup, T. G., Connerney, J. E. P., 1987. A micrometeorite erosion model and the age of Saturn's rings. *Icarus* **70**, 124–137.

Ostro, S. J., Elachi, C., Anderson, Y., Boehmer, R., Callahan, P., Hamilton, G., Janssen, M., Johnson, W., Kelleher, K., Lopes, R., Roth, L., Wall, S., West, R., Allison, M., Kirk, R., Wood, C., Posa, F., Stofan, E., Zebker, H., Lorenz, R., Lunine, J., Francescetti, G., Picardi, G., Seu, R., Muhleman, D., Encrenaz, P., 2004. Cassini RADAR Observations of Phoebe. *Bull. Amer. Astron. Soc.* **36**, 1071 (abstract only).

Parker, E. N., 1987. The dynamo dilemma. *Solar Phys.* 110, 11–21.

Porco, C. C., West, R. A., McEwan, A., Del Genio, A. D., Ingersoll, A. P., Thomas, P., Squyres, S., Dones, L., Murray, C. D., Johnson, T. V., Burns, J. A., Brahic, A., Neukum, G., Veverka, J., Barbara, J. M., Denk, T., Evans, M., Ferrier, J. J., Geissler, P., Helfenstein, P., Roatsch, T., Throop, H., Tiscareno, M., Vasavada, A. R., 2003. Cassini imaging of Jupiter's Atmosphere, Satellites, and Rings. *Science* **299**, 1541–1547.

Porco, C. C., Baker, E., Barbara, J., Beurle, K., Brahic, A., Burns, J. A., Charnoz, S., Cooper, N., Dawson, D. D., Del Genio, A. D., Denk, T., Dones, L., Dyudina, U., Evans, M. W., Giese, B., Grazier, K., Helfenstein, P., Ingersoll, A. P., Jacobson, R. A., Johnson, T. V., McEwen, A., Murray, C. D., Neukum, G., Owen, W. M., Perry, J., Roatsch, T., Spitale, J., Squyres, S., Thomas, P. C., Tiscareno, M., Turtle, E., Vasavada, A. R., Veverka, J., Wagner, R., West, R., 2005a. Cassini Imaging Science: Initial Results on Phoebe and Iapetus. *Science* **307**, 1237–1242.

Porco, C. C., Baker, E., Barbara, J., Beurle, K., Brahic, A., Burns, J. A., Charnoz, S., Cooper, N., Dawson, D. D., Del Genio, A. D., Denk, T., Dones, L., Dyudina, U., Evans, M. W., Giese, B., Grazier, K., Helfenstein, P., Ingersoll, A. P., Jacobson, R. A., Johnson, T. V., McEwen, A., Murray, C. D., Neukum, G., Owen, W. M., Perry, J., Roatsch, T., Spitale, J., Squyres, S., Thomas, P., Tiscareno, M., Turtle, E., Vasavada, A. R., Veverka, J., Wagner, R., West, R., 2005b. Cassini Imaging Science: Initial Results on Saturn's Atmosphere. *Science* **307**, 1243–1247.

Porco, C. C., Baker, E., Barbara, J., Beurle, K., Brahic, A., Burns, J. A., Charnoz, S., Cooper, N., Dawson, D. D., Del Genio, A. D., Denk, T., Dones, L., Dyudina, U., Evans, M. W., Giese, B., Grazier, K., Helfenstein, P., Ingersoll, A. P., Jacobson, R. A., Johnson, T. V., McEwen, A., Murray, C. D., Neukum, G., Owen, W. M., Perry, J., Roatsch, T., Spitale, J., Squyres, S., Thomas, P., Tiscareno, M., Turtle, E., Vasavada, A. R., Veverka, J., Wagner, R., West, R., 2005c. Cassini Imaging Science: Initial Results on Saturn's Rings and Small Satellites. *Science* **307**, 1226–1236.

Porco, C. C., Baker, E., Barbara, J., Beurle, K. Brahic, A., Burns, J. A., Charnoz, S., Cooper, N., Dawson, D. D., Del Genio, A. D., Denk, T., Dones, L., Dyudina, U., Evans, M. W., Fussner, S., Giese, B., Grazier, K., Helfenstein, P., Ingersoll, A. P., Jacobson, R. A., Johnson, T. V.,

McEwen, A., Murray, C. D., Neukum, G., Owen, W. M., Perry, J., Roatsch, T., Spitale, J., Squyres, S., Thomas, P., Tiscareno, M., Turtle, E. P., Vasavada, A. R., Veverka, J., Wagner, R., West, R., 2005d. Imaging of Titan from the Cassini spacecraft. *Nature* **434**, 159–168.

Sánchez-Lavega, A., Pérez-Hoyos, S., 2005. On the Vertical Wind Shear of Saturn's Equatorial Jet at Cloud Level. *Bull. Amer. Astron. Soc.* **37**, 681 (DPS2005 abstract 30.23).

Shemansky, D. E., Stewart, A. I. F., West, R. A., Esposito, L. W., Hallett, J. T., Liu, X., 2005. The Cassini UVIS Stellar Probe of the Titan Atmosphere. *Science* **308**, 978–982.

Smith, B. A.; Soderblom, L. A.; Beebe, R.; Boyce, J.; Briggs, G.; Bunker, A.; Collins, S. A.; Hansen, C. J.; Johnson, T. V.; Mitchell, J. L.; Terrile, R. J.; Carr, M.; Cook, A. F., II; Cuzzi, J. N.; Pollack, J. B.; Danielson, G. E.; Ingersoll, A.; Davies, M. E.; Hunt, G. E.; Masursky, H.; Shoemaker, E.; Morrison, D.; Owen, T.; Sagan, C.; Veverka, J.; Strom, R.; Suomi, V. E., 1982. Encounter With Saturn: Voyager 1 Imaging Science Results. *Science* **212**, 163–191.

Spilker, L. J. (ed.), 1997. *NASA SP-533, Passage to a Ringed World - The Cassini-Huygens Mission to Saturn and Titan,* NASA, Washington D.C., 157 pp.

Spilker, L. J., Pilorz, S. H., Wallis, B. D., Ferrari, C., Altobelli, N., Brooks, S. M., Edgington, S. G., Pearl, J. C., Flasar, F. M., Pollard, B. J., CIRS Team, 2005. Cassini CIRS: Thermal Changes in Saturn's Main Rings with Increasing Phase Angle. *Bull. Amer. Astron. Soc.* **37**, 764 (DPS2005 abstract 62.06).

Wahlund, J.-E., Boström, R., Gustafsson, G., Gurnett, D. A., Kurth, W. S., Pedersen, A., Averkamp, T. F., Hospodarsky, G. F., Persoon, A. M., Canu, P., Neubauer, F. M., Dougherty, M. K., Eriksson, A. I., Morooka, M. W., Gill, R., André, M., Eliasson, L., Müller-Wodarg, I., 2005. Cassini Measurements of Cold Plasma in the Ionosphere of Titan. *Science* **308**, 986–989.

Waite, J. H. Jr., Cravens, T. E., Ip, W.-H., Kasprzak, W. T., Luhmann, J. G., McNutt, R. L., Niemann, H. B., Yelle, R. V., Mueller-Wodarg, I., Ledvina, S. A., Scherer, S., 2005a. Oxygen Ions Observed Near Saturn's A Ring. *Science* **307**, 1260–1262.

Waite, J. H. Jr., Niemann, H., Yelle, R. V., Kasprzak, W. T., Cravens, T. E., Luhmann, J. G., McNutt, R. L., Ip, W.-H., Gell, D., De La Haye, V., Müller-Wodarg, I., Magee, B., Borggren, N., Ledvina, S., Fletcher, G., Walter, E., Miller, R., Scherer, S., Thorpe, R., Xu, J., Block, B., Arnett, K., 2005b. Ion Neutral Mass Spectrometer Results from the First Flyby of Titan. *Science* **308**, 982–986.

Young, D. T., Berthelier, J.-J., Blanc, M., Burch, J. L., Bolton, S., Coates, A. J., Crary, F. J., Goldstein, R., Grande, M., Hill, T. W., Johnson, R. E., Baragiola, R. A., Kelha, V., McComas, D. J., Mursula, K., Sittler, E. C., Svenes, K. R., Szegö, K., Tanskanen, P., Thomsen, M. F., Bakshi, S., Barraclough, B. L., Bebesi, Z., Delapp, D., Dunlop, M. W., Gosling, J. T., Furman, J. D., Gilbert, L. K., Glenn, D., Holmlund, C., Illiano, J.-M., Lewis, G. R., Linder, D. R., Maurice, S., McAndrews, H. J., Narheim, B. T., Pallier, E., Reisenfeld, D., Rymer, A. M., Smith, H. T., Tokar, R. L., Vilppola, J., Zinsmeyer, C., 2005. Composition and Dynamics of Plasma in Saturn's Magnetosphere. *Science* **307**, 1262–1266.

10 The Ice Giant Systems of Uranus and Neptune

Heidi B. Hammel

Abstract. The current state of knowledge of the Ice Giants, Uranus and Neptune, is presented. The changing appearance of the atmosphere of Uranus is discussed, and its current cloud patterns and zonal winds are reviewed. Highlights of recent uranian ring and satellite observations are presented, along with a brief discussion of the ionosphere as deduced from ground-based observations. For the Neptune system, the rapidly evolving atmosphere is assessed, with a discussion of the long-term record to put recent observations into context. Also discussed are advances in characterizing the clumpy ring system of Neptune. Remarkable changes in the atmosphere of Neptune's moon Triton are described, and the ever-growing number of smaller satellites is reported. Concluding remarks include a synopsis of future exploration of these dynamic planetary systems.

10.1 Introduction

Our understanding of Uranus and Neptune has evolved rapidly in recent years due to a combination of improvements in astronomical technology and observational techniques, and to intrinsic changes of the planets themselves. For planets where the length of each season is measured in decades rather than months, understanding planetary variation change requires more than the mere snapshots provided by Voyager 2 of Uranus (in 1986) and Neptune (in 1989). Many years of painstaking telescopic observations are required before we can begin to unravel their secrets. Planetary scientists now appreciate that the planets Uranus and Neptune differ from Jupiter and Saturn in more than just size: their interiors differ in composition and phase; their zonal winds differ in magnitude and direction; and their magnetic fields differ in structure. Hence we refer to these mid-sized planets as "Ice Giants" to distinguish them from their larger "Gas Giant" cousins.

10.2 The Uranus System

Uranus had perhaps the most difficult adolescence of any of the giant planets. Sometime in its early history, after the bulk of its core had formed, Uranus experienced a catastrophic collision: an impact so severe that the planet was twisted completely over on its side, where it spins to this day with a 98° obliquity.

In spite of that dynamic past, Uranus subsequently was typecast as the Solar System's "most boring" planet after presenting a rather bland face to the Voyager 2 spacecraft in 1986. Its cameras discerned less than a dozen discrete clouds, compared with Jupiter's thousands and Saturn's hundreds. Seeing the disappointing atmospheric display, some scientists speculated that the early collision so disrupted the planet's atmosphere that it was, effectively, stirred from the bottom up, leaving the remaining atmosphere dormant.

We suspect now that the dormant appearance of Uranus may have been a fluke. Much like a garden appears dormant in the late winter but then blossoms with the advent of spring, so too might Uranus simply have appeared dormant due to the particular season when Voyager flew by. It was the height of summer in the Southern hemisphere, and that hemisphere had been baking under the glow of the Sun literally for decades. Equinox – when Uranus displays a side view to the Sun and consequently us Earth-dwellers near the Sun – is now rapidly approaching.

The formerly bland atmosphere of Uranus appears to be responding to the changes in illumination more rapidly and perhaps more deeply than scientists expected based on their models. I review here recent Uranus observations that show signs of atmospheric activity, and discuss the implications for theoretical models. I also discuss new observations of the planet's rings and ionospheric activity, and review the current status of its system of moons.

10.2.1 The atmosphere of Uranus at Equinox

In 2007, Uranus will reach equinox. Any planet with an axial tilt will experience equinox, of course. For Uranus, however, an equinox is an extreme event. Due to the planet's severe obliquity, the poles spend decades either basking in constant insolation, or radiating ceaselessly to the cold blackness of outer space (Fig. 10.1). The effect on global radiation balance was nevertheless thought to be negligible, at least in the visible regions of the atmosphere. The blank atmosphere imaged by the Voyager cameras indicated a dynamically dead world, with a scant handful of convecting clouds and no other atmospheric activity beyond subtle latitudinal banding visible only with aggressive imaging processing.

Hubble Space Telescope (HST) images of the South pole of Uranus provided the first evidence of atmospheric change, showing that as the polar region began to receive less direct sunlight, it began to darken (Rages et al. 2004). HST images taken with the NICMOS infrared camera in 1998 revealed incredibly bright Northern cloud features –

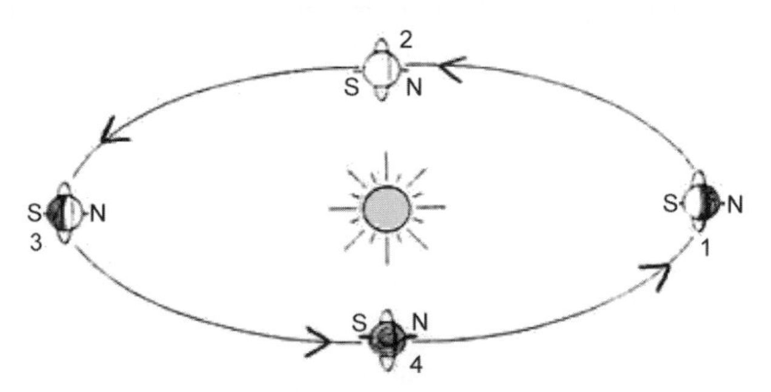

Fig. 10.1. Seasons of Uranus. Uranus takes about 84 years to complete one trip around the sun. (1) Southern-hemisphere solstice occurred in 1985; Voyager flew by Uranus in 1986. (2) The next equinox, or ring-plane crossing, is in 2007. (3) The next solstice occurs in 2028, when the Northern hemisphere of the planet will be fully illuminated. (4) Uranus will not return to an equinoctial position until 2049.

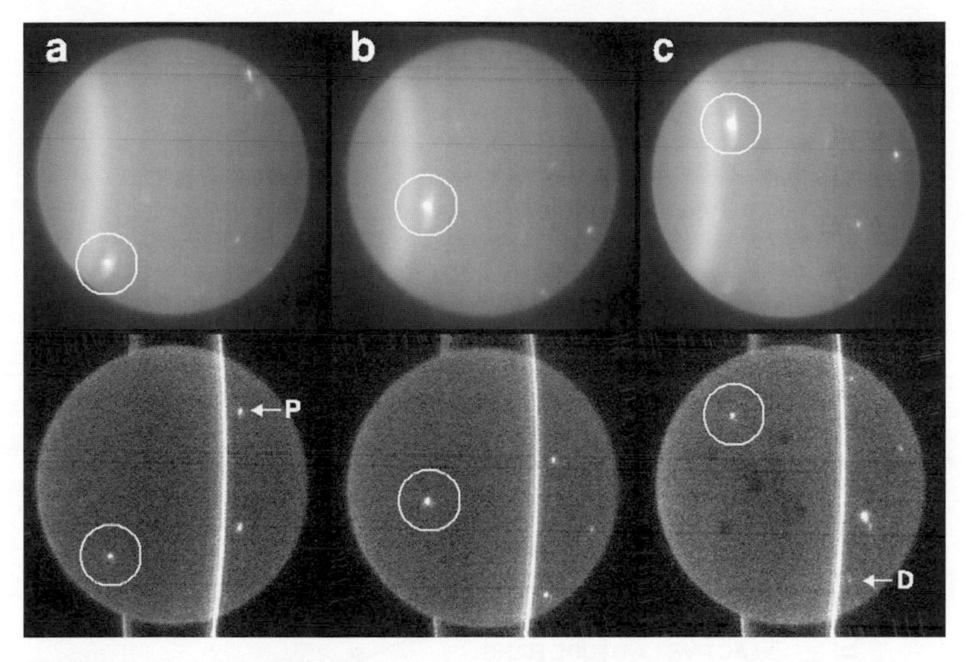

Fig. 10.2. Uranus in July 2004 from Hammel et al. (2005a). These images from the Keck telescope contained an unusual bright feature (circled in each image). For each pair, the upper image is H (1.6 μm) and the lower image is K' (2.2 μm). The South pole of Uranus is oriented to the left. Images from 4 July are in the top row (a, b, and c). At K' images are dominated by ring flux, and uranian satellites often masquerade as Northern cloud features. We indicate the interlopers in these images with arrows and letters: P=Portia and D=Desdemona. The dark splotches in the panel c image at K' are an artifact caused by residual charge in the detector.

one of these clouds had the highest contrast ever recorded for an outer planet (Karkoschka 1998), a major change from Voyager's low-contrast features. The newest images from HST and from the Keck 10-meter telescope in Hawaii (Fig. 10.2) show tremendous cloud activity across the planet, including convective activity on very short time-scales (Hammel et al. 2005a) and the first detection of a planetary-scale wave feature in the clouds wrapping around the equator of Uranus (Hammel et al. 2005b).

The modern observations of clouds in the atmosphere of Uranus permit the nearly complete characterization of its zonal wind profile (it is not fully complete because as of this writing the far Northern hemisphere is still not yet visible). The first comprehensive measurements of Uranus' zonal winds were produced from Voyager images, which tracked a scant handful of Southern hemisphere clouds (Smith et al. 1986). These were supplemented by a single equatorial wind measurement determined via Voyager radio science experiment. When Hubble and Keck images of Uranus revealed discrete features, new measurements were possible – not only of the Southern hemisphere, confirming the Voyager results, but also for the first time of winds in the Northern hemisphere (Fig. 10.3; Hammel et al. 2001; Hammel et al. 2005b). The winds of Uranus, like those of Neptune, show a significant retrograde equatorial wind, along with strong mid-latitude prograde jets in both hemispheres. Evidence suggests that the zonal winds of Uranus

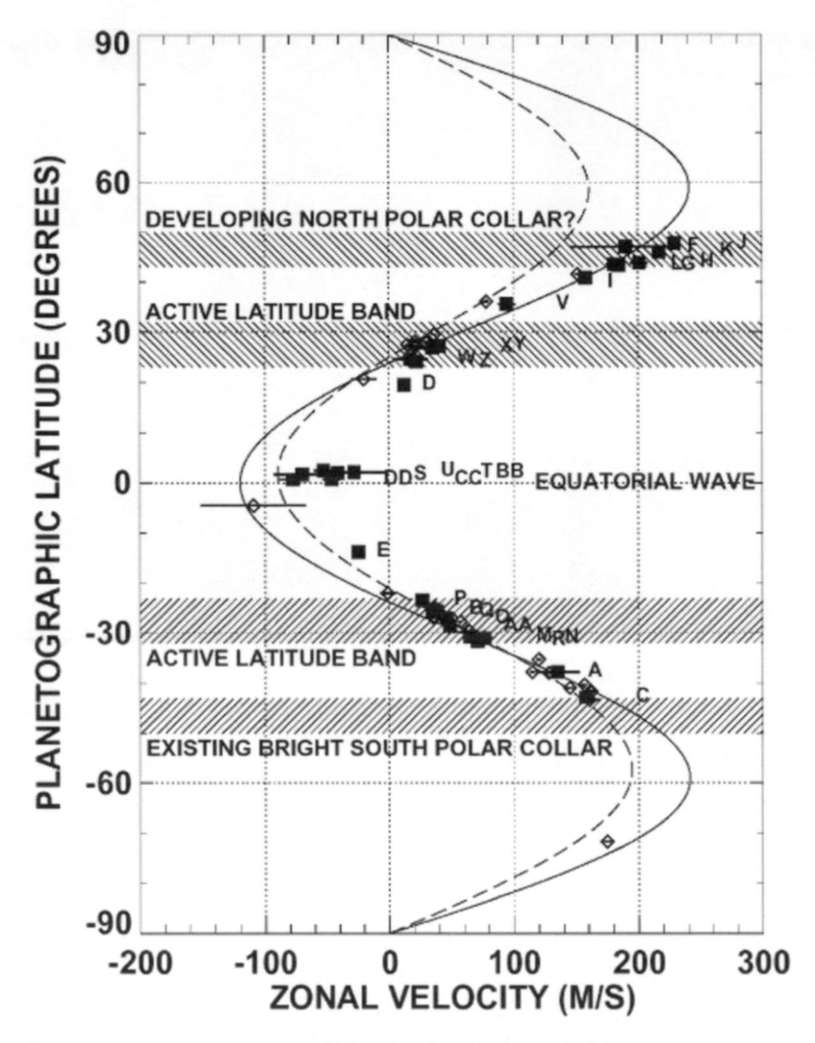

Fig. 10.3. Zonal Winds of Uranus in 2003 from Hammel et al. (2005b). Solid squares are the points measured in 2003 with Keck (letters refer to specific cloud features); open symbols are past measurements from HST, Keck, and Voyager. The 2003 letter labels are offset to the right for clarity. All points are shown with error bars, though the errors are often smaller than the size of the points. The large error bars on the 2003 equatorial features are indicative of their diffuse nature; to within their measured errors, they have similar zonal velocities, and may be a manifestation of a wave pattern. The hatched region from $-50°$ to $-43°$ corresponds to the bright polar collar clearly evident at H (Fig. 10.2). The hatched region from $+43°$ to $+50°$ mirrors the Southern values. Many Northern features are clustered at these latitudes. The hatched region from $-32°$ to $-23°$ was defined to enclose most of the Southern mid-latitude features seen in the 2003 data and the earlier measurements. The symmetric band in the North, from $+23°$ to $+32°$, encloses many Northern features. The lines indicate models: solid – non-weighted fit to Voyager data (Allison et al. 1991); dotted – asymmetric model of Karkoschka (1998).

may vary with time. However, an alternative suggestion may be vertical variation of zonal velocity: when features are closer to the limb they probe higher altitudes; lower altitudes are sensed when such features are further on the disk.

As of this writing, the atmosphere of Uranus is strikingly asymmetric at visible and near-infrared wavelengths (Rages et al. 2004; Hammel et al. 2005a, 2005b). A bright ring surrounds its Southern pole – a South polar 'collar' – but HST and Keck images show no bright counterpart around the Northern pole. Using current images, one can create a computer-generated simulation of Uranus, and rotate it backward in time to earlier seasons to compare with data obtained at earlier Uranus seasons. The comparison reveals Uranus must have had a Northern polar collar in the past (Hammel and Lockwood 2006). Interestingly, maps made at cm wavelengths with the Very Large Array interferometer appear to show symmetry: bright regions covering the polar regions in both the Northern and Southern hemispheres (Hofstadter et al. 2005). These measurements of thermal emission probe the temperature and composition of the atmosphere at pressures between 0.5 and 50 bars.

We do not yet understand why the appearance of Uranus is changing, nor how the deep atmosphere is linked to the upper atmosphere. The varying position of the Sun relative to the equator undoubtedly influences atmospheric circulation patterns, but the insolation changes predicted by models do not appear to be sufficient to drive the observed level of variation. We expect more clues for solving the puzzling variations of Uranus to emerge as the planet moves through equinox and beyond.

10.2.2 Perhaps a new ring and dusty old rings

In 2002, 2003, and 2004, de Pater et al. (2005b) obtained images of the uranian ring system at 2.2 μm with the adaptive optics camera NIRC2 on the Keck II telescope (Fig. 10.4). Using those data, they reported the first detection in backscattered light of a ring – which they call the ζ (" zeta") ring – interior to Uranus' known rings. This ring consists of a generally uniform sheet of dust that extends inward at a gradually decreasing

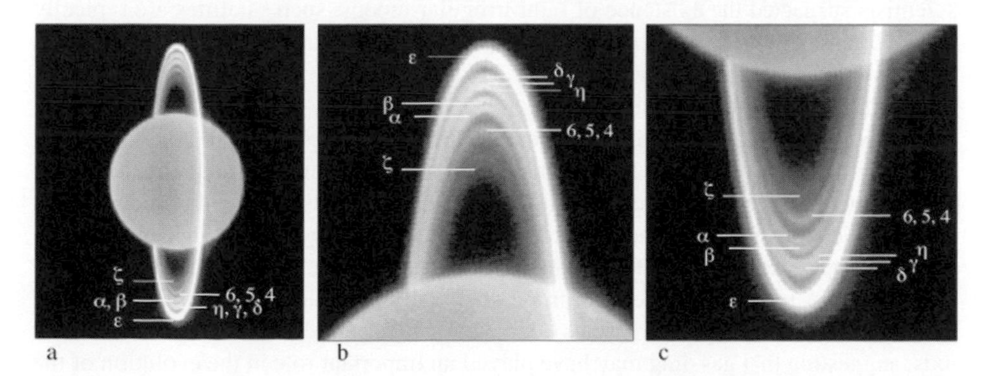

Fig. 10.4. Ring System of Uranus in 2004 from de Pater et al. (2005b). Most of the known major ring components (the ε ring; the α and β pair; the trio η, γ, and δ; and the cluster containing 6, 5, and 4) can be resolved in this combined image, and a new inner dust sheet labeled ζ was revealed (panels b and c are close-up views of the ansae from panel a).

brightness. This ring might be related to the Voyager ring 1986 U2R, although both its location and extent differ, perhaps due to a difference in observing wavelength and solar phase angle, or perhaps due to temporal variation of the ring.

By careful modeling of the brightnesses of the individual rings at each ansa, de Pater et al. (2005b) also detected narrow (few 100 km wide) sheets of dust between the δ and ε rings, and between rings 4 and α. The spatial distribution and relative intensity of these dust sheets differ from that seen in Voyager images taken in forward scattered light, due perhaps to the different observing wavelength and the solar phase angle, to changes over time, or perhaps a combination of both. They may have detected the λ ring in one scan, but not several other scans. However, azimuthal asymmetries are known to exist in this ring. They further demonstrate the presence of azimuthal asymmetries in all rings. They confirm the eccentricity in rings 4, 5, and 6, which in 2004 showed a North ansa about 70 km closer to Uranus than the South ansa.

Showalter et al. (2005) resolved the individual rings of Uranus (except 6, 5 and 4) using the Hubble Space Telescope. Their analysis suggests the presence of subtle clumps and arcs within the ring system, which appear to be primarily periodic variations in the widths of the known rings. The variations may be related to modes of the δ and γ rings, or perhaps to resonant perturbations by Cordelia and Ophelia.

10.2.3 How many moons does Uranus have?

At last count, Uranus had 26 satellites. William Herschel, discoverer of Uranus, identified Titania and Oberon – the two largest moons – shortly after Uranus was recognized as a planet. By the middle of the next century, William Lassell detected Umbriel and Ariel (Fig. 10.5) – the next two largest. Nearly a century went by before G. Kuiper discovered the fifth large moon, Miranda. The Voyager spacecraft added ten smaller moons to the tally, making serious inroads into the list of Shakespearean characters for whom these Uranian moons are named (an 11th moon purportedly identified by Voyager has not been confirmed, and is not included in the final total of 26 moons). All of these moons travel in prograde orbits. Based on experience with Jupiter and Saturn, however, planetary scientists suspected the existence of faint irregular moons; such satellites are typically small, often exhibiting orbits that are highly inclined with large semi-major axes and that sometimes even are retrograde.

From 1997 to 2001, dedicated deep searches in the vicinity of Uranus finally revealed eight tiny retrograde moons (Gladman et al. 1998, 2000; Kavelaars et al. 2004). In 2003, discovery of another three regular (prograde) moons brought the final count of confirmed uranian moons to 26 (Sheppard et al. 2005). The elements of the cadre of uranian moods have implications for the planet's formation and evolutionary history (Kavelaars et al. 2004; Sheppard et al. 2005). For example, the observed size distribution is somewhat shallow compared to that expected for a collisionally-dominated system. Furthermore, the moons' orbital distribution reveals a correlation between eccentricity and semi-major axis, suggesting that gas drag may have played an important role in the evolution of the uranian system.

As equinox nears for Uranus, its satellites yield surprises as the bodies' Northern hemispheres tilt toward to Earth. For example, recent near-infrared spectra of Ariel revealed the signature of CO_2 ice (Fig. 10.5). Not only is this the first time CO_2 ice has

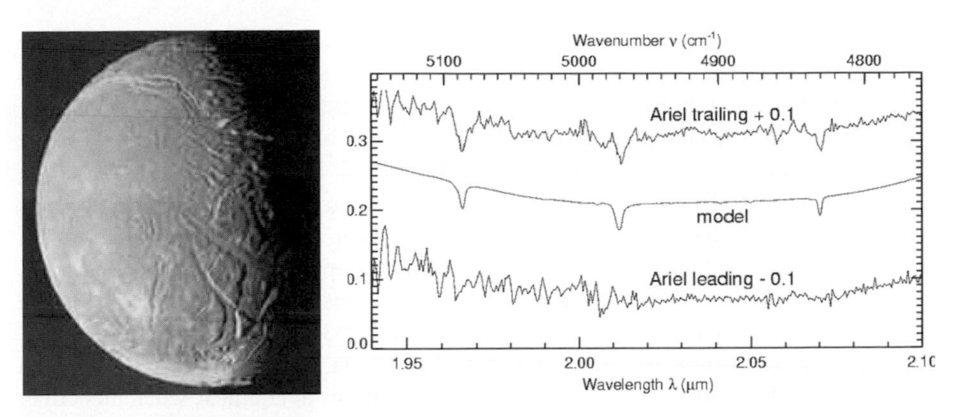

Fig. 10.5. Ariel and near-infrared spectra of its leading and trailing hemispheres from Grundy et al. (2003). The Voyager spacecraft took the image in 1986. CO_2 (labeled "model") is detected on the trailing hemisphere of the satellite, but not on the leading side. The y-axis units are albedo, with the Ariel data offset from the model as indicated for clarity.

been detected in the Uranian system; it is the first new species of ice to be discovered on a Uranian satellite in twenty years, since water ice was detected. Preliminary models indicate that more than 8% of Ariel's trailing hemisphere must have CO_2 ice on it (Grundy et al. 2003). The CO_2 ice may be confined to the Northern hemisphere not mapped by the Voyager spacecraft in 1986, but it could equally well be distributed all over the trailing hemisphere. The presence of CO_2 ice raises many interesting questions, such as: Why is it only seen on Ariel, and not on other uranian moons? How much CO_2 ice is there? How is it distributed? Where did it come from? And how long should it survive? These are just a few of the mysteries which future observations of the uranian satellites will seek to resolve.

10.2.4 The ionosphere of Uranus

Current research on the ionosphere of Uranus centers mainly on the detection and interpretation of near-infrared (2–4 μm) emission from rotationally-vibrationally excited H_3^+ ions. The importance of the H_3^+ ion in giant planet atmospheres was suggested over 40 years ago (Martin et al. 1961). However, decades elapsed before Trafton et al. (1993) reported the first detection of H_3^+ emission features in the ionosphere of Uranus. Because of the strength and high contrast of these H_3^+ emission lines with respect to background emission, and because these lines are observable from the ground, the detection of H_3^+ on Uranus has given researchers a new tool for probing its ionosphere.

The H_3^+ emission on Uranus is, in intensity and computed column abundance of $\sim 10^4\ \mathrm{cm}^{-3}$, a few percent of the jovian emissions (Trafton et al. 1993). Curiously, H_3^+ uranian emission lines appear to be stronger than those observed on Saturn (Geballe et al. 1993). This is puzzling: Saturn's level of UV auroral activity, and the expectation that the H_3^+ emission intensity would decrease with increasing distance from the Sun, together suggest that saturnian emissions should be stronger than those of Uranus. The solution to this conundrum – which requires more accurate information about the temperature, column abundance, spatial distribution, and temporal behavior of the H_3^+ emission on both Saturn and Uranus – awaits further observations.

The upcoming equinox of Uranus is providing investigators with a first opportunity to examine diurnal behavior. At electron densities typical of the Uranian ionosphere, the chemical lifetime of H_3^+ is less than 3 hours (Trafton et al. 1999). That is much less than the planet's rotational period (of order 16–18 hours), suggesting that H_3^+ concentrations on the night side should be low, with significant emission ramping up from the sunrise limb to reach a broad maximum in the late afternoon approaching the evening limb. Trafton and Miller (2004) obtained imaging observations of Uranus' H_3^+ emission in 1999 in the relatively bright Q(1) and Q(3) lines at 3.953 and 3.986 µm, respectively. The emission was symmetric along the central meridian, and exhibited the predicted behavior from the morning to evening limbs. Observations at equinox, when this effect is most observable, will yield invaluable information on the ionospheric processes driving the variability.

10.3 The Neptune System

Through a large telescope, Neptune – the most distant giant planet in our Solar System – appears as a small blue disk, 2.3 seconds of arc in diameter. If Neptune is a twin of Uranus, then it is a fraternal twin: its atmosphere, moons, and ring system all differ markedly from their uranian counterparts. We discuss our current knowledge of the distant yet dynamic Neptune system, highlighting the differences from its Solar System sibling Uranus.

10.3.1 Neptune's evolving atmosphere

The dynamic nature of Uranus' seasonal atmosphere may have been misunderstood until recently (see above), but Neptune's extremely variable atmosphere is simply not understood. Prior to 1989, the best pictures of Neptune from Earth's surface showed discrete bright clouds and a bright haze over the South pole of the planet (Hammel 1989). In 1989, the Voyager 2 spacecraft confirmed these sightings when it reached Neptune, flying less than 5,000 km above the planet's cloud tops. Its cameras revealed many features (Smith et al. 1986), including a large, dark storm system in the Southern hemisphere named the Great Dark Spot, or GDS (Fig. 10.6). The GDS had disappeared by 1994, when the Hubble Space Telescope discovered a new GDS in the Northern hemisphere (Hammel et al. 1995). Voyager 2 also saw numerous other clouds, including wispy cirrus-like clouds that cast shadows on deeper cloud decks below.

The Voyager 2 images and subsequent Hubble Space Telescope observations showed that Neptune's clouds change rapidly, often forming and dissipating over periods of several hours (Smith et al. 1986; Sromovsky et al. 2001). The surprisingly fast changes indicate that the weather of Neptune is as dynamic and variable as that of the Earth. No suitable mechanism has explained Neptune's dynamic atmospheric variation. On longer timescales, the basic banded structure changes. The sequence of Hubble images

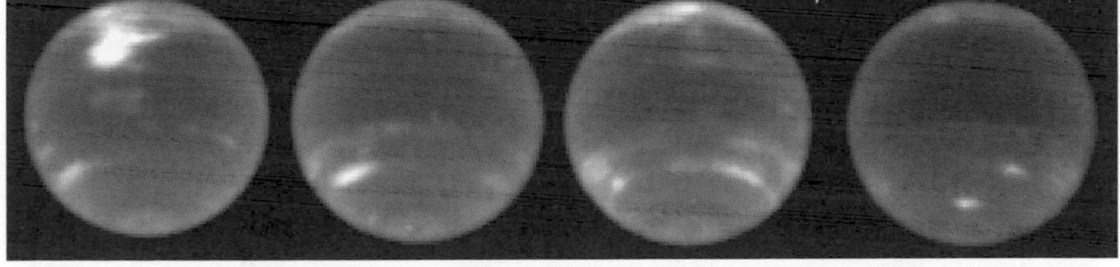

Fig. 10.6. Neptune's evolving atmosphere, adapted from Hammel and Lockwood (2006). The Voyager spacecraft took the far-left image in 1989 through a broad-band blue filter. The other images – taken in the WFPC2 619-nm methane-band filter – were obtained in 1994, 1997, 2001, and 2004 (left to right, respectively). The 1994 image of Neptune shows the companion to the Northern Great Dark Spot (Hammel et al. 1995). Images in subsequent years show Southern mid-latitude activity increasing (a maximum was reached in 2003). The 2004 image shows that much of that mid-latitude activity has now subsided, perhaps heralding a return to a Voyager-era cloud distribution.

in Fig. 10.6 show a steady build-up of Southern hemispheric activity at middle latitudes, which was also seen in ground-based near-infrared Keck images (Hammel and Lockwood 2006). By 2004, much of that activity had subsided.

Observations at longer wavelengths also reveal long-term changes in stratospheric temperature (Fig. 10.7). Both sub-millimeter molecular abundance observations (Marten et al. 2005) and mid-infrared hydrocarbon emission (Hammel et al. 2006) indicate an increasing trend of temperature with time. As with the cloud distribution at deeper levels, the driving mechanism for these upper atmospheric changes is not yet understood. A simple model that seasonal change was driving long-term changes as it does on Uranus (Sromovsky et al. 2003) is not supported by older photometric data (Lockwood and Jerzykiewicz 2006). A suggestion that long-term change is due to a subtle response to innate solar variations is still being vigorously debated (Hammel and Lockwood 2006).

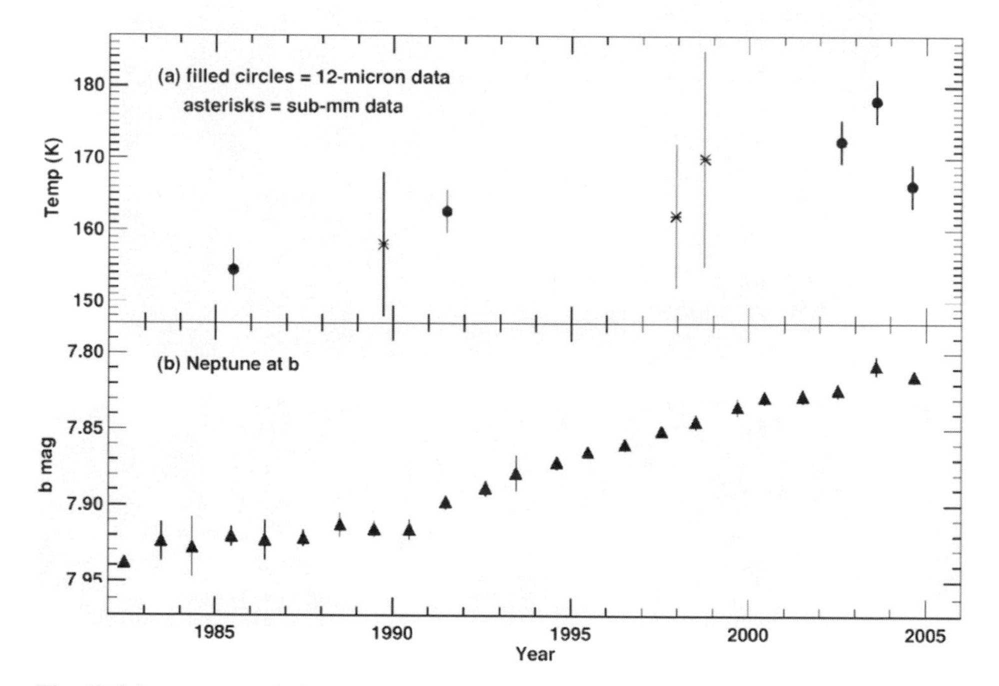

Fig. 10.7. Long-term variations of Neptune over two decades. (a) Solid circles indicate effective stratospheric temperature determined from mid-infrared ethane emission spectra (Hammel et al. 2006). Asterisks are from Marten et al. (2005): their initial nominal model based on Voyager observations (year: 1989); the value used by Bézard et al. (1999) to model ISO observations (year: 1997); and the value Marten et al. deduce from sub-mm CO observations (year: 1998). The sub-mm observations provide an independent measure of temperature change in the atmosphere of Neptune, consistent with our results. (b) The triangles are long-term disk-integrated photometry of Neptune at Strömgren b from Lockwood and Jerzykiewicz (2006).

10.3.2 Clumpy Rings

The presence of rings around Neptune had been a subject of debate prior to the Voyager encounter (see de Pater et al. 2005a and references therein). Several ground-based observations had suggested rings, but others had not. Studies of Voyager images eventually revealed that five rings surround Neptune: two bright, narrow rings and three fainter, fuzzier sheets of orbiting materials (Smith et al. 1989). Some sections of the bright rings have significantly higher densities than others, and it was these arcs of higher density that had been detected occasionally by Earth telescopes. In near-infrared Keck images taken in 2002 and 2003, de Pater et al. (2005a) recovered the full Adams and Le Verrier rings for the first time since their initial imaging by Voyager (Fig. 10.8). Showalter et al. (2005) detected the Adams and Le Verrier rings of Neptune at visual wavelengths with HST. The intensity ratio between Adams and Le Verrier rings, and their azimuthal brightness variations, are similar to those seen in the Voyager data.

The recent ground-based observations revealed fascinating changes in the details of the ring system (de Pater et al. 2005a, Showalter et al. 2005). Keck and HST data suggest both the location and intensity of Neptune's ring arcs changed drastically relative to trailing arc Fraternité. The observed changes suggest that material migrates between resonance sites. For arc Liberté, a "twin-arc" structure seen in 2002 might have been a resonance shift in progress, and by 2003 Liberté appeared to be on the verge of disappearing completely. In another possible manifestation of migration, Voyager (sub-)arcs Egalité 1 and 2 appear to have reversed in relative intensity. In yet another example, leading arc Courage has jumped $\sim 8^\circ$, or, in terms of a theory espoused by Namouni and Porco (2002), this arc advanced by one full corotation potential maximum (the theory predicts 43 possible maxima locations).

Taken together, the HST and Keck observations reveal a system that is surprisingly dynamic. The red color of the ring arcs, as derived both by Dumas et al. (2002) and de Pater et al (2005a), is consistent with ring arcs being composed of dust, the natural product of moon erosion. If indeed the entire arc system is decaying in decadal timescales, as the data suggest, the loss mechanism must be more rapid than the regeneration mecha-

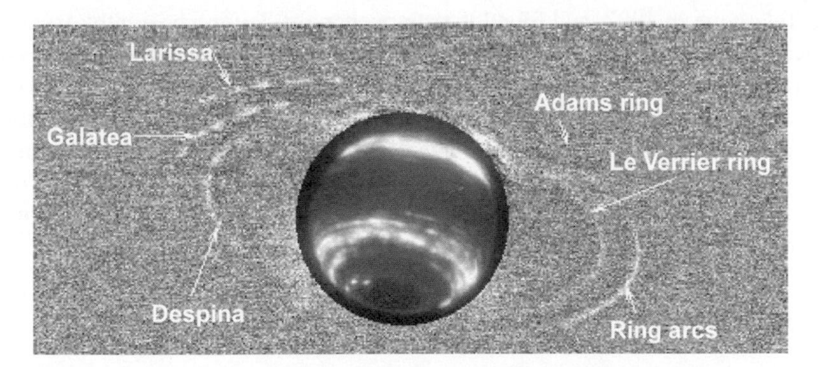

Fig. 10.8. Neptune's rings and moons in 2003 as seen by Keck. In this composite frame, a single Neptune exposure is surrounded by multiple exposures of its rings and moons at 2.2 microns (adapted from de Pater *et al.* 2005a).

nism. No quantitative theory has yet been devised that can describe the observed rapid dynamical evolution.

10.3.3 Triton: an icy body with a warming atmosphere

Triton – the largest satellite of Neptune – travels in a retrograde direction, unlike any other of the Solar System's large satellites. Triton has a diameter of 2,706 km, somewhat smaller than Earth's Moon. The Voyager 2 spacecraft revealed remarkably diversity on Triton's surface (Smith et al. 1989). Some areas were smooth and free of craters, suggesting relative youth by Solar-System standards. Grooves and ridges cut across the cold, icy landscape. Prockter et al. (2005) suggest that Triton's ridges may have formed by diurnal tidal stresses, in a manner similar to that proposed for Europa's ridge formation. The magnitudes of the stresses and heat fluxes required to generate such ridges are comparable to expected values during the latter part of Triton's orbital evolution from its initial highly eccentric state. The greater widths of ridges on Triton (relative to those on Europa) is likely a consequence of a lower surface temperature and greater brittle-ductile transition depth (Prockter et al. 2005). Other unique surface features include a cantaloupe terrain characterized by roughly circular depressions with upraised rims. The depressions are probably not impact craters (too similar in size and too regularly spaced); they more likely are associated with local melting and collapse. Yet other regions have dark gray markings with sharply defined bright borders. The slightly pinkish and highly reflective South polar cap may consist of a slowly evaporating layer of nitrogen ice. Northward the surface is generally darker and redder, perhaps through modification of surface methane ice by ultraviolet and magnetospheric radiation. Triton has a relatively high overall density of 2.1 g/cm^3, suggesting that perhaps up to 75% of its inner core is rocky material.

Triton's atmosphere is composed primarily of nitrogen, with some methane, and may extend as high as 800 km. In addition, a thin haze appears at about 3 km above the surface but may extend upward to 14 km. The haze may be condensed atmospheric gases (nitrogen ice) or complex organic molecules produced by irradiation of methane in the atmosphere. Elliot et al. (1998) observed a stellar occultation by Triton with the Hubble Space Telescope in 1997, and found the atmosphere of Triton apparently experienced significant warming since 1989, based on inferred surface-pressure increases. The inferred rate of temperature and surface-pressure increase suggested that the atmosphere is doubling in bulk every ten years – significantly faster than predicted by published frost models for Triton (Hansen and Paige 1992; Spencer and Moore 1992). In 1997, Triton was approaching its extreme Southern summer (a condition arising only every few hundred years); the observed variation suggest the polar caps on Triton play a dominant role in regulating seasonal atmospheric change (Elliot et al. 1998).

10.3.4 More moons around Neptune

Prior to the Voyager Encounter, only two satellites of Neptune were known: Triton and Nereid. Both satellites had unusual orbits: Triton moves in a retrograde direction around Neptune, and Nereid has the most eccentric orbit of any moon in the Solar System. In 1989, Voyager 2 discovered six new satellites during its passage. Proteus is the largest of these satellites, just slightly larger than Nereid (its dark color and proximity to Neptune hid it from Earth-based searchers). Larissa is an irregularly shaped, dark object. The non-circular shapes of Proteus and Larissa suggest that they remained cold and icy throughout much of their history. Aside from their orbits, little is known about the other four satellites found by Voyager: Despina, Galatea, Thalassa, and Naiad. The census of Neptune's moons, like that of Uranus, has increased in recent years. Holman et al. (2004) reported the discovery of 5 new irregular moons of Neptune based on observations made in 2003. Their extreme faintness had shielded them from view until today's powerful telescopes and sensitive cameras could be employed. The five satellites orbit Neptune at great distances, two in prograde orbits, and three in retrograde orbits. None is larger than about 50 km in diameter. These bodies, like Triton, may be captured objects. Alternatively, they may be the remnants of a collision between an existing moon and a passing comet or asteroid.

10.4 Future Exploration of the Ice Giants

The remoteness of Uranus and Neptune engender significant challenges to planetary exploration. Even with our most optimistic scenarios for standard or nuclear launch capabilities, the travel times are still a decade or longer. Thus a spacecraft and all its components must be long-lived. At the far reaches of the Solar System, little sunlight is available for power, almost certainly requiring nuclear power sources that are both stable and robust.

NASA has begun preliminary designs for missions to Neptune. However, in the most recent NASA "Roadmapping" (planning) exercise, a Neptune mission launch is not even considered until after 2035. Uranus does not appear on any manifest. Given these circumstances, all future exploration of these bodies will be done remotely. Much of the work described in this chapter has been from Earth's surface (ground-based telescopes) and with the Hubble Space Telescope in Earth-based orbit (which should hopefully operate for a few years yet).

A future facility that may yield fascinating results is the James Webb Space Telescope, an infrared-optimized 6.5-m segmented telescope that is planned for launch in 2013 to Earth's L2 point. One potential drawback with this facility is that its plans currently do not include the capability to tracking a moving object such as Neptune or Uranus (or their satellites). Fortunately, there are still many years to devise means to work around this limitation.

More than a decade ago, Hunt and Moore (1989) remarked that despite Voyager 2's remarkable close fly-by encounters with Uranus and Neptune, "the planets guard their secrets well." With a helping hand from the natural world *vis a vis* seasonal change, we are now slowly teasing out some of these guarded secrets. With ever-more powerful telescopes and increasingly sophisticated instrumentation, we can be assured of new discoveries for many decades to come.

References

Allison, M., et al., 1991. Uranus atmospheric dynamics and circulation. In *Uranus*, eds. J.T. Bergstralh, E.D. Miner and M.S. Matthews, Univ. Arizona Press, Tucson, Arizona, pp. 253–295.

Bézard, B., et al., 1999, Detection of the methyl radical on Neptune. *Astrophys. J.* **515**, 868.

de Pater, I., et al., 2005a. The dynamic neptunian ring arcs: evidence for a gradual disappearance of Liberté and resonant jump of courage. *Icarus* **174**, 263.

de Pater, I., Gibbard, S., & Hammel, H.B., 2005b. Evolution of the Dusty Ring System of Uranus. *Icarus* **180**, 186.

Dumas, C., et al., 2002. Astrometry and near-infrared photometry of Neptune's inner satellites and ring arcs. *Astrophys. J.* **123**, 1776.

Elliot, J.L., et al., 1998. Global warming on Triton. *Nature* **393**, 765.

Geballe, T.R., Jagod, M.F., & Oka, T., 1993. Detection of infrared emission lines in Saturn. *Astrophys. J.* **408**, L109.

Gladman, B.J., et al., 1998. Discovery of two distant irregular moons of Uranus. *Nature* **392**, 897.

Gladman, B.J., et al., 2000. The discovery of Uranus XIX, XX, and XXI. *Icarus* **147**, 320.

Grundy, W.M., et al., 2003. Discovery of CO_2 ice and leading-trailing spectral asymmetry on the uranian satellite Ariel. *Icarus* **162**, 222.

Hammel, H.B., 1989. Neptune cloud structure at visible wavelengths. *Science* **244**, 1165

Hammel, H.B., & Lockwood, G. W., 2006. Atmospheric Variability on Uranus and Neptune: Seasonal, Solar-Driven, or Stochastic? *Icarus*, submitted.

Hammel, H.B., et al., 1995. Hubble Space Telescope Imaging of Neptune's Cloud Structure in 1994. *Science* **268**, 1740.

Hammel, H.B., et al., 2001. New measurements of the winds of Uranus. *Icarus* **153**, 229.

Hammel, H.B., et al., 2005a. New Cloud Activity on Uranus in 2004: First Detection of a Southern Feature at 2.2 microns. *Icarus* **175**, 284.

Hammel, H.B., et al., 2005b. Uranus in 2003: zonal winds, banded structure, and discrete features. *Icarus* **175**, 534.

Hammel, H.B., et al., 2006. Mid-infrared ethane emission on Neptune and Uranus. *Astrophys. J.*, submitted.

Hansen, C.J., & Paige, D.A., 1992. A thermal model for the seasonal nitrogen cycle on Triton. *Icarus* **99**, 273.

Hofstadter, M.D., Butler, B.J., & Gurwell, M.A., 2005. Imaging the troposphere of Uranus at millimeter and centimeter wavelengths. *Bull. Amer. Astron. Soc.* **37**, 662.

Holman, M.J., et al., 2004. Discovery of five irregular moons of Neptune. *Nature* **430**, 865.

Hunt, G., & Moore, P., 1989. *Atlas of Uranus*, Cambridge University Press. Karkoschka, E., 1998. Clouds of high contrast on Uranus. *Science* **280**, 570.

Kavelaars, J.J., et al., 2004. The discovery of faint irregular satellites of Uranus. *Icarus* **169**, 474.

Lockwood, G.W., & Jerzykiewicz, M., 2006. Photometric Variability of Uranus and Neptune, 1950 – 2004. *Icarus* **180**, 442.

Marten, A., et al., 2005. Improved constraints on Neptune's atmosphere from submillimetre-wavelength observations. *Astron. & Astrophys.* **429**, 1097. Martin, D.W., McDaniel, E. W., & Meeka, M.L., 1961. On the possible occurrence of H_3^+ in interstellar space. *Astrophys. J.* **134**, 1012.

Namouni, F., & Porco, C., 2002. The confinement of Neptune's ring arcs by the moon Galatea. *Nature* **417**, 45.

Prockter, L.M., Nimmo, F., & Pappalardo, R.T., 2005. A shear heating origin for ridges on Triton. *Geophys. Res. Letters* **32**, L14202.

Rages, K.A., Hammel, H.B., & Friedson, A.J., 2004. Evidence for temporal change at Uranus' South pole. *Icarus* **172**, 548.
Sheppard, S.S., et al., 2005. An ultradeep survey for irregular satellites of Uranus: limits to completeness. *Astron. J.* **129**, 518.
Showalter, M.R., Lissauer, J.J., & de Pater, I., 2005. The Rings of Neptune and Uranus in the Hubble Space Telescope. *Bull. Amer. Astron. Soc.* **37**, 772.
Smith, B.A., et al., 1986. Voyager 2 in the Uranian system: imaging science results. *Science* **233**, 43.
Smith, B.A., et al., 1989. Voyager 2 at Neptune – imaging science results. *Science* **246**, 1422.
Spencer, J.R., & Moore, J.M., 1992. The influence of thermal inertia on temperatures and frost stability on Triton. *Icarus* **99**, 261.
Sromovsky, L.A., et al., 2001. Coordinated 1996 HST and IRTF imaging of Neptune and Triton. III. Neptune's atmospheric circulation and cloud structure. *Icarus* **149**, 459.
Sromovsky, LA., et al., 2003. The nature of Neptune's increasing brightness: evidence for a seasonal response. *Icarus* **163**, 256.
Trafton, L.M., et al., 1993. Detection of H_3^+ from Uranus. *Astrophys. J.* **405**, 761.
Trafton, L.M., et al., 1999. H2 quadrupole and H_3^+ emission from Uranus: the uranian thermosphere, ionosphere, and aurora. *Astrophys. J.* **524**, 1059.
Trafton, L.M., & Miller, S., 2004. Images of Uranus' H_3^+ emission. *Bull. Amer. Astron. Soc.* **36**, 1073.

11 The Solar System Beyond The Planets

Audrey Delsanti and David Jewitt

Abstract. The Kuiper belt contains a vast number of objects in a flattened, ring-like volume beyond the orbit of Neptune. These objects are collisionally processed relics from the accretion disk of the Sun and, as such, they can reveal much about early conditions in the Solar system. At the cryogenic temperatures prevailing beyond Neptune, volatile ices have been able to survive since the formation epoch 4.5 Gyr ago. The Kuiper belt is the source of the Centaurs and the Jupiter-family comets. It is also a local analogue of the dust disks present around some nearby main-sequence stars. While most Kuiper belt objects are small, roughly a dozen known examples have diameters of order 1000 km or more, including Pluto and the recently discovered (and possibly larger) giant Kuiper belt objects 2003 UB_{313}, 2003 EL_{61} (a binary and a triple system, resp.) and 2005 FY_9.

11.1 Introduction

As is well known, Pluto was discovered in 1930 as the result of a search motivated by Percival Lowell's prediction of a 9th planet. Lowell based his prediction on anomalies in the motion of Uranus that could not be explained by the gravitational tug of Neptune. In hindsight, we now know that these anomalies were simply astrometric errors, and that Lowell's prediction of Pluto was baseless. Nevertheless, Clyde Tombaugh found Pluto and he and many others assumed that it was the massive object predicted by Lowell (Tombauch, 1961). Doubts about the correctness of this assumption were raised almost immediately, with the realization that the likely mass of Pluto was too small to measurably perturb Uranus or Neptune. Still, the planetary label stuck.

Pluto's true significance became apparent only in 1992, with the discovery of the trans-Neptunian object 1992 QB_1 (Jewitt & Luu, 1993). Since then, about 1000 Kuiper belt objects (KBOs) have been found, with sizes from a few 10's km to several 1000 km and orbits in a number of dynamically distinct classes. Their defining feature is that their semi-major axes are larger than that of Neptune (30 AU). Most have perihelia beyond 30 AU: the region inside Neptune's orbit tends to be dynamically unstable owing to strong perturbations from the giant planets. The known dynamical sub-types (their average orbital parameters are listed in Table 11.1) in the Kuiper Belt are:

11.1.1 Resonant objects

The resonant objects are those trapped in mean-motion resonance with Neptune. The prime example is Pluto, which sits with a great many other KBOs in the 3:2 mean motion resonance at 39.5 AU. Occupancy in a mean-motion resonance conveys dynamical stability, by limiting the possibilities for close encounters between the trapped KBO and Neptune. Pluto, for example, has a perihelion distance (q = 29.7 AU) inside Neptune's orbit, but never encounters Neptune because the resonance ensures that the longitude of perihelion is separated from that of the planet by about $\pm 90°$. The 3:2 resonance is the most densely populated and, to draw a parallel with the largest trapped object, the residents of this resonance are known as "Plutinos". Numerous other resonances are also occupied, including so-called secular resonances in which there is commensurability between the rates of precession of angular variables describing the orbits of the KBOs and corresponding quantities of the orbit of Neptune.

11.1.2 Classical Kuiper belt

The Classical Kuiper belt is effectively the region between the 3:2 mean motion resonance at 39.5 AU and the 2:1 mean motion resonance at 48 AU, excluding objects which are in those resonances. The first-detected objects in this region, like the prototype 1992 QB_1, have small inclinations, reminiscent of dynamical expectations that the Belt would be a dynamically cold (thin) structure left over from the accretion epoch. Later discoveries, however, have included a large number of highly inclined KBOs in the Classical region, and we now recognize that there are two populations superimposed in this region. The low inclination ("cold") population has $i \leq 4°$ while the high inclination ("hot") population extends to inclinations of 30° to 40°, and possibly higher. As discussed below, there is some evidence for a difference between the physical properties of the hot and cold populations.

11.1.3 Scattered objects

These are objects with perihelia in the ~35 to 40 AU range and characteristically large eccentricities and inclinations. The first example found, 1996 TL_{66}, is typical with a = 82.9 AU, $e = 0.577$ and $i = 24°$. The scattered objects, often called scattered disk objects (although their large inclinations more resemble a torus than a disk) are thought to have been lofted into eccentric orbits by weak scattering from Neptune (Morbidelli et al., 2004). In this scenario they would be survivors of a once much larger population that has been steadily depleted by Neptune perturbations over solar system time. Recent work

Table 11.1. Mean orbital elements of KBOs sub-classes and Centaurs

	a[AU]	$\bar{e}$	$\bar{i}$ [deg]	$N_{objects}$
Classical	$39.4 < a < 47.8$	0.06	6.97	681
3:2 resonance	39.4	0.22	9.93	47
5:3 resonance	42.4	0.20	9.35	4
7:4 resonance	43.6	0.20	3.97	4
2:1 resonance	47.6	0.31	11.5	7
5:2 resonance	55.8	0.41	8.03	6
Scattered	>30	0.29	12.6	384
Centaurs	<35	0.31	12.9	57

has shown that many scattered objects are in fact trapped in high order mean motion resonances, blurring at least the nomenclature and their distinction from the resonant population. The likely mechanism of emplacement of the bodies remains scattering by Neptune, however, so that the label is not entirely without meaning.

11.1.4 Detached objects

Two objects, 2000 CR_{105} (Gladman et al., 2002) and (90377) Sedna (Brown et al., 2004), have perihelia so large that they cannot have been emplaced by gravitational interactions with Neptune, at least not in the same way as were the other scattered objects. These bodies define the class of detached objects (Emel'yanenko et al., 2003) also called "extended scattered disk" objects. Their emplacement might have been due to perturbations associated with a passing star (Morbidelli & Levison, 2004).

11.2 Pluto and Other Large Kuiper Belt Objects

About a dozen KBOs are now known with diameters in the 1000 km range and larger (Table 11.2). In July 2005, three Pluto-sized bodies were announced: 2003 UB_{313}, 2003 EL_{61} (a binary and triple system, respectively) and 2005 FY_9. They are the result of an ongoing survey dedicated to the search of bright KBOs (Trujillo & Brown, 2003). Indeed, results from previous surveys are in agreement to predict the existence of a few tenths of KBOs of the size of Pluto, some of them possibly larger. Their discovery implies a large sky coverage (often limited to the ecliptic region and the Solar system invariable plane, where the high-inclination objects like 2003 UB_{313} spend only a small fraction of their orbit and are therefore more difficult to discover), a task that is time consuming. The Pan-STARRS project, a set of four 1.8m telescopes on top of Mauna Kea that will scan the whole available sky down to V$\sim$24 every couple of days, should significantly contribute to the discovery of new, large KBOs. In this section, we briefly describe the four largest objects known as of October 2005.

11.2.1 Pluto

Observations of the orbital motion of Pluto and its main satellite Charon have given the combined mass of the pair as 1.5×10^{22} kg. This is only $\sim$0.2% of the mass of the Earth and five times less even than the mass of the Moon. "Mutual events", a series of occultations and eclipses visible as Earth passed through the orbital plane of Charon, have revealed many details of Pluto and Charon (see Table 11.3: Pluto) including the density, about 2000 kg m^{-3}, which suggests a composition about 70% rock by mass. The large specific angular momentum of the Pluto-Charon pair suggests that Charon may have formed by a glancing impact, presumably early in the history of the Solar system (Canup, 2005).

Independently of the mutual events, occultations of field stars by Pluto have revealed astounding details of this distant world. An atmosphere is present, with a pressure at 1250 km radius near 1 μbar (1 bar = 10^5 N m^{-2}, see Elliot et al., 1989). The occultation

Table 11.2. Large KBOs

Name	Type[a]	H^b	p[c]	D [km][d]	a [AU]	e	i[deg]	Multiple?
2003 UB_{313}	Scat	−1.2	0.6?	2600?	67.6	0.44	44.2	Yes
Pluto	3:2	−1.0	0.6	2320	39.5	0.25	17.1	Yes
2005 FY_9	Scat	0.3	0.6?	1250?	45.7	0.15	29.0	
2003 EL_{61}	Cla	0.4	0.6?	1200?	43.3	0.19	28.2	Yes
(90377) Sedna	Det	1.6	0.2?	$<$1500?	495	0.85	11.9	
(90482) Orcus	3:2	2.2	0.12?	$\sim$1500	39.4	0.22	20.6	
(50000) Quaoar	Clas	2.6	0.12	1200 ∓ 200	43.5	0.03	8.0	
(28978) Ixion	3:2	3.2	0.09	1065 ∓ 165	39.6	0.24	19.6	
(55565) 2002 AW_{197}	Clas	3.2	0.1	890 ∓ 120	47.4	0.13	24.4	
(20000) Varuna	Clas	3.7	0.07 ± 0.02	900 ∓ 140	43.0	0.05	17.2	

a: Dynamical type: 3:2 = resonant, Clas = Classical, Cen = Centaur, Scat = Scattered, Det = Detached
b: absolute magnitude
c: Geometric Albedo
d: Diameter

lightcurves show a steep drop in intensity near Pluto that cannot be accurately matched by a model of an isothermal atmosphere in hydrostatic equilibrium, leading to some ambiguity in the structure of the atmosphere and even the occultation radius of Pluto. The steep drop in the intensity can be explained by invoking a near-surface haze, or by invoking steep near-surface temperature gradients that would refract light from the occulted star away from the direction to the observer. In either case, the key uncertainty is the distance between the 1250 km reference radius probed by occultations and the surface, and this distance remains unknown. If we assume that the radius derived from mutual events is accurate, then the surface pressure (extrapolated from 1250 km) in the atmosphere must be near 30 μbar, but this value is uncertain by at least a factor of several. Pressures of this magnitude correspond to the vapor pressure above solid N_2 at Pluto's surface temperature $\sim$38 K. Recent observations of occultations of a bright star (Sicardy et al., 2003; Elliot et al., 2003) showed that the pressure at a given height is time-variable: the measured pressures were two times higher than in 1988, despite the greater heliocentric distance of Pluto. Presumably, time variations reflect seasonal variations in the insolation of patches of volatile matter exposed on the surface of Pluto.

Reflectance spectroscopy of Pluto reveals a variegated surface dominated by methane ice (CH_4) incorporated in frozen nitrogen (N_2). Carbon monoxide (CO) and some water ice (H_2O) are also detected. Hubble Space Telescope images evidenced some dark and bright areas over the surface. Bright areas are most probably covered by nitrogen and water ice, whereas the origin of dark areas is less clear. Methane is believed to be destroyed over time under the effect of the steady flux of ions and UV radiations (from solar wind and cosmic rays) that hits the surface. Laboratory work shows that such processes lead to the formation of dark-colored, complex organic compounds. The abundance of methane on the surface therefore indicates that a replenishment mechanism should be acting. At Pluto's surface equilibrium temperature ($\sim$40K), methane can be volatile and mix with more volatile nitrogen to form a tenuous, weakly bound atmosphere. With seasonal variations of pressure, one can expect a future redeposition of frozen methane and nitrogen over the surface. A subsurface methane source is also invoked, that would

Table 11.3. Parameters of the Pluto system

Object	Mass [kg]	Radius [km]	Density[g/cm^3]	a [km][a]	P [days][b]
Pluto	1.32×10^{22}	1160	2.0	—	6.37
Charon	1.6×10^{21}	625	1.7	19,400	6.37
S/2005 P 1	—	25-80?	—	64,700?	38.2?
S/2005 P 2	—	20-70?	—	49,500?	25.5?

a: semi-major axis of the orbit around the barycenter of the Pluto system
b: rotation period (Pluto), orbital period (Charon, S/2005 P 1, S/2005 P 2)

replenish the surface through cryovolcanism, as hypothesized for Neptune's satellite: Triton.

Two faint satellites (visible magnitude ~23 and 23.4) were discovered in May 2005 on Hubble Space Telescope images (Weaver et al., 2005). They both are on a near circular orbit far outside Charon's. Preliminary studies give distances to the Pluto system barycenter of 64,700 km and 49,500 km for the brightest and the faintest satellite resp., corresponding to orbital periods of ~38 and ~25 days. All three satellites (including Charon) are orbiting on approximately the same plane. The project that lead to this discovery was aimed at mapping Pluto's sphere of gravitational influence: no other detection was claimed down to a magnitude of V~27, which means that if any other satellite exists, it should be smaller than 20km in diameter. From the apparent brightness of the discovered satellites, the approximate diameters are estimated in the range 40 km–180 km.

Pluto is now known as a quadruple system. We can expect that several other multiple systems will be discovered in the Kuiper belt in the near future (the first triple system, 2003 EL_{61} was announced in December 2005). Further studies of these two new satellites will lead to improved estimates of the mass and density of Pluto and Charon and bring constraints on the formation scenario and tidal evolution of the whole system (and on the other KBO multiple systems).

11.2.2 2003 UB_{313}

2003 UB_{313} is a scattered-disk object with an extremely high orbital inclination (44°). It is currently located near aphelion at 97 AU from the Sun. With a semi-major axis of ~68 AU, it will reach perihelion at 38 AU in ~250 years: its surface will probably undergo significant seasonal changes on the way. Its albedo is not determined yet, but even with the most constraining assumptions, 2003 UB_{313} is likely to be larger than Pluto. Preliminary spectroscopic studies (Brown et al., 2005a) show that it has a methane dominated surface, very close to that of Pluto. The optical spectral slope, however, is less red. This is the first time methane has been firmly detected on a KBO other than on Pluto. Unlike the methane bands on Pluto, those on 2003 UB_{313} show no shift in wavelength that might be attributed to solution in solid nitrogen. Equilibrium surface temperatures are expected to be ~30K at 97 AU. The corresponding vapor pressure over pure methane ice makes it in-volatile: it is most likely present on the surface as an ice, segregated from

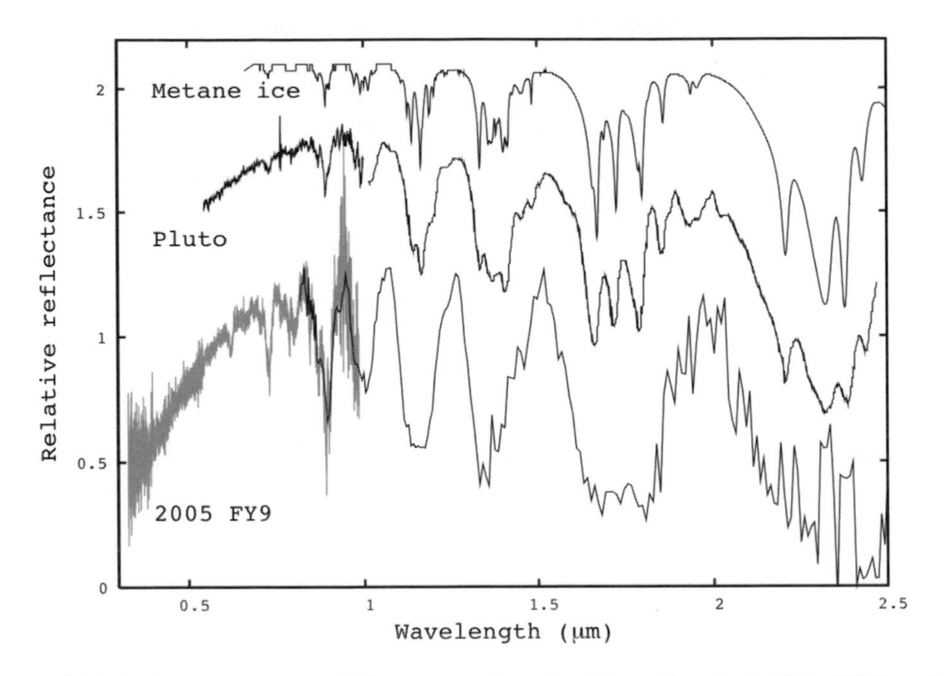

Fig. 11.1. Reflectance spectra of laboratory methane ice, Pluto (Grundy & Fink, 1996), and 2005 FY_9. Spectra are normalized at 0.6μm and vertically shifted for clarity. Pluto and 2005 FY_9 show very similar chemical surface composition. Figure from Licandro et al. (2006).

other components. However, as 2003 UB_{313} approaches the Sun, methane will become volatile again and might mix with (not detected yet) nitrogen in a tenuous atmosphere, as for Pluto. 2003 UB_{313} therefore provides a low-temperature analog to study processes occurring on Pluto's surface. One possibility is that the presence of methane might be a characteristic of the largest KBOs, the smaller bodies being too small to retain it on their surface due to higher escape rates, or being unable to produce it owing to their frigid interiors.

A faint satellite has been reported around 2003 UB_{313} (with a fractional brightness of 2% of the primary). This satellite is much fainter than other KBO satellites and may have a different origin. The preliminary binary system characteristics favor the impact formation scenario (see Section 11.3).

11.2.3 2005 FY_9

2005 FY_9 is the third brightest known KBO (in absolute magnitude, after 2003 UB_{313} and Pluto). Its size may approach that of Pluto. With a semi-major axis of 46 AU, a perihelion of 39 AU and an inclination of $\sim 29^o$, it belongs to the classical KBOs family. Preliminary spectroscopic studies (Licandro et al., 2006) reveal the presence of methane (as in Pluto and 2003 UB_{313}). The methane lines detected are very close to those of laboratory pure methane ice. So far, no other ices such as N_2 or water ice were detected. The visible data show a spectroscopically red surface (similar to Pluto) which is compatible with the presence of complex organic compounds. Its general similarities with Pluto (size, surface composition, heliocentric range) make 2005 FY_9 a good candidate to hold a tenuous, bound atmosphere. This hypothesis can be observationally tested in the future during an occultation of a star by 2005 FY_9. According to Brown et al. (2005b), no satellite was discovered within 0.4 arc-seconds of 2005 FY_9 and with a brightness of more than 0.5% of the object. This KBO will be the object of extensive studies.

11.2.4 2003 EL_{61}

2003 EL_{61} is the fourth brightest known KBO, and its size should be about that of Pluto. Early studies revealed both water ice on its surface and the presence of two satellites. The first satellite is on a near circular orbit at 49,500km from 2003 EL_{61}. Preliminary results (Brown et al., 2005c) show that the orbital period of the system is $\sim$49 days and the satellite brightness is 6% of 2003 EL_{61} (for reference, Pluto-Charon has an orbital period of 6.4 days and a flux ratio of 20). The mass of this system is estimated to 4×10^{21}kg which is 32% the mass of Pluto. The rotation of 2003 EL_{61} ($\sim$4h, Rabinowitz et al., 2005, to be confirmed by further studies), is extremely fast. From the mass of the system and the rapid rotation period of the primary (which should lead to extreme rotational deformations), first estimations give a length of the primary of 1900 to 2500km, a mean density of 2600 to 3300 kg m^{-3} (consistent with Pluto) and a visual albedo greater than 60%. The parameters of this system are in partial agreement with an impact formation scenario (see Sec. 11.3) as for Pluto. In that case, tidal evolution will modify the eccentricity and orbital period of the system over time, with an amplitude determined by the strengthness of both the primary and the satellite. Another possibility is that this pair formed by capture (see Sec. 11.3), and the semi-major axis subsequently shrunk to the current one owing to dynamical friction. The current parameters of the system only partially match this scenario. As these lines are written, a second satellite is detected. Its brightness is 1.5% that of 2003 EL_{61}. Preliminary circular orbit fits (Brown et al., 2005b) give a semi-major axis of 39,300 km and an orbital period of 34 days. Its orbital plane is inclined of $\sim$40 degrees with respect to the other satellite's plane. The presence of multiple satellites around giant KBOs (as for Pluto) might point towards a formation in a disk, although this possibility requires further exploration.

Unfortunately, mutual events are not predicted in the near future and the characterization of 2003 EL_{61} (as regards mass, density, etc) will consequently be much poorer than for Pluto.

11.3 Binaries and Multiple Systems

Pluto's main satellite, Charon, was discovered as late as 1979 in photographic images taken to refine the orbit of this "planet" about the Sun. Pluto-Charon became the first-recognized of many Kuiper belt binaries (see Table 11.3). In fact, the Pluto-Charon system is synchronously locked, with the primary and secondary having the same spin period as the orbital period about the barycenter. This is a natural state for a binary in which gravitational tides are strong. Tides on the primary raised by the secondary (and vice-versa) raise a bulge on which gravitational torques act to bring the orbital and spin

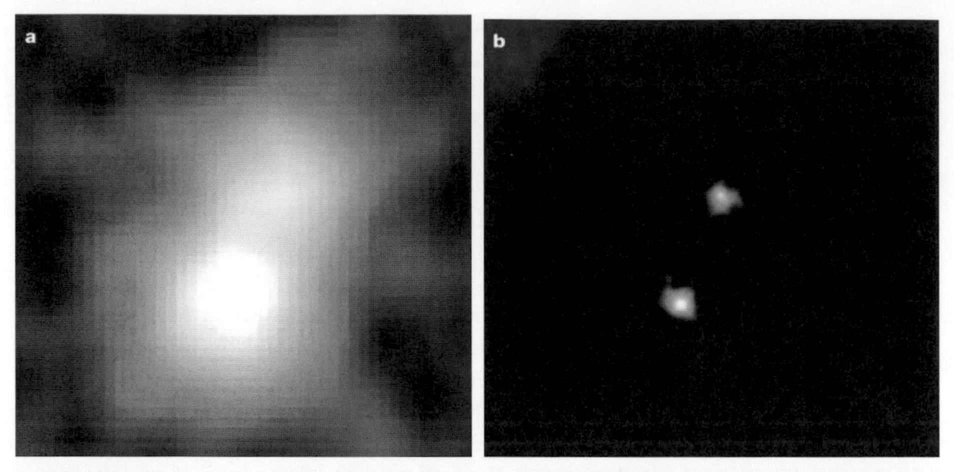

Fig. 11.2. The first discovered binary after Pluto-Charon, 1998 WW_{31}, as seen from the ground at CFHT on 2001 September 12 (left) and from the Hubble Space Telescope on September 9, 2001 (right). Separation is 0.59 arcsec on both images (same scale), semi-major axis of the system is ~20,000 km. Images from Veillet et al. (2002).

Table 11.4. Parameters of the multiple KBOs

Object	a [km][a]	e^b	i[deg][c]	Type[d]	Q[arc-sec][e]	P[days][f]	Δmag
Pluto				3:2			
Charon	19,400	0.00	96	—	0.9	6.4	3.2
S/2005 P1	64,700?	—	—	—	2.2	38.3?	9.0
S/2005 P2	49,400?	—	—	—	1.7	25.5?	9.4
1998 WW_{31}	22,300±800	0.82	42	Cla	1.2	574	0.4
1999 OJ_4	—	—	—	Cla	—	—	—
2000 CF_{105}	—	—	—	Cla	0.8	—	0.9
2000 CQ_{114}	—	—	—	Cla	—	—	—
2000 OJ_{67}	—	—	—	Cla	—	—	—
2001 QC_{298}	3,690±70	—	—	Cla	0.17	19.2	—
2001 QW_{322}	—	—	—	Cla	4.0	—	0.4
2003 EL_{61}				Sca			
S/2005 (2003 EL_{61}) 1	49,500±400	0.050±0.003	234.8±0.3	—	1.3	49.12±0.03	3.3
S/2005 (2003 EL_{61}) 2	39,300?	—	—	—	1.0	34.1?	4.5
2003 QY_{90}	—	—	—	Cla	—	—	—
2003 UB_{313}	36,000	—	—	Sca	0.5	14	4.2
2003 UN_{284}	—	—	—	Cla	—	—	—
2005 EO_{304}	—	—	—	Cla	—	—	—
(26308) 1998 SM_{165}	11,310±110	—	—	2:1	0.2	130	1.9
(42355) 2002 CR_{46}	—	—	—	Sca	0.11	—	1.2
(47171) 1999 TC_{36}	7,640±460	—	—	3:2	0.4	50.4	1.9
(48639) 1995 TL_8	—	—	—	Sca	—	—	—
(58534) 1997 CQ_{29}	8,010±80	0.45	121	Cla	0.2	312±3	0.3
(60458) 2000 CM_{114}	—	—	—	Cla	0.07	—	0.5
(66652) 1999 RZ_{253}	4,460±170	0.46	152	Cla	—	46.3	—
(79360) 1997 CS_{29}	—	—	—	Cla	—	—	—
(80806) 2000 CM_{105}	—	—	—	Cla	—	—	—
(82075) 2000 YW_{134}	—	—	—	Sca	—	—	—
(88611) 2001 QT_{297}	27,300±340	0.31	128	Cla	0.6	825	0.5

a: semi-major axis of the binary system
b: eccentricity
c: inclination
d: Dynamical type: 3:2, 2:1 = resonant, Clas = Classical, Scat = Scattered
e: Angular separation
f: Orbital period

periods to the same value. With the Pluto-Charon separation of ~20,000 km, the tidal dissipation time is small compared to the age of the Solar system, and a synchronously locked system is the result.

The mechanisms by which the Kuiper belt binaries formed remain unidentified. Close binaries in nearly circular orbits, like Pluto-Charon and 2003 EL_{313}, suggest a collisional origin in which the satellite was blasted out of the primary by an ancient impact. Numerical simulations have shown that glancing impacts can produce bound satellites, at least over a range of carefully selected initial conditions (Canup 2005).

The wide, eccentric binaries that are seen elsewhere in the Kuiper belt must have a different origin, and several have already been suggested. In an early, dense Kuiper belt consisting of objects having a wide range of sizes, the process called "dynamical friction" might have acted to stabilize objects in wide binaries. In this process, the collective effects of the smaller objects exert a net force on massive bodies passing through them. One prediction of this model is that the fraction of binaries should increase as the separation decreases (Goldreich et al 2002). It is also possible that a dense early Kuiper belt could have a significant number of three-body interactions. In these, scattering between bodies can result in the ejection of one, which carries excess energy from the system and leaves behind a stable binary (Weidenschilling 2002). Three-body captures result mainly in wide binaries, and the detection of a substantial fraction of close binaries would require another explanation. From lightcurve studies, the fraction of contact binaries (perhaps products of continued dynamical friction) has been estimated as 10% to 20% (Sheppard & Jewitt, 2004).

The common feature of the binary formation models is that they all assume an initial dense phase, with densities ~100 to ~1000 times the present value in the trans-Neptunian region. If these conditions once prevailed, it is reasonable to assume that binaries were both formed and dissociated in an active early epoch, probably in association with or soon after formation. The existing suite of binaries are merely the survivors from this long-gone stage.

11.4 Observed Structure of the Kuiper Belt

11.4.1 Inclination distribution and velocity dispersion

Early models of the Kuiper belt predicted a modest range of inclinations of $\sim 1^\circ$ due to gravitational stirring by Neptune (e.g., Holman & Wisdom, 1993). The observed belt is much thicker, as may be seen in Figure 11.3 which shows an unbiased inclination distribution. The thickness of the Kuiper belt is directly related to the velocity dispersion amongst the objects within it. Careful measurements show that the velocity dispersion is now about 1.6 km s^{-1} (Trujillo et al., 2001). At this speed, KBOs shatter and pulverize each other when they collide instead of sticking and growing. The implication is that the velocity dispersion has been pumped up by some process (or processes) since the epoch of formation, when the velocity dispersion would have been an order of magnitude or more smaller than now. What could have done this? It is known that resonance trapping tends to pump up the inclinations of the objects that become trapped (Malhotra, 1995). This resonance pumping can explain why Pluto has $i \sim 17^\circ$, for example, but it is only a partial

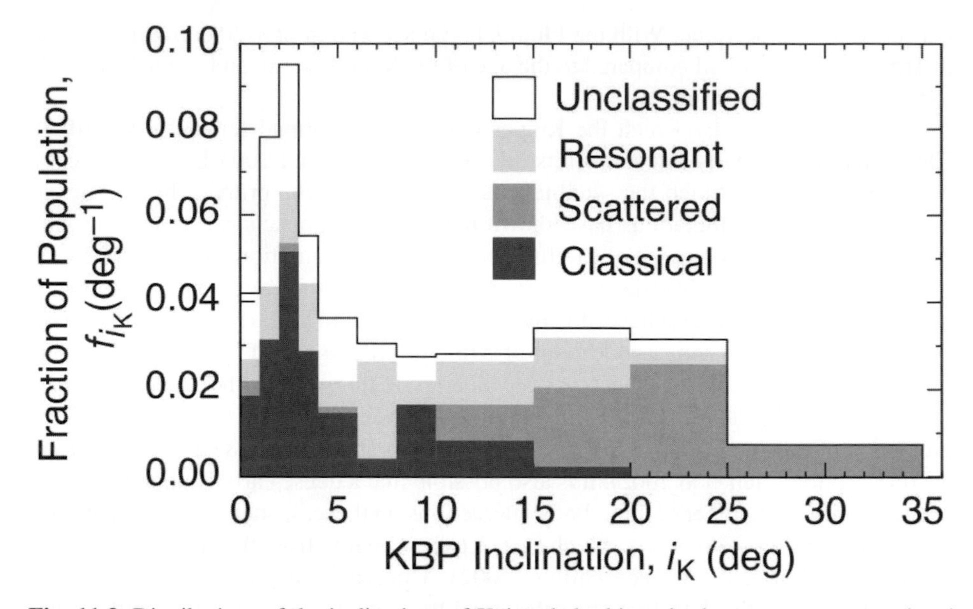

Fig. 11.3. Distributions of the inclinations of Kuiper belt objects in the resonant, scattered and classical populations (dynamically unclassified objects are also shown). The classical objects are bimodally distributed, with the cold and hot populations having peaks near 3° and 8°. The resonant and scattered populations show broad inclination distributions extending to 20° and above. Figure from Elliot et al. (2005).

answer for the belt as whole because the resonant objects are a minority population. For the scattered disk objects, it is likely that the inclinations were amplified by the same scattering events that launched KBOs onto their highly eccentric orbits: numerical simulations of this Neptune scattering produce inclination distributions broadly similar to the one observed (Gomes, 2003b).

More problematic is the inclination distribution of the classical objects which, as shown in Fig. 11.3, is bimodal. This suggests that the classical belt has a composite structure. The cold (with red surfaces, as described in Sect. 11.5.1) classical KBOs maybe objects originally formed beyond Neptune and swept out by resonances during Neptune's migration phase. The hot (with neutral to red surfaces) classical KBOs may have been scattered outwards by the inner giant planets during the planetary migration phase and captured in orbits external to Neptune's (Gomes, 2003a). Why the cold and hot classical KBOs should have different color patterns is unknown and unspecified in this model, but the idea that the differences in the inclination and color distributions is somehow related to a difference in formation locations is attractive.

Still other causes of the broad inclination distribution have been suggested. It is possible that a star, passing ~150 to 250 AU from the Sun, is responsible for observed structures in the Kuiper belt, notably the "edge" to the classical belt at about 47 AU to 48 AU (Ida et al., 2000). Perturbations from this star could stir up the belt, exciting the precursor disk into the puffy distribution we see now (and maybe truncating further growth of the KBOs in the process). The likelihood of a sufficiently close (~150 AU) stellar encounter in the modern epoch is negligible: such interactions might have been common if the Sun were born in a dense cluster of stars.

11.4.2 Radial extent

The classical belt has an edge near the 2:1 mean motion resonance (47 AU to 48 AU) in the sense that the distribution of semi-major axes is truncated at about this radial distance (Jewitt et al., 1998; Allen et al., 2001; Trujillo et al., 2001). It is not known if this is a real edge, or just the inner boundary of a gap, with the surface density of objects increasing to a high value at some larger radius. Orbits of the scattered disk objects extend over a much larger range of distances. The most extreme objects reach aphelia near 1000 AU (the current record is 2000 OO_{67}, which reaches 1014 AU) and others yet to be found must push further out towards the inner Oort Cloud. It is probably misleading to think of a distinct boundary to the Kuiper belt, or a sharp junction between it and the Oort cloud. Dynamical models tend to suggest that the past histories of the Belt and the Cloud are intertwined, so that no absolute distinction is possible.

11.4.3 Mass and size distribution

Optical surveys give the number of objects as a function of apparent brightness. The precise conversion to size distributions and masses requires knowledge of the albedos that we do not yet possess. Under the assumption that the albedo does not vary systematically with size, the survey data indicate that the size distribution of the larger (diameters $\geq$50 km) KBOs are well described by a differential power law with an index near 4.0$\pm$0.5 (Trujillo et al., 2001). Models of agglomeration of KBOs in a cold disk are compatible with this index, suggesting that it might be a relic of the formation process. At small sizes, the role of collisional shattering should become important and the distribution is expected to flatten (slightly) towards the Dohnanyi value, an index of 3.5. Limited evidence from a deep Hubble Space Telescope (HST) survey suggests that this flattening has been detected (Bernstein et al., 2004): earlier and contradictory evidence from HST (Cochran et al., 1995) has apparently been invalidated.

Kuiper belt masses were initially estimated on the assumption that all KBOs have albedos of 0.04, like the nuclei of the Jupiter-family comets. These initial estimates gave masses $\sim$0.1 $M_\oplus$ (Jewitt et al., 1998). Recent measurements of large KBOs suggest albedos on average about three times larger but it is not known if this is generally true or if the high albedos apply only to the largest KBOs (we cannot yet measure the smaller ones directly). In any case, 0.1 $M_\oplus$ is an upper limit to the total mass. With mass scaling as (albedo)$^{-3/2}$, the derived mass is $3^{3/2} \sim 5$ times smaller than first thought, corresponding to a few percent of an Earth mass. These estimates are still quite uncertain because we possess few reliable determinations of the albedos, because we do not know the densities of most KBOs and, especially, because we have meaningfully sampled only the inner regions of the Kuiper belt. For all these reasons, the relative masses of the various components of the Kuiper Belt are uncertain. A reasonably safe conclusion, however, is that the mass of the scattered disk objects is larger than the mass of the other components of the Kuiper belt (Trujillo et al., 2000).

11.4.4 Origin of the structure

Understanding of the origin of dynamical structure in the Kuiper belt is at a juvenile stage and major changes can be expected as new survey data become available (soon, from Pan-STARRS). There are several important parameters that must be fitted by the models. These include the (low) value of the total mass, its distribution amongst the various components (in the order scattered disk > classical belt > resonant population), the bounded distribution of semi-major axes of the classical objects (edge near the 2:1 resonance), the broad distributions of inclination and the bimodal distribution within the classical belt (and probably not elsewhere). While the number of observational constraints is already considerable, the number of degrees of freedom in models designed to fit the observations is much greater. Thus, the subject suffers from the classical problem of non-uniqueness. Interesting models are proposed and published, but many are based on initial conditions or other assumptions that are either arbitrary or untested. Still, it is interesting to speculate on the origin of the dynamical structure of the Kuiper belt.

The most plausible way to populate the resonances is through slow radial migration (Malhotra, 1995; Gomes, 2003b). A fraction of the KBOs encountered by a drifting resonance become captured and their inclinations and eccentricities may become enlarged as a result. Simple models suggest that migration of Neptune over $\sim$7 or 8 AU distances on time-scales near 10 Myr could account for the resonant populations. One puzzle is that the resonant populations are small ($\sim$10% of the total, possibly less) whereas migration models can be much more efficient. An explanation may lie in the jumpy nature of Neptune's migration, with the jumps being caused by individual interactions with massive bodies in the protoplanetary disk.

The cold component of the classical belt suggests that it is most closely related to (but still substantially thicker than) the original protoplanetary accretion disk. The existence of an edge at the 2:1 resonance may suggest that bodies in the cold population were dragged outwards by this resonance as Neptune migrated (Levison & Morbidelli, 2003). Otherwise we would have to suppose that the closeness of the edge and the resonance is merely a coincidence.

The scattered population shows signs of having been emplaced by long-term perihelic interactions with Neptune. While the implications of these belt components look secure, it is unclear how they fitted together in producing the Belt from a much thinner, more uniform and (probably) denser protoplanetary disk. One recent suggestion is particularly fascinating and, to avoid being drawn into a long description of the many models, we mention it here alone. Tsiganis et al. (2005) suppose that the early planetary system started with Jupiter and Saturn close to the 2:1 mean motion resonance (they are now close to, but not in, the 5:2 mean motion resonance with each other). Uranus and Neptune interacted gravitationally with each other and with Jupiter and Saturn. Torques between Neptune and a dense Kuiper belt caused a slow, outward migration of that planet. With the right assumptions of planetary orbit radius and Kuiper belt primordial density, the outward migration of Neptune can be adjusted to continue for hundreds of millions of years, eventually, pulling Jupiter and Saturn into the 2:1 resonance, whereupon strong interactions between these massive planets excite Neptune into a destructive interaction with the primordial belt, scattering its contents throughout the solar system. With the right choice of initial conditions, this model might account for the delayed late-heavy bombardment of the surface of the Moon (at about 3.8–3.9 Gyr) as a result of final destabilization of the trans-Neptunian region (Gomes et al., 2005). It would also produce the scattered disk from an initially confined distribution of inclinations, and it might produce the hot component of the classical belt. The origin of the cold component is not well explained by this scenario.

11.4.5 Problem of the missing mass

The mass of the current Kuiper belt estimated from survey observations and limited numbers of albedo measurements lies in the 0.01 to 0.1 $M_\oplus$ range. Several indirect lines of evidence suggest that this is a small fraction of the initial mass and that, therefore, most of the initial mass has been lost. First, the surface density of the protoplanetary disk can be estimated from the distribution of the refractory materials in the giant planets. An extrapolation of this surface density across the region of the Kuiper belt predicts a mass $\sim$100 times larger than observed. Second, models in which the KBOs grow by binary accretion fail to produce KBOs of the sizes observed unless the protoplanetary disk density in the Kuiper belt is increased by factors of $\sim$100 or more (Kenyon & Luu, 1999). Thirdly, while the mechanism for the formation of binaries remains unclear, it is reasonably certain that binaries cannot form in the present-day low density environment. The binary formation models that have been suggested require formation in a much higher density Kuiper belt, perhaps 100 to 1000 times denser than now. Likewise, massive impacts capable of producing Pluto-Charon type binaries are now vanishingly rare. If they occurred at all, it must have been in an earlier, denser phase of the Kuiper belt.

If these indirect arguments are valid, how might the mass have been lost? The answer to this question is not known, but there are several conjectures in the literature. For example, some part of the initial mass might have been lost by dynamical erosion. Neptune cannot do the job, depleting the mass of the belt by only a factor of a few over the age of the solar system except for orbits in its immediate vicinity. A passing star might substantially erode the belt, but it is not obvious that the remnant left behind would be like the Kuiper belt in detail. Perhaps the mass was lost through collisional grinding. In this, the KBOs collide and shatter, producing a cascade of particles of sizes all the way down to dust. At the smallest sizes, these particles become responsive to radiation drag forces, and would spiral into the Sun. However, bodies larger than $\sim$50 km to 100 km are not easily shattered and the collisional grinding model only works if the initial size distribution in the Kuiper belt were so steep that most of the mass was held by the smallest (most destructable) bodies. It is unclear whether this size distribution ever prevailed.

Another possibility is that the lines of evidence for mass loss are misleading. The KBOs could have formed in a denser environment, perhaps at smaller heliocentric distances, perhaps aided by concentration of the condensible materials in the protoplanetary disk under the action of aerodynamic forces (Youdin & Chiang, 2004), followed by wholescale movement of this population out to the observed location. Understanding the reality and cause of the inferred mass loss is perhaps the biggest problem of the Kuiper belt.

11.5 Surface Properties of KBOs

The study of the surface properties of KBOs is in its infancy: the field is probably at the same level of knowledge as for the main belt asteroids in the late 1970s. The main reason is that the KBOs are extremely faint, even for telescopes in the 8-10m class. Also, secure orbits are needed to place the studied KBOs in the small fields of view of the instruments. This implies repeated observations of newly discovered objects, and recovery programs are generally not able to handle in real time the volume of discoveries. Of the thousand objects discovered, only half have well established orbits. Visible colors have been reported for ~200, while only ~50 objects have near infrared colors, and useful reflectance spectra are available for a mere handful.

Albedos and sizes are known only for a few objects (see Table 11.2 and Grundy et al., 2005). The visible flux measured from the object is proportional to the product of the geometric albedo and the square of the diameter. The thermally emitted flux, on the other hand, is proportional to (1 – albedo) times the diameter squared, because the fraction of the incident sunlight that is absorbed and thermally re-radiated is (1 – albedo). With surface temperatures near 40 K to 50 K, the Wien peak lies near 60μm to 75μm and is inaccessible from Earth. Observations of thermal emission have been made through sub-millimeter wavelength windows in the Earth's atmosphere but thermal observations nearer the emission peak require space-based data. NASA's Spitzer satellite is taking such data now, but with a sensitivity less than originally planned for this mission, only a few dozen KBOs can be attempted. Observations in both the optical and thermal or sub-millimeter range are therefore required for one object to assess its albedo (and size).

11.5.1 Colors

The surface properties of KBO are assessed from the optical and near infrared light they reflect from the Sun, through broadband photometry and reflectance spectroscopy. Ideally, we would use spectra to study the surface compositions of the KBOs and related bodies. In practise, however, most KBOs are too faint for meaningful spectra to be acquired, and most investigators have resorted to broadband colors in the optical and near infrared wavelength regimes. Colors provide only very weak constraints on composition, but they are nevertheless useful in classifying the KBOs, in comparing them with other types of small solar system body, and in searching for correlations (e.g. with size, with orbital parameters) that might be physically revealing. Indeed, in the 1970s, surface color studies of main belt asteroids (orbiting between Mars and Jupiter) soon revealed a color pattern with heliocentric distance (McCord & Chapman, 1975; Zellner et al., 1977, 1985), unveiling a radial compositional structure of the primordial nebula at these distances (Gradie & Tedesco, 1982). It is therefore very tempting to try to reveal such patterns for the Kuiper belt population.

Quantitatively, the color is conveniently measured by the slope of the spectrum after division by the spectrum of the Sun, a quantity conventionally expressed as S' [%/1000Å] and known as the reflectivity gradient. This quantity can be derived from visible photometry. The least contested observational result is that the colors of the KBOs occupy a wide range, from neutral ($S' \sim 0\%/1000$Å) to "very red" ($S' \sim 60\%/1000$Å), indicating a wide diversity of surface types on KBOs (see Fig. 11.4). The question of the

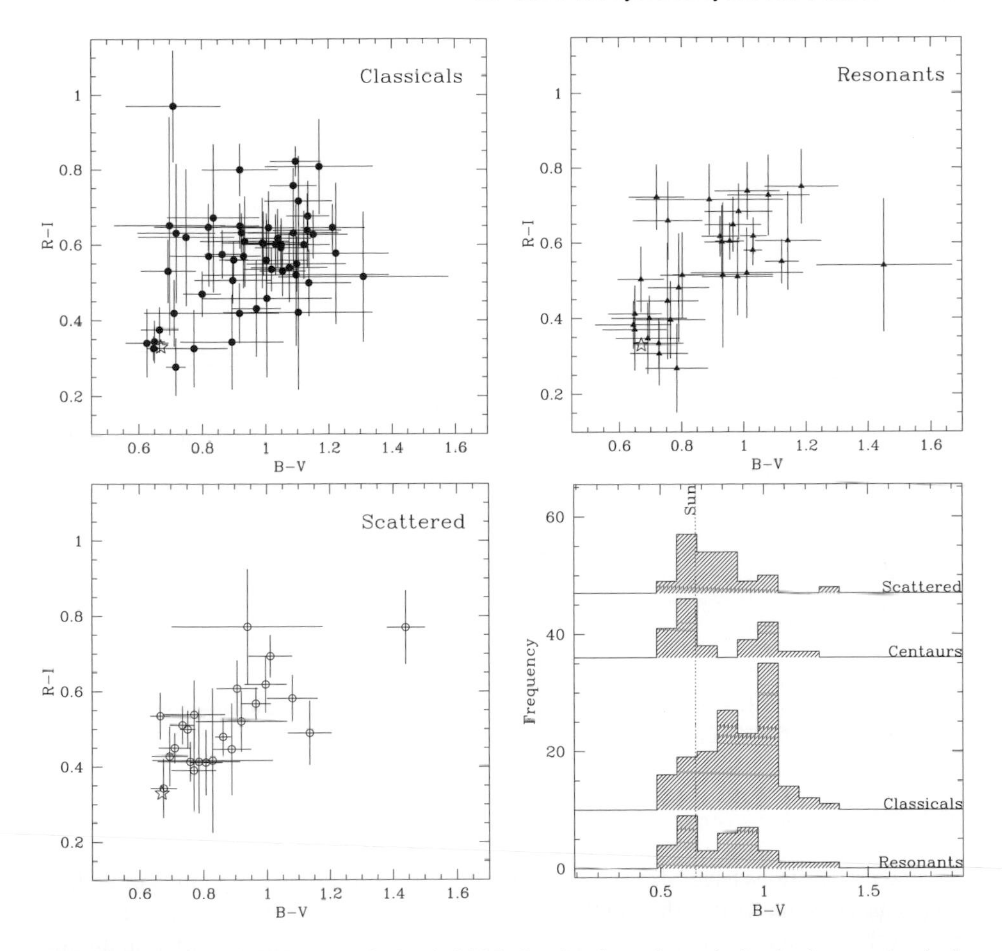

Fig. 11.4. Colors of KBOs (as of March 2005) for the three dynamical sub-classes: classical, resonant and scattered disk objects. Solar colors are represented by an open star. The bottom right plot represents a histogram of the B−V colors. Centaurs have been included for comparison. See also Fig. 11.5.

shape of the distribution of S' has received considerable attention. Most investigators report that S' is unimodally distributed while Tegler & Romanishin (1998, 2003) have reported a bimodal distribution, with KBOs being either nearly neutral, or very red and with very few in between. Later, Peixinho et al. (2003) reported that Centaur colors are bimodally distributed while those of KBOs are not, and suggested that Tegler and Romanishin's finding of bimodal KBO colors was caused by their mixing of these two types of object into a single sample. The sum total of published data (as of 2005) show unimodal color distributions in the Kuiper belt but favor a bimodal distribution of Centaur surface colors (see Figs. 11.4 and 11.5). The unimodal vs. bimodal color question is important because it limits the range of options available for explaining the existence of the color diversity in the Kuiper belt.

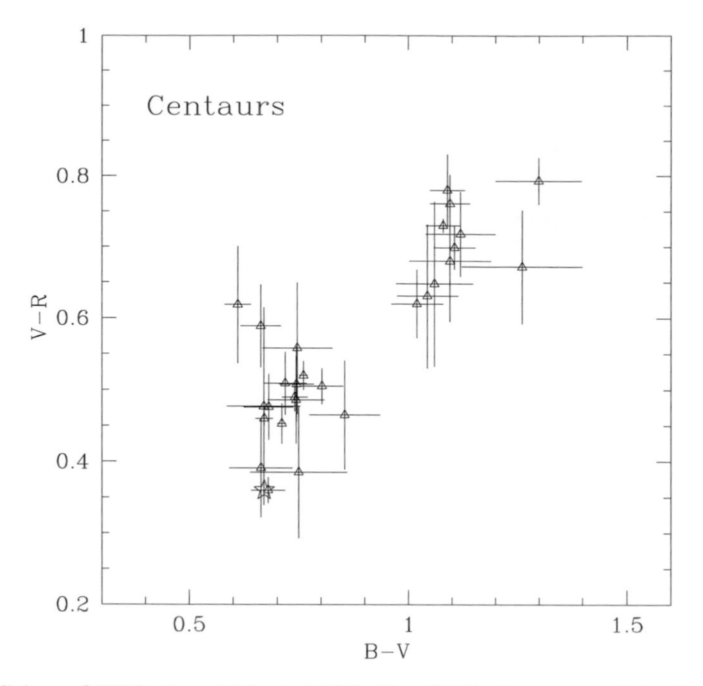

Fig. 11.5. Colors of KBOs (as of March 2005). The distribution appears bimodal for the 28 measured Centaurs. The star symbol represents the solar colors.

The origin of color diversity

Several explanations have been suggested for the origin of the amazing range of surface colors observed on KBOs. The simplest invokes a wide variety of chemical compositions. However, the temperature gradient across the Kuiper belt is about 30–50K, (assuming a blackbody in radiative equilibrium with the Sun), which is too low to produce a wide diversity in chemical compositions. Furthermore, there is no correlation of color with distance or semi-major axis, as might be expected if formation location and temperature were important. Gomes (2003a) describes how a population of highly inclined classical objects (the so-called "hot population") formed much closer to the Sun and was subsequently transported to the Kuiper belt by dynamical mechanisms. This could increase the range of formation temperatures of the KBOs and explain the hot/cold neutral/red color variations, but no specific relation between color and a potential formation location has been identified yet.

The second possibility is that the colors are evolutionary artifacts. Perhaps all KBOs had the same initial composition, a mix of solid organics, silicates and volatile ice, constrained by the cosmochemical abundances in the primordial nebula. Over time, external processing by irradiation from high energy particles (cosmic rays, solar wind and solar UV), by non disruptive collisions between KBOs and by the sublimation of volatiles could change the surface compositions.

The surface processes

Laboratory work shows that fresh (neutral colored) carbon-containing ices become gradually darker under irradiation by high energy particles, due to the progressive loss of hydrogen atoms and carbonization of the surface layers (Strazzulla & Johnson, 1991). Complete processing of a meter-thick crust of organic-rich material occurs on a time-scale of 10^{7-9} yrs (Shul'Man, 1972). In that context, the red colors of some of the KBOs are generally interpreted as the presence of an aged, irradiated organic-rich surface. However, while darkening is a general result, the effect of irradiation on color is less clear, being a complicated function of the material composition, particle fluency and even grain size distribution (Thompson et al., 1987; Moroz et al., 2004). Subject to this ambiguity, it is likely that high albedo neutral color surfaces refer to fresh icy material, while low albedo neutral surfaces should belong to highly irradiated objects. As a KBO surface ages, it should become darker and may change in color.

Non disruptive collisions between KBOs also may play a role as well in removing the dark red irradiation mantle at the location of the impact, revealing fresh neutrally colored underlying material. However, collision rates are not well constrained across the Kuiper belt, and big collisions that would result in an important rejuvenating of the KBO surfaces are statistically very unlikely.

Cometary activity is potentially efficient at creating uniform, neutral colored surfaces through sublimation (for example CO, and N_2 are both volatile at KBO temperatures) and the deposition of sub-orbital debris. Just as on comets, particles on the surface may be lifted by the outgassing and some fall back to create a uniform neutral surface. A burst of cometary-type activity might be triggered either by a collision impact, or by the increasing solar heat while orbiting near perihelion. Unfortunately, there is no observational evidence for routine cometary activity amongst KBOs and, except for occasional bursts triggered by local impacts, none is expected on most KBOs. The very largest objects may be an exception to this statement. We know that Pluto has a μbar atmosphere of N_2 and similarly large KBOs could likewise sustain sublimation-driven atmospheres and associated deposits of seasonally migrating frost.

Several numerical models implying the competition of two or three of these physical processes acting on KBOs surfaces over time succeeded in generally describing the color distribution observed (Jewitt & Luu, 2001; Delsanti et al., 2004). However, these models are incompatible with the Centaur color bimodality (see Fig. 11.5) and with the color-inclination dependency of classical KBOs (see below).

The observed color trends

Apart from the wide spread of surface colors that prevail for the Kuiper belt as a whole (whether it is in the form of a continuous or bimodal distribution), more subtle trends were looked for. With the increasing number of color measurements, it became possible to look closer into the photometric properties of some of the dynamical subclasses. For classical KBOs, objects with perihelia inside about 40 AU show a wide range of optical colors while those beyond 40 AU tend to display only the reddest colors (Fig. 11.6). This trend might be caused by an observational bias (the neutral colored objects with larger perihelion being for some reasons undetectable with the current observing tools),

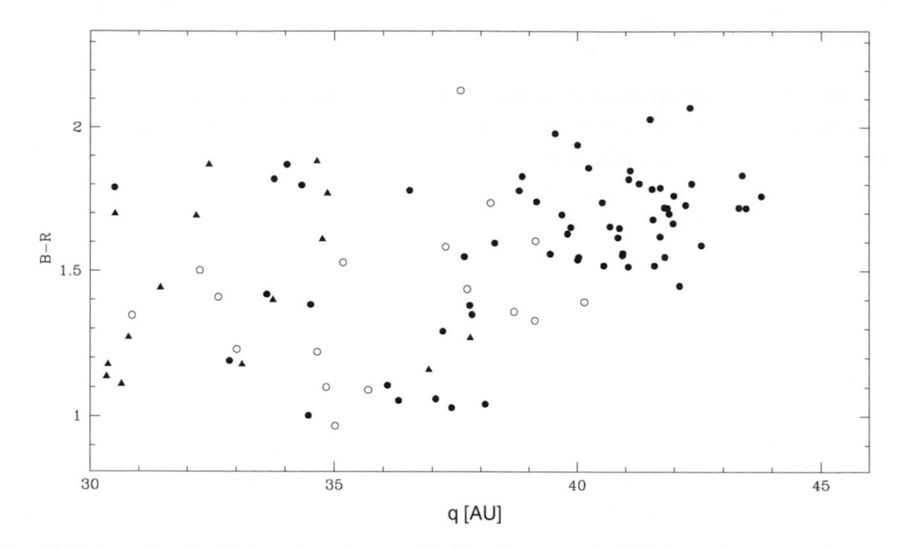

Fig. 11.6. $B - R$ color index plotted vs. perihelion distance, q [AU] for KBOs. At $q \leq 40$ AU, the KBOs show a wide scatter in $B - R$, from neutral to very red. For $q > 40$ AU, only red colors ($B - R \geq 1.5$), are found. The red, large q objects are also low inclinations members of the classical Kuiper belt population. Figure from Delsanti et al. (2004).

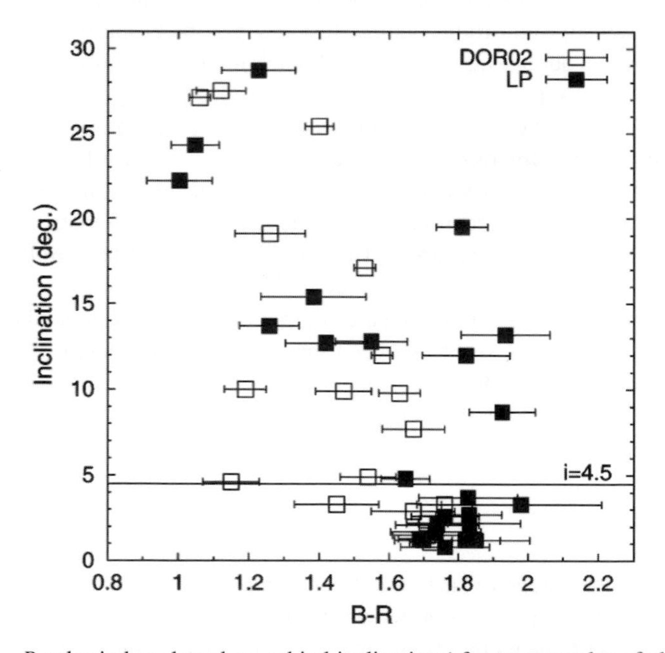

Fig. 11.7. $B - R$ color index plotted vs. orbital inclination i for two samples of classical KBOs from Peixinho et al. (2004). At $i \leq 4.5°$, the classical KBOs show only the reddest colors. For $i > 4.5°$, the whole color range from neutral to very red is displayed.

Table 11.5. KBOs and Centaurs with Water Ice Firmly Detected

Name	Type[a]	H^b	Depth[c]	Reference[d]
(2060) Chiron	Cen	6.5	~10%[e]	Foster et al. (1999), Luu et al. (2000)
(5145) Pholus	Cen	7.0	~16%	Cruikshank et al. (1998)
(10199) Chariklo	Cen	6.4	20%	Brown & Koresko (1998)
(19308) 1996 TO_{66}	Cla	4.5	~20%	Brown et al. (1999)
(31824) Elatus	Cen	10.1	24%?	Bauer et al. (2002)
(50000) Quaoar	Cla	2.6	~22%	Jewitt and Luu (2004)
(83982) 2002 GO_9	Cen	9.1	~16%	Doressoundiram et al. (2006)
(90482) Orcus	3:2	2.3	~30%	de Bergh et al. (2005)

a: Dynamical type: 3:2 = resonant, Clas = Classical, Cen = Centaur
b: Absolute magnitude
c: 2 μm band depth as a fraction of the continuum
d: Prime reference: for brevity we list only one reference where several exist
e: The band depth is variable because of coma dilution. We list the maximum value.

or simply these objects are absent. In both cases, a change of physical environment at larger distances should be invoked. In the second case, this trend might support the idea of cometary activity amongst some KBOs, responsible for the blue objects at shorter perihelion distances, while for objects with larger perihelia, cometary activity is not expected. Collisional resurfacing model simulations by Thébault & Doressoundiram (2003) predicts this color-perihelion trend in the case of a disk truncated at 48 AU. The red, large perihelion classical KBOs also have the smallest inclinations (as displayed in Fig. 11.7). The inclination cut-off (~4°) of the difference of color behavior (Fig. 11.7) is compatible with the concept of a primordial dynamically "cold" population (with small orbital inclinations) superimposed to a dynamically "hot" population (Levison & Stern, 2001) that formed elsewhere in the Solar System (Gomes, 2003a). However, there is currently no clear physical explanation why the cold population should display only the red colors, while the hot population should display the whole range of observed colors.

Other less significant trends scarcely show in the different photometry projects, but they are generally based on small number statistics and most of the time they fail to be found on other samples. As for the main asteroid belt, the most important and physically meaningful trends will strengthen with the increasing number of measured colors.

11.5.2 Spectra

The overtones and combination bands of common molecular bonds such as O-H, C-N, N-H occur in the near-infrared (1 to 2.5 μm) wavelength range, and the terrestrial atmosphere is also comparatively transparent there, at least from mountain sites where the atmospheric water column density is low. Information gleaned from spectra cannot in general be uniquely interpreted, because the reflection spectrum depends on many poorly-known factors in addition to the composition of the target material. For example, the physical state of the surface (whether it be solid, or particulate) plays a role in determining the scattering characteristics. In particulate surfaces, the grain size distribution,

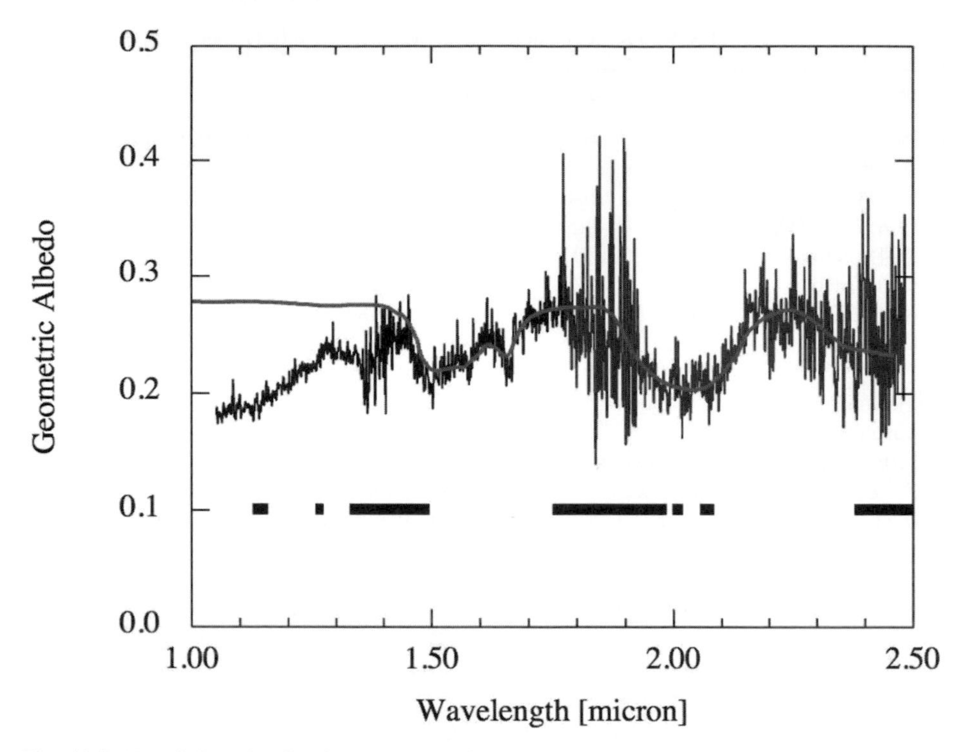

Fig. 11.8. Near infrared reflection spectrum of KBO (50000) Quaoar from the Subaru 8-m telescope. In addition to the broad bands due to water at 1.5 μm and 2.0 μm, a narrow feature at 1.65 μm proves that the surface ice is crystalline, not amorphous. Horizontal bars mark regions of strong absorption by the Earth's atmosphere. The solid line is a water ice spectrum overplotted on the data. From Jewitt and Luu (2004).

the porosity and even the grain shapes and temperature can be important. Nevertheless, near infrared absorptions are quite diagnostic of particular molecular bonds, providing a basis from which to make conjectures about molecular composition. The main practical problem is that the KBOs are faint and the resulting signal-to-noise ratios in KBO spectra are typically low. The discovery of the giant KBOs 2003 UB_{313} and 2005 FY_9, (see Sect. 11.2) lead to the detection of the first methane bands on a KBO besides Pluto. However, the most readily detected specie is water, which has been evidenced by its 2μm absorption in 3 KBOs and 5 Centaurs (see Table 11.5). As these lines are written, water was announced to be present at the surface of the giant KBO 2003 EL_{61}.

Observations of (50000) Quaoar show an absorption band at 1.65 μm that is found in crystalline (but not in amorphous) water ice (see Fig. 11.8). The surface temperature of Quaoar is near ~50 K or less, and at these low temperatures water should be stable indefinitely in its amorphous form. The presence of crystalline ice shows that the ice has been heated above the 100 to 110 K temperature at which amorphous ice transforms to the crystalline (cubic) form. The source of the heat needed to effect this transformation is unclear, but internal heating by the decay of radioactive elements followed by eruption onto the surface is possible. Impact heating could also transform ice in the upper layers although how this happens without vaporizing the ice leading to its deposition as amorphous frost elsewhere on the frigid surface has not been demonstrated. A more serious question concerns the effect of high energy particles on the surface, from the cosmic rays and from particles in the solar wind. The effect of particle bombardment is to break the bonds in crystalline ice, leading to a transformation to the amorphous state. The timescale for the amorphization of ice is uncertain but short compared to the age of Quaoar: estimates range from 10^6 to 10^7 yrs. The fact that the ice is predominantly crystalline argues for comparatively recent emplacement, consistent with on-going endogenic activity.

Other species like ice of N2, NH_3, CO_2 etc, and other organic molecules are from cosmochemical arguments likely to be present at the surface of KBOs. To date, they fail to be detected; we suspect it mainly for technological limitation reasons.

11.6 Related Objects

11.6.1 Centaurs

Centaurs are objects whose orbits bring them into close interaction with the giant planets: a practical (but not unique) definition is that the Centaurs have perihelia *and* semimajor axes between the orbits of Jupiter and Neptune. So defined, about 50 Centaurs are known; some like (2060) Chiron and 29P/Schwassmann-Wachmann 1 are cometary in nature, while others like (5145) Pholus appear completely devoid of coma. These bodies have short dynamical lives owing to strong interactions with the giant planets. Characteristic lifetimes are 10^6 to 10^7 yrs, with a wide scatter (Tiscareno & Malhotra, 2003). Their source is most likely in the Kuiper belt: prevailing evidence suggests that the Centaurs are escaped members of the Scattered disk part of the Kuiper Belt population but other regions may also contribute to the Centaur populations. The main sinks of the Centaurs include ejection to the interstellar medium, capture by Jupiter followed by strong sublimation that leads these objects to be relabelled as comets of the Jupiter family, and collision with the planets or with the Sun. For these reasons, the Centaurs hold special interest in the study of the small body populations of the Solar system. They are closer, brighter counterparts to the KBOs, and they are precursors to the active nuclei of the Jupiter family comets.

In terms of their colors and surface compositions, the Centaurs appear similar to the KBOs: water ice is seen more commonly than other species. The great diversity of spectral types indicated by the wide range of colors of the KBOs is also present in the Centaur population in a bimodal version (see Fig. 11.5). Characteristics from these two groups of surface colors are well represented by (2060) Chiron and (5145) Pholus. The former is spectrally neutral or slightly blue, and shows absorption at 2 μm due to water ice. No other features are detected in its spectrum. The latter is one of the reddest objects in the solar system (in the optical) and shows a structured near-infrared spectrum with bands due to water, a hydrocarbon possibly close to methanol, and olivine (Cruikshank et al., 1998). The very red optical spectrum is commonly attributed to an organic surface (most likely containing N_2 and CH_4) although, lacking discrete optical features, no unique diagnosis of the organic matter has been made.

11.6.2 Comets

Comets were initially envisioned as the most pristine material in the Solar System, as expected from relics from the accretion disk of the Sun. The nuclei of long period comets (that most likely formed in the Oort cloud and remained there for a long period of time) are still considered as true fossils. However, it is now believed that the nuclei of short period comets (the "Jupiter Family") have originated in the Kuiper belt, in the form of collisional fragments. In that case, short period comet nuclei have been somewhat processed, either by the heat of their parent body (if the latter is large enough to sustain heat from radiogenic decay), or by the physical processes that occured during the collision that generated them. To test this scenario, surface colors of short period comets are compared to those of KBOs (Hainaut & Delsanti, 2002; Jewitt, 2002; Doressoundiram et al., 2005). Colors of short period comets are not compatible with the observed colors of KBOs and Centaurs. The "ultrared matter" that is a unique characteristic of KBOs is rare or absent on the surface of short period comets nuclei. The most likely explanation is short period comets nuclei underwent different resurfacing processes during their journey from the Kuiper belt to their present location. For example, cometary activity is a short timescale process that resurfaces the nuclei with neutral colored suborbital particles that were lifted by the sublimation of the volatiles inside the orbit of Jupiter. This process might also be responsible for the lack of intermediate colors amongst the Centaurs. As a conclusion, short period comets, due to their most probable origin in the Kuiper belt and the subsequent thermal and physical processing they underwent, might not be the fossils the planetary scientists long dreamed of.

11.6.3 Irregular satellites

The irregular satellites of the giant planets have large, eccentric and highly inclined orbits. Most known examples are retrograde, in fact, and the only plausible explanation is that they were captured by their respective planets probably long before the present epoch and perhaps in association with planet formation. The source region from which the irregular satellites were captured is not known. One possibility is that the irregular satellites are planetesimals that were initially in similar orbits but which narrowly escaped incorporation into the growing giant planets. In this case, it would be appropriate to think of the irregulars as samples of the material that agglomerated to form the massive, dense cores of the giant planets. Another possibility, currently growing in fashion, is that the irregulars were captured from distant locations in the Kuiper Belt. In this case, the irregulars could be considered as (more) local samples of material from the outer reaches of the Solar system.

Saturn's largest irregular satellite, Phoebe, has been studied close-up by NASA's Cassini spacecraft (Figure 11.9). The surface is heavily cratered but, otherwise, not remarkable compared to the surfaces of other small bodies that have been imaged at similar resolution. The density of Phoebe measured from gravitational deflections of the Cassini spacecraft (Porco et al., 2005) is 1630 ± 45 kg m^{-3}, consistent with a rock/ice mixture as the bulk composition. Johnson & Lunine (2005) assert that this high density points to a Kuiper belt origin, but the connection between density and formation location is unclear. Clark et al. (2005) and Esposito et al. (2005) detected numerous ices and

Fig. 11.9. Saturn satellite Phoebe imaged by the NASA Cassini spacecraft. Image courtesy NASA.

other compounds on Phoebe that they believe point to a Kuiper belt source. However, the possibility that the surface is coated by ices delivered by cometary impact cannot be discounted. Therefore, while we possess data of staggering resolution and quality on Phoebe, its origin is unclear.

The capture mechanism for irregular satellites is unknown. The favored hypothesis – gas drag in the bloated atmospheres of the young giant planets – might apply at Jupiter and Saturn where the total gas mass is large (the gas giants are >90% hydrogen and helium by mass). Uranus and Neptune also possess irregulars, however, and these ice giant planets are comparatively gas free, and formed by a process different from that which formed the gas giants. Three -body interactions (two small bodies in an encounter within the Hill sphere of the larger planet) might be capable of producing capture around both gas and ice giants, but the details of this mechanism have not been fully worked out. An intriguing new observation is that, measured down to a given limiting size, the four giant planets all possess about the same number of irregular satellites. To within a factor of ~2 they all have ~100 irregulars larger than 1 km in radius (Jewitt & Sheppard, 2005). This strange result, which is not predicted by any of the suggested satellite capture models, surely tells us something of importance about the origin of the irregulars, whether in the Kuiper belt or not.

Acknowledgments

This work was supported by a grant to DJ from the NASA Planetary Astronomy Program and by the NASA Astrobiology Institute under Cooperative Agreement No. NNA04CC08A issued through the Office of Space Science

References

Allen, R. L., Bernstein, G. M., & Malhotra, R. 2001, The Edge of the Solar System. *ApJ Letters*, **549**, L241

Bauer, J. M., Meech, K. J., Fernández, Y. R., Farnham, T. L., & Roush, T. L. 2002, Observations of the Centaur 1999 UG5: Evidence of a Unique Outer Solar System Surface. *PASP*, **114**, 1309

de Bergh, C., Delsanti, A., Tozzi, G. P., Dotto, E., Doressoundiram, A., & Barucci, M. A. 2005, The surface of the transneptunian object 90482 Orcus. *A&A*, **437**, 1115

Bernstein, G. M., Trilling, D. E., Allen, R. L., Brown, M. E., Holman, M., & Malhotra, R. 2004, The Size Distribution of Trans-Neptunian Bodies. *AJ*, **128**, 1364

Brown, R. H., Cruikshank, D. P., & Pendleton, Y. 1999, Water Ice on Kuiper Belt Object 1996 TO_{66}. *ApJ Letters*, **519**, L101

Brown, M. E., Trujillo, C., & Rabinowitz, D. 2004, Discovery of a Candidate Inner Oort Cloud Planetoid. *ApJ*, **617**, 645

Brown, M.E., Trujillo, C.A., Rabinowitz, D.L. 2005, Discovery of a planetary-sized object in the scattered Kuiper belt, submitted to *ApJ Letters*

Brown, M.E, van Dam, M.A, Bouchez, A.H., Le Mignant D., Campbell, R.D., Chin, A., Conrad, A. et al. 2005, Satellites of the largest Kuiper belt objects, submitted to *ApJ Letters*

Brown, M.E, Bouchez, A.H., Rabinowitz, D., Sari, R., Trujillo, C.A., van Dam, M.A, Campbell, R.D. et al 2005, Keck Observatory Laser Guide Star Adaptive Optics Discovery and Characterization of a Satellite to the Large Kuiper Belt Object 2003 EL61. *ApJ Letters*, **632**, L45

Brown, M. E., & Koresko, C. C. 1998, Detection of Water Ice on the Centaur 1997 CU 26. *ApJ Letters*, **505**, L65

Canup, R. M. 2005, A Giant Impact Origin of Pluto-Charon. *Science*, **307**, 546

Clark, R. N., et al. 2005, Compositional maps of Saturn's moon Phoebe from imaging spectroscopy. *Nature*, **435**, 66

Cochran, A. L., Levison, H. F., Stern, S. A., & Duncan, M. J. 1995, The Discovery of Halley-sized Kuiper Belt Objects Using the Hubble Space Telescope. *ApJ*, **455**, 342

Cruikshank, D. P., et al. 1998, The Composition of Centaur 5145 Pholus. *Icarus*, **135**, 389

Delsanti, A., Hainaut, O., Jourdeuil, E., Meech, K. J., Boehnhardt, H., & Barrera, L. 2004, Simultaneous visible-near IR photometric study of Kuiper Belt Object surfaces with the ESO/Very Large Telescopes. *A&A*, **417**, 1145

Doressoundiram, A., Peixinho, N., Doucet, C., Mousis, O., Barucci, M. A., Petit, J. M., & Veillet, C. 2005, The Meudon Multicolor Survey (2MS) of Centaurs and trans-neptunian objects: extended dataset and status on the correlations reported. *Icarus*, **174**, 90

Doressoundiram, A., Barucci, M.A., Tozzi, G.P., Poulet, F., Boehnhardt, H., de Bergh, C. and Peixinho, N. 2006, Spectral characteristics and modeling of the trans-neptunian object (55565) 2002 AW197 and the Centaurs (55576) 2002 GB10 and (83982) 2002 GO9. PSS in press

Elliot, J. L., Dunham, E. W., Bosh, A. S., Slivan, S. M., Young, L. A., Wasserman, L. H., & Millis, R. L. 1989, Pluto's atmosphere. *Icarus*, **77**, 148

Elliot, J. L., Ates, A., Babcock, B. A., Bosh, A. S., Buie, M. W., Clancy, K. B., Dunham, E. W., et al. 2003, The recent expansion of Pluto's atmospher, *Nature*, **424**, 165

Elliot, J. L., et al. 2005, The Deep Ecliptic Survey: A Search for Kuiper Belt Objects and Centaurs. II. Dynamical Classification, the Kuiper Belt Plane, and the Core Population. *AJ*, **129**, 1117

Emel'yanenko, V. V., Asher, D. J., & Bailey, M. E. 2003, A new class of trans-Neptunian objects in high-eccentricity orbits. *MNRAS*, **338**, 443

Esposito, L. W., et al. 2005, Ultraviolet Imaging Spectroscopy Shows an Active Saturnian System. *Science*, **307**, 1251

Foster, M. J., Green, S. F., McBride, N., & Davies, J. K. 1999, NOTE: Detection of Water Ice on 2060 Chiron. *Icarus*, **141**, 408

Gladman, B., Holman, M., Grav, T., Kavelaars, J., Nicholson, P., Aksnes, K., & Petit, J.-M. 2002, Evidence for an Extended Scattered Disk. *Icarus*, **157**, 269

Goldreich, P., Lithwick, Y., & Sari, R. 2002, Formation of Kuiper-belt binaries by dynamical friction and three-body encounters. *Nature*, **420**, 643

Gomes, R. 2003, Planetary science: Conveyed to the Kuiper belt. *Nature*, **426**, 393

Gomes, R. S. 2003, The origin of the Kuiper Belt high-inclination population. *Icarus*, **161**, 404

Gomes, R., Levison, H. F., Tsiganis, K., & Morbidelli, A. 2005, Origin of the cataclysmic Late Heavy Bombardment period of the terrestrial planets. *Nature*, **435**, 466

Gradie, J. & Tedesco, E. 1982, Compositional structure of the asteroid belt. *Science*, **216**, 1405.

Grundy, W. M., & Fink, U. 1996, Synoptic CCD Spectrophotometry of Pluto Over the Past 15 Years. *Icarus*, **124**, 329

Grundy, W. M., Noll, K. S., & Stephens, D. C. 2005, Diverse albedos of small trans-neptunian objects. *Icarus*, **176**, 184

Hainaut, O. R., & Delsanti, A. C. 2002, Colors of Minor Bodies in the Outer Solar System. A statistical analysis. *A&A*, **389**, 641

Holman, M. J., & Wisdom, J. 1993, Dynamical stability in the outer solar system and the delivery of short period comets. *AJ*, **105**, 1987

Ida, S., Larwood, J., & Burkert, A. 2000, Evidence for Early Stellar Encounters in the Orbital Distribution of Edgeworth-Kuiper Belt Objects. *ApJ*, **528**, 351

Jewitt, D., & Luu, J. 1993, Discovery of the candidate Kuiper belt object 1992 QB1. *Nature*, **362**, 730

Jewitt, D., Luu, J., & Trujillo, C. 1998, Large Kuiper Belt Objects: The Mauna Kea 8K CCD Survey. *AJ*, **115**, 2125

Jewitt, D. C. 2002, From Kuiper Belt Object to Cometary Nucleus: The Missing Ultrared Matter. *AJ*, **123**, 1039

Jewitt, D. C., & Luu, J. X. 2001, Colors and Spectra of Kuiper Belt Objects. *AJ*, **122**, 2099

Jewitt, D. C., and Luu, J. 2004, Crystalline water ice on the Kuiper belt object (50000) Quaoar. *Nature*, **432**, 731

Jewitt, D., & Sheppard, S. 2005, Irregular Satellites in the Context of Planet Formation. *Space Science Reviews*, **116**, 441

Johnson, T. V., & Lunine, J. I. 2005, aturn's moon Phoebe as a captured body from the outer Solar System. *Nature*, **435**, 69

Kenyon, S. J., & Luu, J. X. 1999, ApJ, Accretion in the Early Outer Solar System. **526**, 465

Levison, H. F. & Morbidelli, A. 2003, The formation of the Kuiper belt by the outward transport of bodies during Neptune's migration. *Nature*, **426**, 419.

Levison, H. F., & Stern, S. A. 2001, On the Size Dependence of the Inclination Distribution of the Main Kuiper Belt. *AJ*, **121**, 1730

Licandro, J. Pinilla-Alonso, N., Pedani, M., Oliva. E., Tozzi, G.P. and Grundy, W. 2006, The methane ice rich surface of large TNO 2005 FY9: a Pluto-twin in the trans-neptunian belt?, *A&A Letters*, in Press

Luu, J. X., Jewitt, D. C., & Trujillo, C. 2000, Water Ice in 2060 Chiron and Its Implications for Centaurs and Kuiper Belt Objects. *ApJ Letters*, **531**, L151

Malhotra, R. 1995, The Origin of Pluto's Orbit: Implications for the Solar System Beyond Neptune. *AJ*, **110**, 420

McCord, T. B., & Chapman, C. R. 1975, Asteroids – Spectral reflectance and color characteristics. *ApJ*, **195**, 553

Morbidelli, A., & Levison, H. F. 2004, Scenarios for the Origin of the Orbits of the Trans-Neptunian Objects 2000 CR105 and 2003 VB12 (Sedna) *AJ*, **128**, 2564

Morbidelli, A., Emel'yanenko, V. V., & Levison, H. F. 2004, Origin and orbital distribution of the trans-Neptunian scattered disc. *MNRAS*, **355**, 935

Moroz, L., Baratta, G., Strazzulla, G., Starukhina, L., Dotto, E., Barucci, M. A., Arnold, G., & Distefano, E. 2004, Optical alteration of complex organics induced by ion irradiation: 1. Laboratory experiments suggest unusual space weathering trend. *Icarus*, **170**, 214

Peixinho, N., Doressoundiram, A., Delsanti, A., Boehnhardt, H., Barucci, M. A., & Belskaya, I. 2003, Reopening the TNOs color controversy: Centaurs bimodality and TNOs unimodality. *A&A*, **410**, L29

Peixinho, N., Boehnhardt, H., Belskaya, I., Doressoundiram, A., Barucci, M. A., & Delsanti, A. 2004, ESO large program on Centaurs and TNOs: visible colors-final results. *Icarus*, **170**, 153

Porco, C. C., et al. 2005, Cassini Imaging Science: Initial Results on Phoebe and Iapetus. *Science*, **307**, 1237

Rabinowitz, D.L, Barkume, K., Brown, M.E., Roe, H., Schwartz, M., Tourtellotte, S., Trujillo, C. 2005, Photometric observations constraining the size, shape and albedo of 2003 EL61, a rapidly rotating, Pluto-sized object in the Kuiper Belt, submitted to *ApJ*

Sheppard, S. S., & Jewitt, D. 2004, Extreme Kuiper Belt Object 2001 QG298 and the Fraction of Contact Binaries. *AJ*, **127**, 3023

Shul'Man, L. M. 1972, The Chemical Composition of Cometary Nuclei, IAU Symp. 45: The Motion, Evolution of Orbits, and Origin of Comets, **45**, 265

Sicardy, B., et al. 2003, Large changes in Pluto's atmosphere as revealed by recent stellar occultations. *Nature*, **424**, 168

Strazzulla, G., & Johnson, R. E. 1991, ASSL Vol. 167: IAU Colloq. 116: Comets in the post-Halley era, 243

Tegler, S. C., & Romanishin, W. 1998, Two distinct populations of Kuiper-belt objects. *Nature*, **392**, 49

Tegler, S. C., & Romanishin, W. 2003, Resolution of the kuiper belt object color controversy: two distinct color populations. *Icarus*, **161**, 181

Thébault, P., & Doressoundiram, A. 2003, Colors and collision rates within the Kuiper belt: problems with the collisional resurfacing scenario. *Icarus*, **162**, 27

Thompson, W. R., Murray, B. G. J. P. T., Khare, B. N., & Sagan, C. 1987, Coloration and darkening of methane clathrate and other ices by charged particle irradiation – Applications to the outer solar system. *JGR*, **92**, 14933

Tiscareno, M. S., & Malhotra, R. 2003, The Dynamics of Known Centaurs. *AJ*, **126**, 3122

Tombaugh, C. W. 1961, The Trans-Neptunian Planet Search. Planets and Satellites, edited by Gerard P. Kuiper and Barbara M. Middlehurst Chicago: University of Chicago Press, 1961, p.12

Trujillo, C. A., Jewitt, D. C., & Luu, J. X. 2000, Population of the Scattered Kuiper Belt. *ApJ Letters*, **529**, L103

Trujillo, C. A., Jewitt, D. C., & Luu, J. X. 2001, Properties of the Trans-Neptunian Belt. *AJ*, **122**, 457

Trujillo, C. A., & Brown, M. E. 2002, A Correlation between Inclination and Color in the Classical Kuiper Belt *ApJ Letters*, **566**, L125

Trujillo, C. A., & Brown, M. E. 2003, The Caltech Wide Area Sky Survey, E*arth Moon and Planets*, **92**, 99

Tsiganis, K., Gomes, R., Morbidelli, A., & Levison, H. F. 2005, Origin of the orbital architecture of the giant planets of the Solar System. *Nature*, **435**, 459

Veillet, C., et al. 2002, The binary Kuiper-belt object 1998 WW31. *Nature*, **416**, 711

Weaver, H. A., et al. 2005, S/2005 P 1 and S/2005 P 2. *IAU circular*, 8625, 1

Weidenschilling, S. J. 2002, On the Origin of Binary Transneptunian Objects. *Icarus*, **160**, 212

Youdin, A. N., & Chiang, E. I. 2004, ApJ, Particle Pileups and Planetesimal Formation. **601**, 1109

Zellner, B., Andersson, L., & Gradie, J. 1977, UBV photometry of small and distant asteroids. *Icarus*, **31**, 447

Zellner, B., Tholen, D. J., & Tedesco, E. F. 1985, The eight-color asteroid survey – Results for 589 minor planets. *Icarus*, **61**, 355

12 The Nature of Comets

David W. Hughes

Abstract. Our knowledge of comets has advanced hugely since the 1986 apparition of Comet Halley provided the spur to spacecraft investigation. But our understanding is still very much 'skin deep'. The mass of the nucleus, the density of the nucleus and its internal structure and strength are poorly understood. Whether the physical and chemical properties vary from place to place in a specific cometary nucleus, or from comet to comet is still a mater of conjecture. Recent low-resolution images of nuclei surfaces are improving our understanding, but the real breakthrough will probably have to wait until a probe lands on a cometary surface in 2014, and until future probes return material to Earth some decades later. In this chapter we summarise our knowledge of the physical properties and chemical composition of the cometary nucleus paying special attention to the four comets that have had their nuclei imaged by spacecraft, these being comets Halley, Borrelly, Wild-2 and Tempel-1. Special attention is paid to the rate at which the cometary nucleus decays and the effect this has on cometary activity and the thickness of the all-encompassing surface dust layer. The results of recent cometary space mission are summarised, special attention being paid to the Deep Impact mission to comet Tempel-1.

12.1 Introduction

Comets have been observed at irregular intervals since the dawn of history, 'great' comets such as Hale-Bopp and Hyakutake appearing at the rate of about three per century. 'Great' comets are unmissable; they briefly dominate the sky; and in ancient time caused considerable consternation. Before the days of Isaac Newton (1642–1727) and Edmond Halley (1656–1742) no one knew where comets came from or where they went. In those days they were unexpected. Each comet was only seen for a few weeks during which time they slowly moved across the constellations from night to night.

Comets are bright when close to the Sun and so are generally evening or pre-dawn objects. Their tails point in the anti-solar direction, usually shooting up from the horizon. Comets looked threatening and were regarded with great superstition.

Cometary science started with orbital analysis. Isaac Newton applied his gravitational theory to the observations of the great comet of 1680 and he calculated five of its six orbital parameters. Unfortunately he had to assume an eccentricity of unity i.e. a parabolic orbit. His comet came from infinity and returned thence. Newton's orbit calculation technique was taken up by Edmond Halley and in 1696 he told the Royal Society that the comets of 1607 and 1682 had similar orbits. By 1705 and the publication of his paper *A Synopsis of the Astronomy of Comets* he had added the comet of 1531 to the list. Halley concluded that the comets of 1531, 1607 and 1682 were the same comet returning to the Sun every 75 years after travelling out to a distance of about 35 AU. This comet was periodic and a permanent member of our planetary system. Today we know of about 180 comets with short periods ($P < 20$ years), and we have recorded well over 1000 comets with $P > 20$ years, about two dozen of these being intermediate period ($20 < P < 200$ years) Halley types. (60 to 80 new comets are added to the list each year).

The investigation of the chemistry and physics of comets started in earnest in the last half of the nineteenth century. The 1860s saw Sir William Huggins examining comets spectroscopically and comparing these spectra with laboratory sources. The relationship between comets and meteoroid streams was established in 1866 when it was realised that comet 55P/Tempel-Tuttle had a similar orbit to the November Leonids and the orbit of 109P/Swift-Tuttle was the same as the August Perseids. Cometary decay produced dusty meteoroids.

In the early twentieth century there were three main cometary mysteries. The first concerned the source of the gas and dust that produced cometary comae and tails. The fact that comets such as Halley and Encke had been seen at about twenty to forty historical returns, and their brightness at perihelion had not varied greatly from one apparition to the next, meant that this source was abundant. The second concerned the cause of the so-called non-gravitational effects. Some comets seemed to be gradually speeding up, and others were slowing down. The final mystery was mass. Renaissance scientists assumed that comets were as massive as planets. This suggestion was discredited when comet Lexell (D1770/L1) passed through the satellite system of Jupiter (in 1779) and caused no discernable changes in the satellite orbits.

Modern cometary science really started in 1950 when the American, Harvard University astronomer Fred L. Whipple (1950, 1972) introduced the dirty snowball nucleus model. Here the fount of all cometary activity (the coma, tail, gas and dust), was a single central kilometric 'dirty snowball'. Other astronomers disagreed (see for example Lyttleton, 1953), their suggested comet being an orbiting swarm of dust particles, this swarm being condensed at perihelion and extended at aphelion. Cometary activity was supposedly produced by the inter-particle collisions that occur when all the particles return to their places of production (either by nucleus emission or particle break-up) these production points, in the main, being close to perihelion. The disagreement between the two suggested models was resolved in March 1986 when the European Space Agency's Giotto spacecraft flew within 600 km of the nucleus of comet 1P/Halley. A camera was trained on the brightest region at the centre of the coma. And there (much to the relief of many) was a single potato-shaped, slowly spinning, cometary nucleus about 15.3 km long and 7 km wide (see Fig. 12.1). 1986 saw the first cometary nucleus images being returned to Earth. The nucleus surface was extremely dark, reflecting only about 4% of the light that fell on it. In Halley's case only about 10% of the surface was actively releasing jets of gas and dust. The composition of the gas and dust was measured. Most of the gas molecules were H_2O. The dust was a simple phylosilicate (similar in bulk composition to the terrestrial planets). The first two questions posed above could be answered. The abundant cometary snows provided an ample source of gas, and one that could provide a new coma around perihelion time after time. A spinning nucleus plus a time lag between solar heating and subsurface gas emission leads to a jet effect. This exerts a tangential force on the nucleus which, dependent on whether the spin is direct or retrograde, accelerates or decelerates the comet producing the so-called non-gravitational effects. Cometary mass was (and still is) much of a mystery.

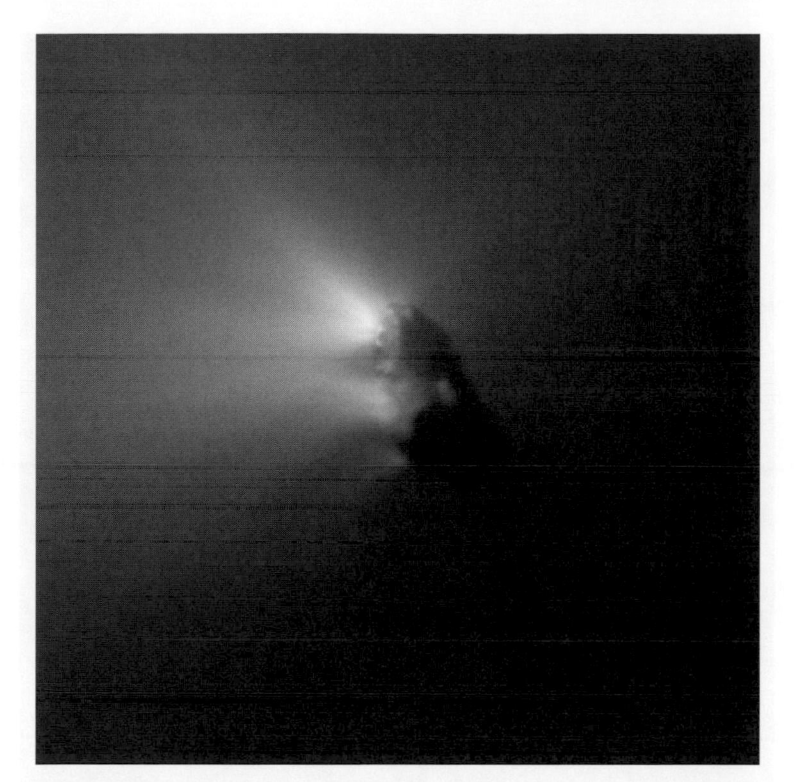

Fig. 12.1. On 14 March 1986 images of the nucleus of Halley's Comet were taken by the Halley Multicolour Camera on board the Giotto spacecraft. The comet had passed perihelion just over three weeks before and was about 0.89 AU from the Sun. The spacecraft was approaching at $65\,\mathrm{km\,s^{-1}}$ (145,000 m.p.h.) The figure shows a 68-image composite, the resolution ranging from 60 m to 320 m. The surface of the nucleus is rough but on a rather large scale, the height difference between the shallow depressions and the higher regions being typically 500 m. The 'day' and 'night' portions of the nucleus can be clearly differentiated and the general shape of the nucleus can be seen against the faint background light of the cometary coma. The surface is dark and dusty and infra-red measurements indicated that the average temperature was about 320 K. The nucleus is 'peanut–shaped', and about 15 km long, and about 7 km wide. Only about 10 percent of its $400 \pm 80\,\mathrm{km}^2$ surface was actively emitting gas and dust. Jets can be clearly seen in the upper portion of the image. This was the first image ever taken of a cometary nucleus, the image that proved that Harvard astronomer Fred Whipple's concept of a 'dirty-snowball nucleus' was correct.

12.2 Cometary Characteristics

For fifteen years scientists had to make do with only one set of low-resolution pictures of a single cometary nucleus. But in September 2001, the NASA Deep Space 1 craft flew past comet 19P/Borrelly. An 8×3.2 km nucleus was revealed (see Fig. 12.2). In January 2004 NASA's Stardust mission passed the $5.5 \times 4.0 \times 3.3$ km nucleus of 81P/Wild-2 (see Figs. 12.3 and 12.4). And in July 2005 Deep Impact treated us to images of the 7.6×4.9 km nucleus of 9P/Tempel-1 (see Fig. 12.5). Halley, Borrelly, Tempel-1 and

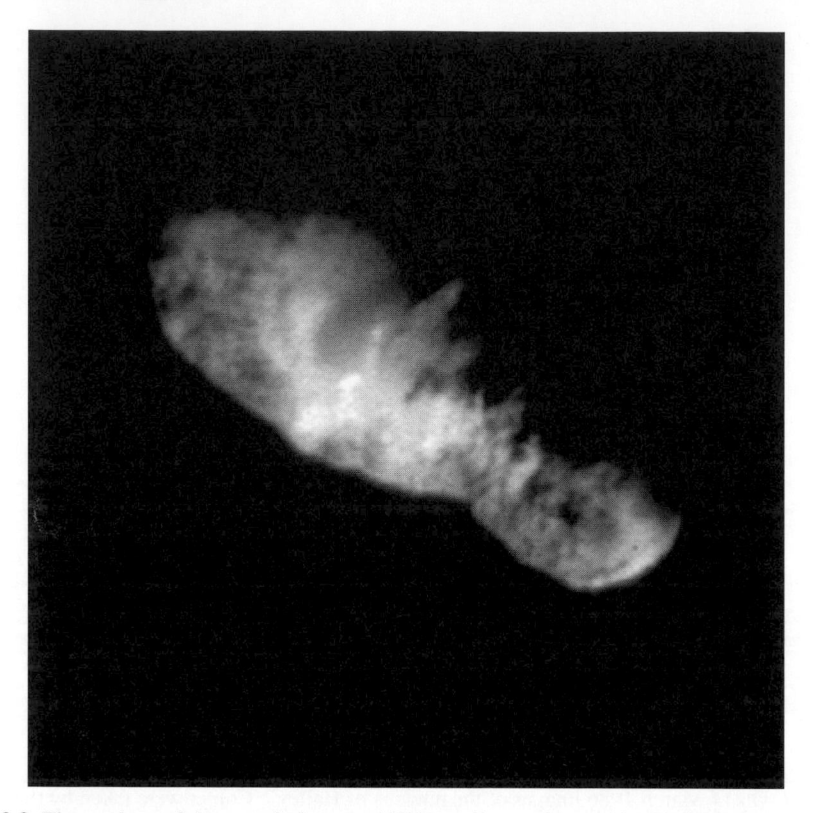

Fig. 12.2. The nucleus of short-period comet 19P/Borrelly was imaged by NASA's Deep Space 1 spacecraft which flew past on 22 September 2001, revealing the nucleus to be shaped rather like a bowling-pin with dimensions 8.0 × 3.2 × 3.2 km. The surface is a dark grey colour and only reflects about 4 percent of the light that hits it. The Sun-comet-spacecraft angle is about 52°. The lower edge of the image shows the illuminated comet silhouetted against the dark background sky and the variability of this edge gives a true impression of the topography of the nucleus surface. The upper part of the image is the terminator. As the angle between the comet surface and the sunlight varies considerably due to the micro-topography, much of the nucleus is in shadow. This affects the perceived brightness. Several jets of gas and dust are being emitted by the nucleus. As the exposure time of the image was chosen to reveal surface details these jets cannot be seen.

Wild-2 are reasonably bright comets with larger than average nuclei. The nucleus of the median known short-period comet is thought to be about 1 km across. Extremely bright comets like the great comet Hale-Bopp (April 1997) has a nucleus estimated to be about 35 km across. The size distribution index of cometary nuclei indicate that they have been formed by accretion and not by fragmentation (as is the caser for the asteroid population.)

The nucleus is thought to be a deep-freeze mixture of 'snow' and 'dirt'. Let me stress, however, that what is to follow is full of caveats. Words such as 'possibly', 'maybe' and 'probably' will be liberally scattered. We have no certainty. Today we have images of four cometary nuclei. The best resolution is 4 m. We have no knowledge of finer detail.

And the interior is a complete mystery. Our understanding is at best skin deep. The nuclei seem to be irregular triaxial ellipsoids, with a long axis usually about 2 to 3 times greater than the small axis. The surfaces are also relatively smooth.

Most scientists suggest that the cometary interior has a temperature below 170 K and has been very close to that temperature since formation about 4.5×10^9 years ago. The fact that cometary nuclei are pristine deep-freeze remnants of the dawn of the solar system makes them prime targets for sample return missions. Unlike planets they have not been differentiated or stirred up during their lifetime.

The snow to dirt mass ratio is typically about 2.2. This estimate comes from the spectroscopic analysis of the solar surface and the mass spectroscopic investigation of the remnants of the asteroid belt that fall to the surface of Earth as meteorites. These provide us with a list of the cosmic abundance of the elements (see, for example Anders and Ebihara, 1982, Hughes, 2000). Consider these elements forming molecules and simple compounds as they cool in the primordial solar nebula. Then divide these compounds into four 'basic' cosmic materials – metal, rock, ice and gas. The metal component contains all the nickel and half the iron. The rest of the iron together with other common siderophile elements such magnesium, silicon and aluminium are combined with oxygen to form rocks by crudely assigning three oxygen atoms to every two 'rocky' atoms. The remaining oxygen is combined with hydrogen or carbon to form the common ices of water, carbon dioxide and carbon monoxide. Other ices such as methane and ammonia are formed out of combinations of carbon, nitrogen and hydrogen. The remaining hydrogen, together with inert elements such as helium, argon and neon, stays in the gaseous form (see for example Whipple 1972).

This primordial cosmo-chemistry leads to typical mass percentiles of 98.40% gas, 1.10% ice, 0.40% rock and 0.09% metal. In order to overcome the escape velocity the accretion of gas requires large (nearly Mars sized) planetesimals with surface temperatures under about 150 K. Comets are thus far too small to have permanent atmospheres. In fact comets simply consist of a combination of the 1.10% ice, 0.40% rock and 0.09% metal mentioned above, the $1.1/(0.4 + 0.09)$ giving the 2.2 snow/dirt mass ratio.

The chemical composition of the cometary snow is reasonably well known. Boice and Huebner (1999) analysed *in situ* data from the Giotto mission to Halley and also general spectroscopic observations of comets and found that for every 1000 H_2O molecules leaving the nucleus there were about 47 CO, 35 CO_2, 24 H_2CO, 24 CH_3OH, 11 N_2 and 35 H_2S, HCN, NH_3, CS_2, H_2CO_2 etc. The physics of the cometary 'snow' is much more problematic. The fact that it is mainly H_2O helps. The solid H_2O that we encounter in everyday life usually comes in two forms. Water ice is compact and strong and has a density of about 920 kg m^{-3}. Water snow is fluffy, vesicular and weak, its density being around 80 kg m^{-3}.

The 2.2 snow/dirt mass ratio is thought to apply throughout the vast majority of the body of the nucleus. The surface is the exception. An upper layer of thickness somewhere between 0.1 and 1 m will have been warmed, and in some instances 'baked' during the successive passages of the comet through the inner solar system. Some volatile components could have been lost. The very top of this layer is expected to be friable, vesicular dirt with a typical albedo of about 0.04.

Cometary activity is extremely intermittent and only occurs in the inner solar system where low albedo surfaces have temperatures above the sublimation temperatures of solid water, carbon dioxide and carbon monoxide. The temperature of a unit area of albedo 0.04, perpendicular to the solar flux, is about $404\,r^{-0.5}$ K, where r is the heliocentric distance in AU. In the vacuum of space, near-surface water, carbon dioxide and carbon monoxide will be sublimating at heliocentric distance less than 2.8, 8 and 55 AU respectively (see Crovisier and Encrenaz, 2000). Think, for example, of a comet on a near-parabolic orbit with a perihelion distance of 0.7 AU. This comet takes only about 36, 69, 103 and 142 days to get from perihelion to heliocentric distances of 1, 1.5, 2 and 2.5 AU respectively. At heliocentric distances less than 1.0 AU it can be assumed that the majority of the incident radiation hitting an active region of the nucleus will be used to sublimate the underlying H_2O snow.

The radiation hitting a 1 m^2 area perpendicular to the Sun-comet line is $1368/r^2$ W, where 1368 W m^{-2} is the solar constant. As the latent heat of sublimation of H_2O is about 2.6×10^6 J kg^{-1}, subsurface snow is sublimating and producing gas at the rate of about $5.2 \times 10^{-4}/r^2\,kg\,m^{-2}\,s^{-1}$. Without the interstitial snow to 'stick' the dust to the surface of the nucleus, the loose dust gets blown away by the gas pressure at the rate of $5.2 \times 10^{-4}/2.2r^2 = 2.6 \times 10^{-4}/r^2\,kg\,m^{-2}\,s^{-1}$. The total mass loss is $7.9 \times 10^{-4}/r^2\,kg\,m^{-2}\,s^{-1}$, and this causes the surface of the active region to retreat at the rate of:

$$\Delta R = \frac{7.9 \times 10^{-4}}{\rho r^2}\,m\,s^{-1}, \tag{12.1}$$

where ρ is the mean density of the nucleus. Here we have a problem. Density is an extremely poorly constrained cometary characteristic. The fragility of cometary nuclei, exhibited by the break up of Comet Shoemaker-Levy 9 in 1992, and the fragmenting of a host of comets at different times, points many cometary scientists toward the paradigm of a fragile, low strength, low density, dirty-snow-ball nucleus, as opposed to a solid, strong, dirty-ice-ball nucleus. This paradigm is supported by cometary origin theories and notions of the primordial accretion of fluffy snow and fluffy dust particles under conditions of very low gravitational field and very low temperatures. Bearing these ideas in mind we will follow Rickman et al. (1987) who concluded that a mean density of about 150 kg m^{-3} reasonably fits the available cometary data. A nucleus ten times this density would consist of solid ice; a nucleus ten times less dense would have huge voids in its interior. Just think in terms of the soft fresh snow that falls to the surface of Earth, with its density of about 80 kg m^{-3}. Then add some dust. In the cometary context density might be decreased due to the formation under condition of zero gravity and zero pressure in the outer solar system regions between Jupiter and Neptune, but increased because the nucleus is a few kilometres in diameter and there might be some compaction, even though the weight of cometary material will be very low due to the small value of the acceleration of gravity. Unfortunately there is as yet no accurate ways of measuring cometary densities. And we must note that other researchers assume a different value. Belton et al. (2005) used a 9P/Tempel-1 nucleus density of 500 kg m^{-3}.

The brief calculation that led to equation (12.1) underlines two of the major problems of modern cometary astronomy. 'Reasonable' values for the mean density of the nucleus lie anywhere in the range of 100 to 900 kg m^{-3}. Also we do not know what fraction of the surface area of a typical cometary nucleus is actively emitting gas and dust. Most researchers estimate that this fraction is less than about 0.15 (see, for example Hughes 1988, Crifo and Rodionov, 1997 and Groussin and Lamy, 2004.)

12.3 A Model Of A Cometary Nucleus

Figure 12.6 is a schematic model of a cometary nucleus and its surface layer (see Hughes 2001). The nuclei in Figs. 12.1, 12.2, 12.3, 12.4 and 12.5 have irregular shapes. Comparisons were made with potatoes and avocado pears in the case of Comet Halley. The shape of the nucleus will vary as a function of time, as the comet progressively loses mass. As the nucleus becomes more elongated the propensity for splitting increases. Usually the components formed by this splitting will be more spherical than their parent.

The nucleus is spinning and the spin axis is precessing. Typical spin and precession periods are in the range 20 to 60 hours. Comets are therefore 'tumbling' as they move along their orbits, unlike planet Earth, where the spin period of 24 hours is very much lower than the precession period of about 25700 y. Spin and precession periods will vary due to the jet effects produced by mass loss. If comets spin too quickly they will break up due to the centrifugal force at the equator exceeding the gravitational attraction.

On average only half the surface is illuminated by sunlight at any one time. Due to the frequent visits to the inner solar system the temperature of the interior of a short-period

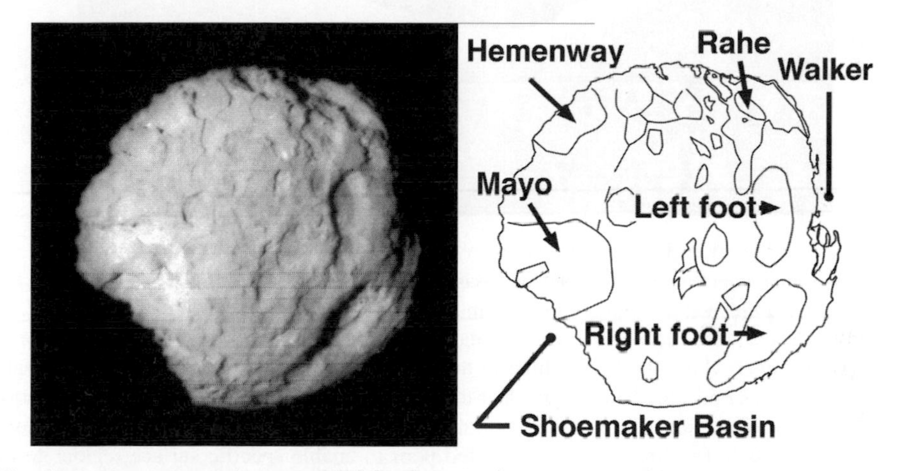

Fig. 12.3. The surface of comet Wild-2 taken was imaged by NASA's Stardust comet sample return mission. The spacecraft flew within 236 km of the nucleus on 2 January 2004, and the comet is seen at a phase angle of 11.4° (this being the Sun-comet-camera angle). The left-hand profile shows the cometary topography silhouetted against the sky background. The nucleus is shaped rather like an oblate spheroid $5.5 \times 4.0 \times 3.3$ km in dimensions, and is much more rounded than either Halley or Borrelly. The terminator is to the right, and the dark shadows emphasise the surface roughness. The pockmark, flat-floored depressions labelled Left Foot and Right Foot are about 1000 m across, and between 140 and 200 m deep. These depressions are bounded by steep (>70°) cliffs. Both the depth and the steepness of the cliffs indicate that the cometary material has sufficient strength to be self-supporting in the weak cometary gravitational field. The depressions are probably formed by mass wastage due to the sublimation of underlying snow, and their shapes probably echo deep structures inside the cometary nucleus. Mass loss keeps the cometary surface fresh and 'new'. As this surface has only been exposed to space for a short time it is very unlikely that these features have been produced by impacts with other orbiting minor bodies in the solar system.

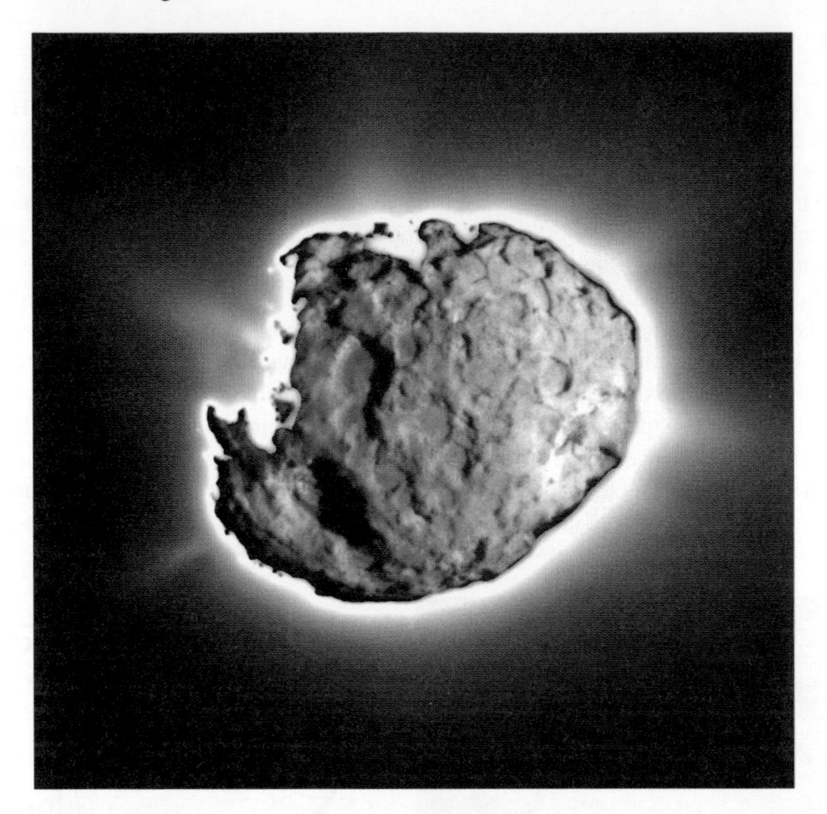

Fig. 12.4. The important features of this long exposure (0.1 s) image of Wild 2 are the gaseous inner coma and the jet features; the nucleus has been greatly overexposed. The jets on the right are emanating from the Walker region and jets can also be seen coming from Mayo (see Fig. 12.3). About twenty jets can be recognised and the majority of these originate from the near-equatorial areas of the comet where the Sun is high in the sky. About 2×10^{28} molecules of water are being lost each second, this being equivalent to about 600 kg of H_2O, around 0.2 m^3 of snow. The narrowness of the jets and their high collimation indicates that many of the source regions are relatively small. The image resolution is too poor to enable specific surface regions to be recognised as jet sources. The jets can be traced outwards until they are about 1 km above the nucleus surface. It would take micrometer-sized dust grains about 60 seconds to travel this far.

(Jupiter family) comet will be about 170 K. Intermediate period comets and Oort-Öpik Cloud comets will have interior temperatures of around 120 and 25 K respectively.

The cross-section on the right of Fig. 12.6 shows the surface and sub-surface region at a time when the comet is in the inner solar system. The interior of the comet contains three things – dust, snow and voids. The dust particles and the snow particles come in a range of sizes and they are also intermixed. It has been established that the snow-to-dust mass ratio is typically about 2.2. What is of more importance in the modelling process, is the snow-to-dust volume ratio.

Observations of the meteors produced when cometary dust particles impact the Earth's upper atmosphere indicate that typical cometary dust particles with masses in the

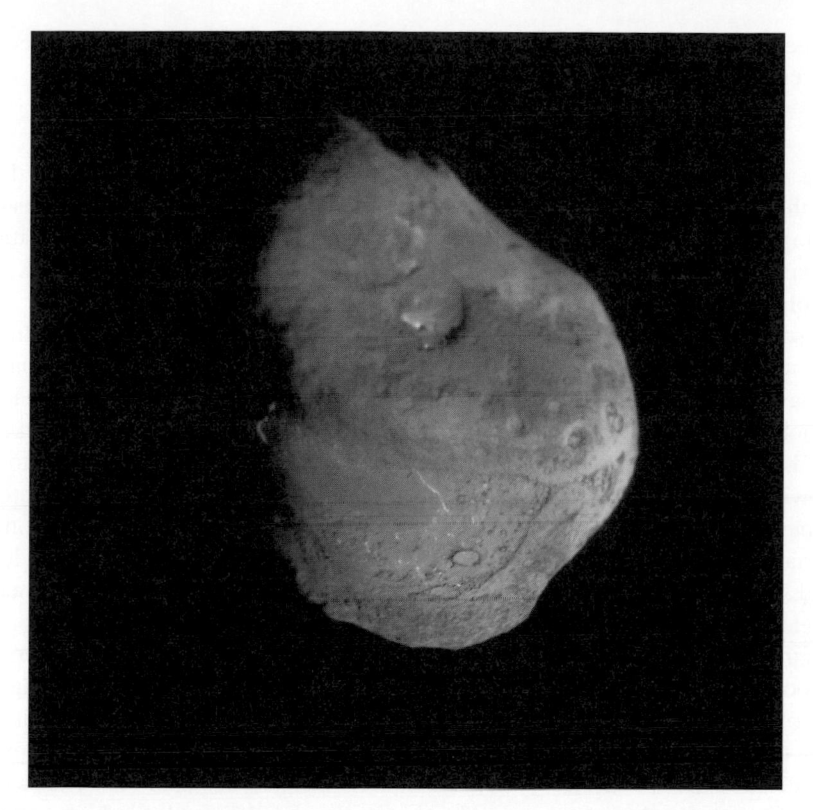

Fig. 12.5. A composite image of the nucleus of Tempel 1 obtained by the impactor probe of NASA's Deep Impact mission on 4 July 2005 as it flew at 10.3 km s^{-1} towards the lower right corner of the nucleus. The longest dimension of the nucleus is about 7.6 km and the shortest 4.9 km. Sunlight streams in from the lower right, so the right hand profile of the image gives a true impression of the surface topography. The surface is an enigmatic mixture of rough terrain, 20-m high cliff faces and unexpected smooth areas reminiscent of flat frozen lake surfaces. The gravitational field above the smooth regions is lower than normal. Maybe they are intruded layers of material, that were formed inside the cometary nucleus, that have subsequently become exposed. The variability in the appearance of the nucleus surface possibly indicates a combination of differing chemical compositions, physical characteristics, ages, and geological processing. A few dozen circular features can be seen on the image, these ranging in diameter from 40 to 400 m. Some of these are depressed below the average surface level; others are not. The size distribution of the Tempel 1 circular features differs from those seen on comet Wild 2. The origin of these features is unknown. The colour and reflectivity of the surface was remarkably homogeneous. No regions of exposed ice could be seen.

10^{-3} to 10^{-8} kg range have densities of around 800 kg m^{-3}, this being about 10 times the density of typical Earth snow. Larger dust particles are probably more vesicular and thus of lower density. The smaller dust particles contain fewer interstitial holes and thus have a larger density. As the dust density is ten times the snow density, a mass ratio of 2.2 converts to a volume ratio of 22. Thus a cursory glance at the interior of a comet will reveal a scene dominated by snow. Dust particles will be rare.

The maximum (i.e. sub-solar) surface temperature, T_s, of a nucleus is expected to be given approximately by:

$$T_s = \frac{404}{\sqrt{r}}\,\mathrm{K}\,, \tag{12.2}$$

where r is the heliocentric distance in AU. The constant in this equation is derived from the Deep Impact spacecraft results (see A'Hearn et al., 2005). The analysis of the imaging infrared (1.05 to 4.8 μm) spectrometer results indicated that the temperature on the sunlit side of 9P/Tempel-1 varied from 260 ± 6 K to a maximum of 329 ± 8 K, at the sub-solar point. The heliocentric distance of the comet was 1.51 AU at the time. In the inner solar system the vast majority of the surface cometary snow will have sublimated. So the top surface layer is dominated by very loosely packed dust. Individual particles will have gently accumulated as the inter-particle snow sublimates. Some heat from the 'hot' cometary surface will be conducted inwards through the dust contact points towards the cold interior. The remainder of the heat is radiated either outwards into space or downwards. When this heat reaches the underlying snow it starts to sublimate. Gas then percolates upwards (in the direction of the pressure gradient) through the cometary material. At the surface, momentum transfer between this gas and the extremely loosely bound surface dust pushes (i.e. blows) dust particles away from the cometary nucleus. The gas pressure is opposed by the very weak cometary gravitational field and any strength that the dust layer might have due to the dust particles sticking together.

Soon the surface of the cometary nucleus is covered with a layer of dust (see Fig. 12.6) some of this dust having fallen back to the surface because the gas pressure was insufficient to blow it away. Imagine that the surface temperature is approximately constant. The nucleus surface is heated by solar radiation whenever the spin of the comet turns that surface into the daylight section. Heat passes through the dust layer down to the cold underlying snows. Due to sublimation the snow temperature in the inner solar system is constant (just like the temperature of boiling water is constant at the surface of Earth). In the vacuum of space the H_2O molecules will be sublimating at a temperature of about 220 K. Escaping gas removes dust from the surface. The thickness of the dust layer will be a function of the temperature difference between the surface and the sublimating snow. Under equilibrium conditions the rate of retreat of the underlying snow will be exactly compensated by the rate of removal of the surface dust. If the snow retreated faster than the dust escaped from the surface then the layer would thicken and the heat transfer would decrease and the sublimation rate would go down. If the snow retreated slower than the dust escaped then the thinner dust layer would transmit more heat and the sublimation rate would go up.

As the comet moves around its orbit the temperature of the nucleus surface changes gradually, maximising at perihelion. The simple model described above indicates that the dust layer will be at its thinnest when the surface is at its hottest. As one moves away from perihelion the layer thickens and will reach a maximum value when the heat being transmitted through the layer is insufficient to the raise the temperature of the underlying snow above the sublimation point, and the pressure of the escaping gas is insufficient to push dust particles away from the nucleus.

The thickness of the dust layer is expected to be in the millimetre to centimetre range. This figure is supported by expected values of the thermal transmission of dust layer coupled with observations of the sublimation rate. Also the dust mass spectrometer

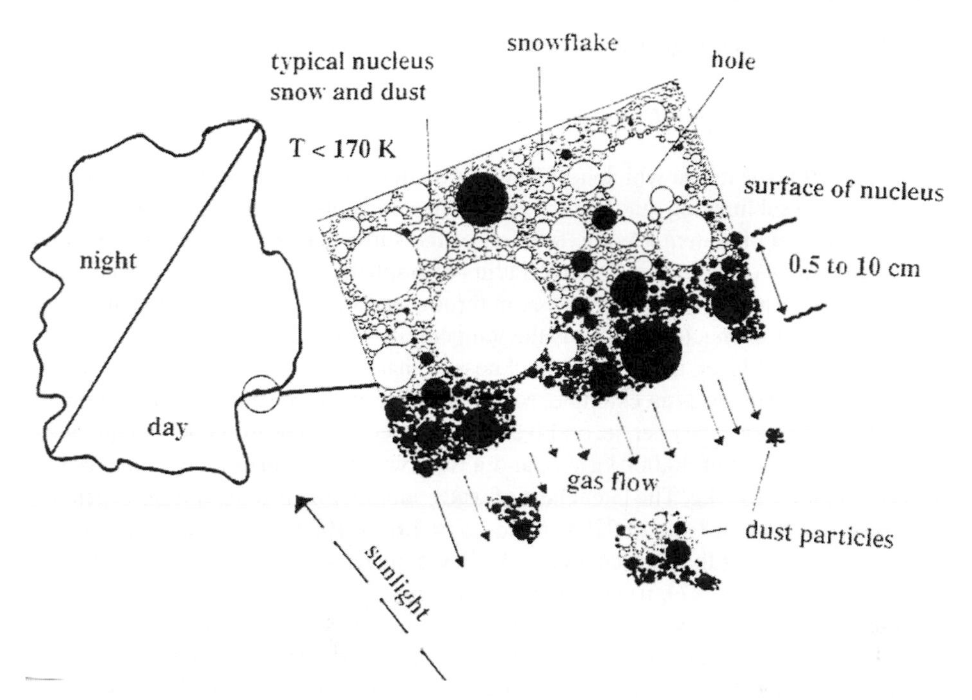

Fig. 12.6. A schematic model of a cometary nucleus and its surface layer. The nucleus, on the left, has an irregular shape (comparisons were made with potatoes and avocado pears in the case of Comet Halley). It spins about every day or so, usually about an axis that maximises the moment of inertia. On average half the surface area is illuminated by sunlight at any one time and the remainder is in darkness. Due to the frequent visits to the inner solar system the temperature of the interior of a short-period (Jupiter family) comet will be about 170 K. Intermediate period comets and Oort-Öpik cloud comets will have interior temperatures of around 120 and 25 K respectively. The cross-section on the left shows the surface of the cometary nucleus surface at a time when the comet is in the inner solar system. The black spherical regions represent dust and the clear spherical regions represent snow. Notice that twenty to forty percent of the cometary interior is probably empty. The lower, sun facing surface of the comet will be heated by solar radiation. Heat will then be transported through the thin dusty surface region by a combination of conduction and radiation. In the inner solar system the dust-snow interface will be about the 240 K isotherm. Gas from the sublimating snows percolates speedily through the surface dust layer and detaches and then blows dust particles away from that surface. The strength of this layer will be very low. Some of the dust particles will still have interstitial snows inside them, as they leave the cometary surface, but this interstitial snow will soon sublimate once the dust is illuminated by the Sun.

on-board the Giotto spacecraft observed some particles that consisted of both dust and sections of subsurface volatiles. These particles were referred to as CHON particles because Carbon, Hydrogen, Oxygen and Nitrogen made up a significant fraction of their mass. So not only was Comet Halley emitting gas and dust, but also some of the escaping dust particles were taking bits of the underlying snow with them. As these particle moved into the sunlight, the interstitial snow sublimated and some of the particles even broke up as the snowy ‘glue’ disappeared. Maybe the CHON particles are only emitted when the rate of mass loss is high. Consider the relevant dust layer equation:

$$\frac{\mathrm{d}H}{\mathrm{d}t} = \frac{\mathrm{d}m}{\mathrm{d}t} \times L_{\mathrm{water}} = \frac{f \times 4\pi R^2 \times \kappa \times (T_{\mathrm{surface}} - T_{\mathrm{snow}})}{d}, \tag{12.3}$$

where dH/dt is the rate at which heat flows through the dust layer, dm/dt is the rate at which snow is sublimating, and gas is leaving the comet and entering the coma, L_{water} is the latent heat of water, f is the fraction of the surface area that is active, R is the equivalent radius of the comet (i.e. the radius of a sphere with the same surface area), κ is the coefficient of thermal transmission through the dust, T_{surface} is the temperature of the surface of the nucleus, T_{snow} is the temperature of the dust-snow interface at the bottom of the dust layer, and d is the thickness of that layer.

Take Comet Halley as an example. Near perihelion it was losing water at the rate of about 1×10^{30} molecules per second (see Feldman et al., 1987), this being equivalent to a water mass loss of 30,000 kg s^{-1}, and a total water loss during the whole 1985/6 apparition of 3×10^{11} kg. The latent heat of water sublimation is temperature dependent but a typical value at a T_{snow} of 220 K is $L_{\mathrm{water}} = 2.61 \times 10^6$ J kg^{-1}. The Giotto camera images indicated that the surface area of Halley's nucleus was about 400 km^2 (i.e. the equivalent radius was 5.7 km). There were three main areas of activity on the nucleus and the active fraction f was found to be about 0.1. This value is non typical; pre and post perihelion f values can vary, and f can change considerably from apparition to apparition. Hughes (1988) analysed the brightnesses of Halley over the last 2000 years and found that the median f value was 0.030. Over this time the value had varied between 0.0076 and 0.10, but 64% of the values were in the 0.015 to 0.05 range.

The coefficient of thermal transmission of the dust layer is unknown. It clearly depends on both the degree of dust compaction, the areas of contact between the dust particles and the particle size distribution. The only material that is vaguely comparable, and known, is the top few centimetres of the fine-grain lunar regolith. At the temperatures encountered by the Apollo astronauts (some 350 K) the heat transfer in this layer was divided in the ratio 70 : 30 between radiative transfer and conduction. The coefficient of heat transmission was 1×10^{-2}W m^{-1} K^{-1}, (a figure similar to the thermal conductivity of Styrofoam.) Near perihelion ($q = 0.587$ AU) Halley would have had a sub-solar surface temperature of about 530 K. Substituting all the above figures into equation (12.3) gives $d = 0.0015$ m. So an active region of a cometary nucleus consists of about 0.0015 m of loose fluffy dust, covering underlying cold snow. The word 'about' sums up our uncertain values for quantities such as f and v. We expect the calculated value of d to be correct within a factor of about 5. So we conclude that the dust layer on top of a cometary active region is between 0.0005 and 0.01 m thick. This layer would be opaque, and could easily have an albedo in the 0.02 to 0.05 range, as observed by Giotto. The layer is so thin that the production of CHON particles would be commonplace.

Observations of the meteoroid streams produced by decaying comets indicate that the meteoroid mass distribution is dominated by particles in the 10^{-5} to 10^{-3} kg range these having diameters of between 0.004 to 0.02 m. These diameters again agree reasonably well with the estimate made above of the dust layer thickness.

Interestingly, there is probably only a small variation in the thickness of this dust layer over the surface of the nucleus. If the dust layer was very much thinner than the 0.0005 to 0.01 m range indicated above the heat transmitted through it would be so large that it would produce very active sublimation of the underlying snows and the snow region would retreat quickly into the nucleus, speedily building up a thicker dust layer. A much thicker layer than that indicated by the above range would act as an efficient insulating blanket. It would transmit a negligible amount of heat, there would be no sublimation of the underlying snows and that region of the cometary surface would become inactive and dormant. In fact a very thick dust layer cannot be produced, simply because the heat transmission is far too small to remove snow from underneath it.

The observation of jets of material in cometary comae indicate that the nucleus surface can be divided into 'active' and 'inactive' areas. Only 40 km^2 of the 400 km^2 of Comet Halley were active in March 1986, there being three active zones, each some 4 km across.

Structurally the cometary volume can be divided into two components. There is the very thin, small volume, dust layer covering the whole surface. This surface appears to be grey and of low albedo. Then from a depth of about 0.01 m right to the centre of the nucleus (some 5700 m below in the case of Halley) we enter the relative unknown. Expectations are of a snow-to-dust mass and volume ratio of around 2.2 and 22 respectively, and a mean density of around 150 kg m^{-3}. The fact that comet Shoemaker-Levy 9 was so easily broken up into twenty pieces (as opposed to 2 or 200 or 2000) when it travelled inside the Roche limit of Jupiter on 7 July 1992 indicates that the strength of the nucleus is very low, and that it is probably extremely porous, and might even consist of a loose fragile agglomeration of smaller individual elements. There is every possibility that there are some large voids inside the nucleus. The 150 kg m^{-3} 'mean' density might be a result of some regions having zero density (and contain nothing) and others have compensating densities of 300 kg m^{-3}. The sizes of these individual void and occupied regions are unknown. Maybe surface features on the cometary nuclei imaged by flyby spacecraft might provide a clue to their extent. Maybe the fact that Shoemaker-Levy 9 was about 1.5 km in diameter before break up and fragmented into about twenty 300 m diameter bits is a hint.

Turning to cometary origin, it is thought that a specific dirty-snowball nucleus was produced by the accretion of a collection of smaller dirty-snowball nuclei in the primordial pre-planetary nebula. These components could have been formed anywhere throughout the Jupiter to Neptune region. The collisional growth was extremely gentle. As the total mass and size of a comet is so small, negligible alteration has occurred post-accretion. There is every chance that the original components have maintained their physical and chemical characteristics, and these might vary slightly depending on the heliocentric distance of their origin.

12.4 Cometary Activity

One of the characteristics that distinguishes comets from other solar system minor bodies (such as asteroids and Edgeworth-Kuiper Disc Objects) is their potential for activity. Comets do far more than orbit around the Sun merely scattering sunlight. On entering the inner solar system comets spring into life and not only surround themselves with huge dusty and gaseous comae but also sport long anti-solar micro-dust and plasma tails.

Take Great Comet Hale-Bopp. At 1 AU from the Sun it was losing water at the rate of about 10^{31} molecules s^{-1}. This dropped to 10^{30}, 10^{29} and 10^{28} molecules s^{-1} at 2.3, 4.0 and 5.0 AU respectively. With an activity variation of over 10,000 fold as the comet speeds around the inner solar system portion of its orbit it is no wonder that great comet apparitions such as this startled and frightened the ancient sky watchers.

Let us briefly estimate the effect of mass loss on a comet like Halley. At maximum activity Halley was losing about 1×10^{30} water molecules per second (i.e. 30,000 kg s^{-1}.) The fact that this is about 10 times less than the loss from Hale-Bopp at 1 AU partially explains why Halley was the fainter of the two comets. Huebner and Boice (1997) estimated that about 78% of the snow mass in a cometary nucleus consisted of H_2O molecules and Hughes (2000) estimated that a typical cometary nucleus had 69% of its mass in the form of snow. So an H_2O mass loss of 30,000 kg s^{-1} is equivalent to a total snow/dust mass loss of about $30{,}000/0.78 \times 0.69$, i.e. 56,600 kg s^{-1}, and a volume loss of 380 $m^3\,s^{-1}$ (assuming a density 150 kg m^{-3}). Halley has an active area of 40 km^2, so the mass loss rate is equivalent to the active area retreating by 9.5×10^{-6} m s^{-1} (i.e. 3.4 cm hr^{-1}). Observers standing on the nucleus would see the comet eroding away before their eyes.

Consider a whole Halley orbit. The total water loss is about 3×10^{11} kg, equivalent to a total mass loss of 5.6×10^{11} kg and a total volume loss of 3.7×10^{9} m^3. The active area of 40 km^2 will therefore retreat about 90 m per orbit. Extremely deep pits would be produced if the active areas only occurred at specific places on the cometary surface. These very deep pits have not been observed by the visiting spacecraft. So we can assume that a specific active region has a short lifetime. Activity quickly stops and then breaks out somewhere else, this change in position possibly being influenced by impacts or by the existence of sub-surface features, such as voids. Averaging over a reasonable number of orbits the mass loss is equivalent to the whole surface retreating by around 9 m per apparition. As the effective radius of Comet Halley is 5700 m this means that in $5700/9 = 600$ apparitions (about 50,000 years) the comet will have decayed away completely. Now these simple calculations depend on the assumed cometary density. But a change in density by a factor of four, from 150 to say 600 kg m^{-3}, will simply quadruple the mass and change the remaining life of the comet to 2400 apparitions (200,000 years).

The comets that we have seen in recent centuries are thus fleeting members of our inner solar system. They are prominently visible because of their mass loss, and this attrition will quickly cause their demise.

We can also go back in time. Comet Halley is orbiting inside a large meteoroid stream this having been produced by mass loss over many previous apparitions. The Halley stream is observed from Earth as the October Orionids and the May Eta Aquarid meteor showers. The mass of dust in this stream has been estimated to be 5×10^{14} kg (see Hughes, 1985) this indicating that the total cometary mass loss required to generate the stream is about 1.60×10^{15} kg. This calculation assumes that the comet remains on or very close to its present orbit for a considerable time. Today the comet has a mass of 1.2×10^{14} kg (assuming density = 150 kg m^{-3} and mean radius = 5700 m). When it was first captured into its present orbit, and started forming a meteoroid stream, the comet had a total mass of 1.7×10^{15} kg and a radius of 14,000 m. Assuming a constant layer thickness loss per orbit, it takes about 900 apparitions (70,000 years) for the radius to decrease from 14,000 to 5,700 m.

There are three important consequences of this material loss. First the dust seen on the surface of the nucleus at the present time would have originally been well inside the comet; in fact in Halley's case over half way down towards the centre. If, as expected, the whole nucleus interior has a similar physical and chemical composition, there should be no variation as a function of depth, and so the surface looks the same as it always did. Looking at a cometary surface gives no indication of age. Secondly the surface of the nucleus that we see today is relatively 'fresh'. Say we punch a crater of diameter 100 m and depth 20 m into the surface of Comet Halley, this crater being similar to the one expected to have been produced in the surface of 9P/Temple-1 by NASA's Deep Impact mission. At an average surface removal rate of 9 m per apparition it would only take a few apparitions for this crater to be eroded away. If the crater produced a new active area it would erode away in a month or so.

To obtain an upper limit to the expected crater density on a nucleus surface let us assume that the impact rate to a comet is similar to that to the Moon (it is probably slightly less because, even though short period comets often traverse the asteroid belt, the velocity of impact decreases as the inverse square root of the heliocentric distance). Present-day lunar cratering is such that ε craters per year bigger than D m are being formed on every 100 km^2 where:

$$\log \varepsilon = -2.58 - 2.5 \log D(\text{m}) , \tag{12.4}$$

(see, for example, Morrison and Owen, 1988.) A 100 m crater would thus be formed on the 400 km^2 surface of Halley every 10^7 years. For craters bigger than 10 m and 1 m one has to wait only 30,000 and 100 years. Large craters are formed very infrequently. So as a consequence of the continuous erosion of the nucleus we can conclude that impact craters larger than a few metres in diameter are not present on cometary surfaces. Centimetre-sized craters, these being deeper than the very thin surface dust layer, will, however, be formed frequently. These are produced both by sporadic meteoroids and Halley stream meteoroids. This 'impact gardening' will reactivate dormant parts of the cometary surface and lead to the production of new active regions.

The third consequence of the high rate of cometary mass loss is that integrating short-period cometary orbits into the future over more than a few tens of thousand apparitions is of little value. Even the large comets will have physically disappeared long before then unless they have been perturbed by the giant planets such that their perihelia have been moved into the outer reaches of the solar system.

Cometary nuclei have two other unusual characteristics. They have a propensity to split apart and they also have sudden and short-lasting outbursts of activity. Mass loss can lead to a change both in the overall shape of a nucleus and its spin and precession state. Interior stresses can be produced by jet effects during mass loss these leading to fracture. Hughes and McBride (1992) concluded that a typical short-period comet breaks into two or more pieces at least once in its inner solar system lifetime. These fragmentations are expected to convert the single rather elongated nuclei (such as Comet Borrelly in Fig. 12.2) into a few more spherically shaped nuclei (such as comet Wild-2 in Figs. 12.3 and 12.4).

A cometary outburst is an unexpected flare-up in the brightness of a comet, with the brightness normally increasing by 2 to 3 magnitudes, i.e. a factor of 5 to 20 (see Hughes, 1991). As stated above, comets are only active over a small fraction of their surface. A typical value of f is about 0.01. Any mechanism that can speedily change this to say 0.02, 0.04, 0.08 or 0.16 would lead to an outburst of magnitude 0.8, 1.5, 2.3 and 3.0. Two possibilities spring to mind. The gradual retreat of the surface could break through into an underlying void thus revealing a large area of fresh snow. The void would have to be big. Doubling the active area changes the magnitude by 0.8. On the other hand there is the possibility that the outburst is due to chemistry. The comet snow might have a range of H_2O/CO_2 mass ratios. If the retreating surface suddenly revealed a region of high CO_2 concentration this would promote a steep increase in sublimation rate. Normality would return when the surface of the hole became dusty, or a more usual H_2O/CO_2 ratio was encountered.

12.5 Cometary Missions And Early Results

Recently we have entered the space-age of cometary investigation. Instead of meekly standing on Earth and waiting for comets to travel past us, we have rocketed off to visit them. The return of Comet Halley thirty years after the 1957 start of the space age triggered the opening of certain government purse strings.

There have been a handful of successful missions so far. The ESA spacecraft Giotto (see Reinhard, 1986) flew past Comet 1P/Halley at a speed of about 65 km s^{-1} on 13 March 1986. From its miss-distance of 600 km it imaged the nucleus at a resolution of about 200 m $pixel^{-1}$. The comet was about 0.9 AU from the Sun at the time, having passed perihelion just over a month before. The dirty snow-ball nucleus was revealed to be potato-shaped and about $15.3 \times 7.2 \times 7.2$ km in size. Major surprises were the unusual spin mode of the comet, the fact that only 10% of the surface was active, and the detection of snowy CHON particles leaving the nucleus. Prior to the Giotto mission scientists had no idea what a cometary nucleus looked like. The USSR's Vega-1 and Vega-2 probes (see Grard, Gombosi and Sagdeev, 1986) also visited Halley and it was observed by Japan's first deep space missions Sakigake and Suisei (see Hirao, 1986). Many on-board experiments investigated the interaction between the plasma and magnetic fields in the solar wind and the cometary coma and tail. In fact the first spacecraft encounter with a comet occurred on 11 September 1985 when NASA's International Sun-Earth Explorer spacecraft (ISEE-3, renamed ICE, International Cometary Explorer) passed through the tail of Comet Giacobini-Zinner (see Bame et al., 1986).

Sixteen years later NASA's Deep Space 1 craft flew within 2200 km of Comet 19P/Borrelly, on 22 September 2001 (see Soderblom et al., 2002 and Rayman, 2003). The best nucleus surface resolution was an improved 47 m $pixel^{-1}$. Borrelly was about half the size of Halley. Its more elongated shape ($8.0 \times 3.2 \times 3.2$ km) reminded researchers of a bowling-pin. It was also less active and only a single jet of dust and gas could be easily discerned.

NASA's Stardust mission imaged Comet 81P/Wild-2 on 2 January 2004 (see Brownlee et al., 2004). This nucleus was less elongated than either Halley or Borrelly, the axes being $5.5 \times 4.0 \times 3.3$ km. From a miss-distance of only 236 km the best resolution was a respectable 14 m pixel^{-1}. The nucleus surface was found to be littered with steep sided depressions. The mission's primary objective was to collect minute dust particles from the inner coma. During the low velocity, 6.1 km s^{-1}, flyby the dust was been captured in aerogel, a low-density porous silicon matrix. This encapsulated dust was returned to Earth on 15 January 2006, being safely parachuted to the US Air Force Utah Test and Training Range southwest of Salt Lake City.

The Deep Impact mission (see A'Hearn et al., 2005) did not just fly by Comet 9P/Tempel-1 but, on 4 July 2005, successfully fired a 370 kg (1-m diameter) self-guided impactor into the surface at a speed of 10.3 km s^{-1}. The expectation was that a 100 m diameter, 20m deep crater would be produced. Unfortunately the damage inflicted by the impactor pulverised the surface dust and snow releasing a huge obscuring gas/dust cloud. Nothing could be seen through this cloud and the details of the new crater are unknown. Apart from this problem, the mission was a huge success, this success being indicated by the smallness of the standard deviations of many of the results quoted below. The nucleus of 9P/Tempel-1 was found to have a radius of 3.0 ± 0.1 km (the longest dimension was 7.6 km and the shortest 4.9 km) and was spinning every 40.832 ± 0.33 days. The albedo and colour of the nucleus surface was remarkably homogeneous. The existence of a complete surface dust layer was confirmed. No exposed regions of ice or frost were visible, even though dust was being lost continuously, the comet having an estimated mass loss of 10^9 kg per orbit (very much less than Halley). Pre-impact observations of the comet showed that outbursts of activity were relatively common, one occurring every 10 days or so. Most of these started when a flat 'shoulder' of the nucleus experienced dawn. This underlines the probability that outbursts are related to sub-surface conditions and are not triggered by impacts. The brightness change during the outburst had a swift onset (less than 10 minutes). Decay was more gradual, the intensity returning to normal after about 18 hours (a time that is considerably shorter period than the cometary day.)

Even though the crater was invisible, the cone of ejected material could be seen clearly. Measurements of the rate of expansion of the base of this cone could be used to estimate the gravitational field, gc, at the impact site. The result was 50 mgal. Unfortunately the uncertainty range is $+34/-25$ mgal. Knowing the size and shape of the nucleus enables this result to be converted into a nucleus mass. (Remember that

$$g_c = \frac{GM_c}{R_c^2}, \tag{12.5}$$

where G is Newton's constant of gravitation, and M_c and R_c are the mass and radius of the comet.)

The value obtained for M_c was 7.2×10^{13} kg (about 70000 times the mass lost per orbit). Unfortunately the uncertainty range was $+4.8 \times 10^{13}$ kg/-3.8×10^{13} kg so the mass estimate was only accurate to within a factor of about 2. Coupled to the uncertainty concerning the volume of the comet this means that the estimated density of 620 kg m^{-3} had an uncertainty range of $+470$ kg m^{-3}/-330 kg m^{-3}. The material blasted from the crater was much hotter (1000 to 2000 K) than the normal material ejected from comets. Not only did this contain expected gasses like H_2O, HCN and CO_2 but the very strong organic features observed in the spectrum indicated that the untypical impact had vaporised carbon compounds that usually stay in the solid state. The fact that the volatiles appeared very quickly after the impact indicates that the snows are very close to the surface.

There is some debate about the physical form of the cometary dust. The solid material leaving the impact site was in the form of microscopic dust particles. Large cometary dust particles, those that produce shootings stars when they burn up in the Earth's upper atmosphere, are not like this. Maybe the energetic impact pulverised the dust in the surface layer of the nucleus. Maybe the pre-existing dust and the dust emitted after the impact have completely different size distributions. Under normal conditions comet dust is lifted very gently from the surface and not explosively ejected. Ground based observations of the plume of ejector indicated that the ratio between radiation pressure and gravity for the typical impact-ejected dust grain was 0.3 (see Meech et al., 2005). According to Dohnanyi (1978), this would make the typical particle size in the ejected dust about 10^{-6} m.

Many observatories observed the ejector cloud after the July 4 impact. The comet, however quickly retuned to normal, the gas production returning to pre-impact levels by 9 July. Ground-based observation of such characteristics as $^{12}C/^{13}C$ and $^{14}N/^{15}N$ indicated that the former was 89 (very similar to that of the Sun), and that the later was about 140, half that of the Earth. Chemically 9P/Tempel-1 seemed to be very similar to other Jupiter-family comets.

The pre-impact comet was faint, the visual magnitude being around 17. About 30 minutes after impact it had brightened by about 2.3 magnitudes. The outbursts seen by the spacecraft were also detected by ground-based telescopes. Dust ejected by the outburst left the nucleus at a velocity of about 160 to 200 m s^{-1}, this being very similar to the mean gas molecular velocity at the sublimation temperature.

Detailed spectroscopic analysis (see Mumma et al, 2005) indicated that 9P/Tempel-1 had a similar chemistry to Oort Cloud comets. (There are estimated to be about 10^{13} of these comets in a near-spherical shell around the Sun, extending from about 10,000 to 50,000 AU.) It is suggested that 9P/Tempel-1 was formed originally in the outer regions of the proto-planetary disc. It was then perturbed into the Oort Cloud and then back into the Kuiper-Edgeworth disc (an ecliptic belt of a few times 10^9 comets orbiting at heliocentric distances between 40 and several hundred AU). Subsequently it was captured onto its present inner solar system orbit.

Infrared observations of the comet (see Sugita et al., 2005 and Harker et al., 2005) coupled with models of cometary dust scattering properties indicated that somewhere between 10^4 and 10^6 kg of dust was ejected by the impact. Interestingly it was noticed that the expansion of this dust was not influenced entirely by gravity. Maybe some of the dust contained embedded snows, and the subsequent vaporisation of this snow acted like a small rocket. The 7.8 to 13 μm spectral energy distribution indicated that the dust had a crystalline silicate composition, similar to the amorphous pyroxene, amorphous olivine and magnesium-rich crystalline olivine that were formed at a temperature less than 1000 K as the early solar nebula cooled.

For future excitement we rely on ESA's Rosetta mission (see for example http://www.esa.int/SPECIALS/Rosetta/index.html and Verdant and Schwehm, 1998). This spacecraft was launched on 2 March 2004. The target is comet 67P/Churyumov-Gerasimenko, and in May 2014 Rosetta will go into a 25 km high orbit around that comet when it is about 4 AU from the Sun. Then, instead of just taking a 'one-moment' snap-shot during a brief fly-by, we will have a mission that will eventually produce an extremely detailed map of the whole comet surface. And the cometary activity will be monitored as the comet travels in towards perihelion and then away from the Sun, so temporal changes of the surface can be recorded, hopefully over a few years, an interval that is a large fraction of the cometary orbital period. In November 2014, a lander (named Philae) will gently touch down (at walking speed) onto the nucleus surface. After securing itself with a harpoon it will start a detailed analysis of the cometary composition and dust structure. The orbiter will be imaging the surface at centimetric resolution. Philae will produce even more detailed images.

The major findings of space missions so far concern the nature of the surface of the nucleus. Let us use the high-resolution images of Comet Wild-2 (see Figs. 12.3, 12.4 and Brownlee et al. 2004) as an example. Wild-2 has an orbit that is controlled by Jupiter and the gravitational influence of this giant planet sporadically 'flips' the comet such that it moves from a large semi-major axis (intermediate P = 40 y) orbit with a perihelion near the jovian orbit and an aphelion near the orbit of Uranus to a small semi-major axis (short-period) orbit with an aphelion near the jovian orbit and a perihelion near the orbit of Mars (see for example Levison and Duncan, 1997). The last flip occurred in September 1974 when Wild-2 was captured onto its present $q = 1.55$ AU, $Q = 5.25$ AU, $i = 3.25°$, $P = 6.2$ year orbit, this bringing it to perihelion in June 1978, August 1984, December 1990, May 1997 and September 2003. In 1974 Wild-2 got to within 0.006 AU of Jupiter, this being only ten times further away than the fragment-inducing flyby of Shoemaker-Levy 9. Many of these orbital 'flips' will have occurred in the past. Wild-2 will have long intervals of inner solar system activity followed by even longer period of outer solar system dormancy.

Figure 12.3 was taken at a phase angle of 11.4°, the cometary illumination being similar to our Moon at a time when it is one day after being 'full'. The terminator is on the right-hand side of the image, this being the region with areas in deep shadow. The right-hand portion of the image is enlivened by two elongated depressions rather unprosaically named by the NASA team 'Left foot' (this being at about 3 o'clock) and 'Right foot' (at about 4:30). These depressions are much more circular than they look, the perspective of the image having them angled away from the viewer. They are around 1400 m across and their bases are about 140–200 m below their rims.

The left hand side of the image shows the cometary surface in profile, the topography being highlighted against the very much fainter background of the inner cometary coma. The 'bite' that seems to have been taken out of the lower left region (at about clock hour-hand direction 7:30) has been named the Shoemaker Basin. The term 'basin' is a misnomer as no suggestion has been made that this region was formed by an impact process (as was, for example, the Caloris Basin on Mercury). This profile is extremely informative. The cometary surface has a 'rolling' topography; it undulates up and down, the 'hills being a mere 170–200 m above the 'dales'. In the main the sides of the hills are at angles of about 35o to the horizontal. Occasionally one encounters features that have been referred to by the image team as 'mesa', 'pinnacles' and 'canyons'. This terminology is reminiscent of USA cowboy movies, and the Monument Valley scenery of Arizona and Utah. Mesa is Spanish for 'table', and in the western USA is an isolated hill with a flattish top and relatively steep (about 70°) sides. A pinnacle is just a mesa with a very small top surface.

On the comet the depressions between the mesa and pinnacles are referred to as canyons. The pinnacles and mesa are typically 100 meters high. The steep sides seeming defy gravity, but this is a misleading concept. Wild-2 has a mean radius of 2090 m and, assuming a density of $150\,\mathrm{kg\,m^{-3}}$, a mass of 5.7×10^{12} kg. This gives the surface a mean acceleration of gravity of about $9 \times 10^{-5}\,\mathrm{m\,s^{-2}}$, this being 100,000 less than the acceleration of gravity at the surface of Earth. So the pinnacles have very little weight and can be relatively easily supported by low strength cometary material. A cliff (or scarp) some 2 km long has also been recognised on the cometary surface. There is only very limited evidence of down-slope mass movement. The cliff debris seems to be very weak. This fine-grained material disintegrates when it falls and is then blown away.

The fact that pinnacles only reach a height of 100 m provides a vital clue to the strength of the underlying cometary material. There is a well known formula in geophysics that relates the maximum height of a planetary mountain, $H_{\max}$, to the maximum compressive change in volume per unit volume $(^{\mathrm{TM}}\mathrm{V/V})_{\max}$ that the underlying material will withstand before giving way. Considering a spherical body of radius R, density ρ, bulk modulus κ, we have

$$\left(\frac{\mathrm{d}V}{V}\right)_{\max} = \frac{\text{base pressure}}{\kappa} = \frac{4G\pi\rho^2 R H_{\max}}{3\kappa}, \tag{12.6}$$

where G is the constant of gravity. For the terrestrial planets it is reasonable to suggest that $R\,H_{\max}$ is a constant and this equation can be calibrated using the radius of Earth (6.375×10^6 m) and the height of Mount Everest (8850 m). Considering all solar system bodies we suggest that $(\rho^2 R H_{\max}/\kappa)$ is a constant, and thus conclude that the bulk modulus of the snowy cometary material is a factor of about 9×10^{-9} less than that of terrestrial rock. Comets cannot, however, be entirely strength-less. If this was the case, their self-gravitation would pull them into a spherical shape.

The fact that pinnacles and mesa exist at all on cometary surfaces suggests (as in Monument Valley!) that there is some lateral surface heterogeneity and that the strength of the material at the top of the pinnacle and mesa is slightly greater than that of the immediate surroundings and is thus slightly more resistant to the erosive decay processes.

Three factors are important in cometary mass loss. The first is the temperature of the surface, the second the thermal transmissibility of the thin dust layer that covers the cometary surface and third is the strength of that layer and specifically its ability to withstand the pressure exerted on it by the wind of sublimating gasses from beneath. The first is by far the most difficult to estimate and will play a dominant role in sculpting the cometary topography. As with the sunlit side of the Moon, the temperature will vary as a function of $\sqrt{\cos\beta}$, where β is the angle between the normal to the surface and the comet-sun vector. If the comet started life as a perfectly spherical object with a specific spin axis perpendicular to its orbital plane and with no precession, the surface temperature would be a simple function of latitude and longitude. But comets are far from spherical; the spin axis is randomly orientated with respect to the normal to the orbital plane and the spin period and precession periods are comparable. We therefore expect the temperature and concomitantly the erosion rate, to vary drastically from place to place. It is this variability that leads to the observed variable topography and it is the strength of the cometary snow/dust mixture that leads to the observed 170–200 m height differential between the hills and dales. The importance of the subsurface strength hints that surface features echo the characteristics of the material inside the comet and will give a clue as to the way in which the comet was built up by the accretion of smaller units.

12.6 Conclusions

We are at an extremely exciting stage in cometary exploration. Modern computers have enabled us to probe cometary orbital evolution in great detail and we have been able to build up a realistic picture of the primordial movement of comets into the Oort cloud and Kuiper-Edgeworth Disc and then back again to the short-period and long-period populations that we observe from Earth. Modern astronomical photometry and spectroscopy have also revealed much about the chemistry and evolution of cometary gas and dust and also much about the way in which these components interacts with the inner solar system environment. The space age has shown us that cometary nuclei are single kilometric-sized entities and has revealed impressive details of their surface topography. But our knowledge of the nucleus is not only skin deep, it is also hampered by the rather poor resolution of these images. Many mysteries remain. Mass, density and interior structure are poorly constrained. Internal chemistry and snow/dust ratios are only intelligent guesses. Hopefully in about 50 years time we will have not only landed on a few cometary nuclei but also been able to return to Earth with refrigerated samples of their material. At the present time we only picking up very minor clues that the basic chemical and physical characteristics might differ slightly from comet to comet. As we learn more, these differences should become clearer. Comets were formed having a range of sizes, and at a range of heliocentric distances. Orbital evolution over the lifetime of the solar system leads to considerable variations in the cometary perihelion value and thus to considerable differences between the rate of mass loss. Some comets are now seen as small eroded remnants of their former glory and other have been hardly altered since formation.

The great beauty of the visible comet in the sky ensures that they have fascinated mankind ever since humans started to gaze upwards from the swamp. The great scientific fascination of comets is due to the fact that they have kept their original components locked up in a low gravity deep freeze ever since formation. Comets are pristine, and represent our only chance of handling the original life-producing building blocks of our planetary home.

References

A'Hearn, M. F. at al., 2005. *Science* **310**, 257.

Anders E. and Ebihara M., 1982. *Geochim. Cosmochim Acta* **46**, 2363.

Bame, S. J., et al., 1986. *Science* **232**, 356.

Belton, M.J.S., Meech, K.J., A'Hearn, M.F., Groussin, O., McFadden, L., Lisse, C., Fernández, Y.R., Pittichová, J., Hseih, H., Kissel, J., Klassen, K., Lamy, P., Prialnik, D., Sunshine, J., Thomas, P., Toth, I., 2005. *Space Science Reviews* **117**, 137.

Boice, D.C. and Huebner W.F., 1999. Physics and chemistry of comets. In *Encyclopaedia of the Solar System*, eds. P.R. Weissman, L.-A. McFadden and T.V. Johnson, Academic Press, San Diego, pp 519–536.

Brownlee, D. E., Horz, F., Newburn, R. L., Zolensky, M., Duxbury, T. C., Sandford, S., Sekanina, Z., Tsou, P., Hanner, M. S., Clark, B. C., Green, S. F., Kissel, J., 2004. *Science* **304**, 1764.

Crifo, J. F., and Rodionov, A. V., 1997. *Icarus* **127**, 319.

Crovisier, J., and Encrenaz, T., 2000. *Comet Science*, Cambridge University Press, p. 98.

Dohnanyi, J. S., 1978. In *Cosmic Dust*, ed. J.A.M. McDonnell, John Wiley and Sons, p. 559.

Feldman, P.D., Festou, M.C., A'Hearn, M. F., Arpigny, C., Butterworth, P. S., Cosmovici, C. B., Danks, A. C., Gilmozzi, R., Jackson, W. M., McFadden, L. A., Patriachi, P., Schleicher, D. G., Tozzi, G. P., Wallis, M. K., Weaver, H. A., Woods, T. N., 1987. *Astron. and Astrophys.* **187**, 325.

Grard, R., Gombosi, T. I., and Sagdeev, R. Z., 1986. In *ESA SP-1066, Space Missions to Halley's Comet*, eds. R. Reinhard and B. Battrick, ESTEC, Noordwijk, Netherlands, p. 49.

Groussin, O., and Lamy, P., 2003. *Astron. and Astrophys.* **412**, 879.

Harker, D.E., et al., 2005. *Science* **310**, 278.

Hirao, K., 1986. In *ESA SP-1066, Space Missions to Halley's Comet*, eds. R. Reinhard and B. Battrick, ESTEC, Noordwijk, Netherlands, p. 71.

Huebner, W.F., Boice, D.C., 1999. In *Encyclopaedia of the Solar System*, eds. P.R. Weissman, L-A. McFadden and T.V. Johnson, Academic Press, San Diego, pp. 519–536.

Hughes, D.W., 1985. *Mon. Not. R. Astr. Soc.* **213**, 103.

Hughes, D.W., 1988. *Mon. Not. R. Astr. Soc.* **234**, 173.

Hughes, D.W., 1991. In *Comets in the Post-Halley Era*, Vol. 2, eds. R.L. Newburn, Jr., M. Neugebauer and J. Rahe, Kluwer Academic Publishers, Dordrecht, p. 825.

Hughes, D.W., 2000. *Mon. Not. R. Astr. Soc.* **316**, 642.

Hughes, D.W., 2001. *J. Brit Interplanetary Soc.* **54**, 169.

Hughes, D.W., and McBride, N., 1992. *J. Brit. Astron. Assoc.* **102**, 5.

Levison, H.F., and Duncan, M.J., 1997. *Icarus* **127**, 13.

Lyttleton, R.A., 1953. *The Comets*, Cambridge University Press. Meech, K.J. et al., 2005. *Science* **310**, 265.

Morrison, D., and Owen T., 1988. *The Planetary System*, Addison Wesley, p. 143.

Mumma, M.J., et al.. 2005, *Science* **310**, 270.

Rayman, M,D., 2003. *Space Technology* **23,** 185.

Reinhard, R., 1986. In *ESA SP-1066, Space Missions to Halley's Comet*, eds. R. Reinhard and B. Battrick, ESTEC, Noordwijk, Netherlands, p. 25.

Rickman, H., Kamél L., Festou, M.C., Froeschlé, C., 1987. In *ESA SP-278, Diversity and Similarity of Comets*, eds. E.J. Rolfe and B. Battrick, ESTEC, Noordwijk, p. 471.

Soderblom, L. A., et al., 2002. *Science* **296**, 1087.

Sugita, et al., 2005. *Science* **310**, 274.

Verdant, M. and Schwehm, G.H., 1998. *ESA Bulletin* **93**, 39.

Whipple, F.L., 1950a. *Astrophysical Journal* **111**, 375 and **113**, 464.

Whipple, F.L., 1972. Cometary nuclei models. In *NASA-CR-129110, Comets: Scientific Data and Missions*, eds. G.P. Kuiper and E. Roemer, NASA, Washington D.C, pp. 12–15.

Index

Printing: Mercedes-Druck, Berlin
Binding: Stein+Lehmann, Berlin

WITHDRAWN